Algorithms for Large Scale Linear Algebraic Systems

NATO ASI Series

Advanced Science Institutes Series

A Series presenting the results of activities sponsored by the NATO Science Committee, which aims at the dissemination of advanced scientific and technological knowledge, with a view to strengthening links between scientific communities.

The Series is published by an international board of publishers in conjunction with the NATO Scientific Affairs Division

A Life Sciences **B Physics**	Plenum Publishing Corporation London and New York
C Mathematical and Physical Sciences **D Behavioural and Social Sciences** **E Applied Sciences**	Kluwer Academic Publishers Dordrecht, Boston and London
F Computer and Systems Sciences **G Ecological Sciences** **H Cell Biology** **I Global Environmental Change**	Springer-Verlag Berlin, Heidelberg, New York, London, Paris and Tokyo

PARTNERSHIP SUB-SERIES

1. Disarmament Technologies	Kluwer Academic Publishers
2. Environment	Springer-Verlag / Kluwer Academic Publishers
3. High Technology	Kluwer Academic Publishers
4. Science and Technology Policy	Kluwer Academic Publishers
5. Computer Networking	Kluwer Academic Publishers

The Partnership Sub-Series incorporates activities undertaken in collaboration with NATO's Cooperation Partners, the countries of the CIS and Central and Eastern Europe, in Priority Areas of concern to those countries.

NATO-PCO-DATA BASE

The electronic index to the NATO ASI Series provides full bibliographical references (with keywords and/or abstracts) to more than 50000 contributions from international scientists published in all sections of the NATO ASI Series.
Access to the NATO-PCO-DATA BASE is possible in two ways:

– via online FILE 128 (NATO-PCO-DATA BASE) hosted by ESRIN, Via Galileo Galilei, I-00044 Frascati, Italy.

– via CD-ROM "NATO-PCO-DATA BASE" with user-friendly retrieval software in English, French and German (© WTV GmbH and DATAWARE Technologies Inc. 1989).

The CD-ROM can be ordered through any member of the Board of Publishers or through NATO-PCO, Overijse, Belgium.

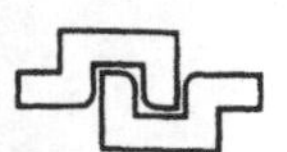

Series C: Mathematical and Physical Sciences – Vol. 508

Algorithms for Large Scale Linear Algebraic Systems

Applications in Science and Engineering

edited by

G. Winter Althaus
University of Las Palmas de Gran Canaria,
Las Palmas, Spain

and

E. Spedicato
Department of Mathematics,
University of Bergamo,
Bergamo, Italy

Kluwer Academic Publishers

Dordrecht / Boston / London

Published in cooperation with NATO Scientific Affairs Division

Proceedings of the NATO Advanced Study Institute on
Algorithms for Large Scale Linear Algebraic Systems:
State of the Art and Applications in Science and Engineering
Gran Canaria, Spain
June 23–July 6, 1996

A C.I.P. Catalogue record for this book is available from the Library of Congress.

Published by Kluwer Academic Publishers,
P.O. Box 17, 3300 AA Dordrecht, The Netherlands.

Sold and distributed in the U.S.A. and Canada
by Kluwer Academic Publishers,
101 Philip Drive, Norwell, MA 02061, U.S.A.

In all other countries, sold and distributed
by Kluwer Academic Publishers,
P.O. Box 322, 3300 AH Dordrecht, The Netherlands.

Printed on acid-free paper

ISBN 978-90-481-5004-5

TABLE OF CONTENTS

Preface

This book was motivated by the NATO Advanced Study Institute (ASI) on Algorithms for Sparse Large Scale Linear Algebraic Systems: State of the Art and Applications in Science and Engineering, which was held at Gran Canaria, Canary Islands, Spain, during June 23-July 6, 1996.

The aim of the book is to provide an overview of the most successful algorithms and techniques to solve large sparse systems of equations and some algorithms and strategies for solving optimisation problems.

The book is tutorial in nature and is addressed to mathematicians, computer scientists, engineers and postgraduate students interested in examining basic tools, progress in recent years, and to know and understand problems which remain open and current trends. The most important topics dealt with in the present book concern iterative methods, specially Krylov methods, ordering techniques and some innovative optimisation tools.

Many contributors are internationally recognised outstanding experts and all of them have been lecturers in the NATO ASI meeting. We would like to thank all the speakers who cooperated and who made the ASI a successful event.

Several simultaneous objectives are pursued in this book:

- To provide a compendium of state of the art on theoretical and numerical aspects of methods for solving large algebraic systems, with special emphasis to convergence and numerical behaviour as affected by rounding error, accuracy in the computed solution for ill-conditioned matrices, improvement of preconditioners effectiveness, ordering procedures, factors affecting stability, hybrid procedures and stopping criteria for iterative methods.

- To present recent advances in numerical matrix calculations, specially efficient strategies to accelerate the computational cost of the numerical solution, with special attention to the most relevant methods for solving symmetric and unsymmetric linear systems.

- As phenomena are often modelled as a set of differential equations leading, after discretization, to algebraic systems, some contributions solving these systems are given with appropriate information to evaluate the error in the solution. In this context, convergence analysis of the multigrid method using a posteriori error estimation in problems formulated as second order elliptic equations are presented. Some inverse problems that appear in science and engineering applications are also included.

- To introduce us to the existence and efficiency of evolution based software-Genetic Algorithms and Evolution Strategies, also relations and Class

Hierarchising, to improve the exploration into large search spaces and finding near-global optima easier than traditional methods in complex optimisation problems. Also recent developments of Messy Genetic Algorithms are described.

Acknowledgments:

The NATO Advanced Study Institute was mainly supported by NATO. Some expenses were supported by the Department of Mathematics and Research Centre of Numerical Applications in Engineering (CEANI) of the University of Las Palmas of Gran Canaria. Some national organisations provided travel expenses for participants. To all these organizations, especially to the NATO Science Committee, we express special gratitude.

Gabriel Winter and Emilio Spedicato.
June 1997

Computational Complexity of Solving Large Sparse and Large Special Linear Systems of Equations *

Victor Y. Pan
Department of Mathematics and Computer Science
Lehman College, City University of New York
Bronx, NY 10468, USA
Internet: vpan@lcvax.lehman.cuny.edu

Abstract

We review the sequential and parallel computational complexity of solving the sparse and special (structured) linear systems of equations and include some recent results. We revisit the known successful algorithms and techniques of their design that enable us to decrease the computational cost of the solution. Dealing with large linear systems, we apply asymptotic analysis and estimates, but also include nonasymptotic study and comment on some practical aspects of the application of the proposed algorithms. We are mostly concerned with numerical computations with round-off but also cite some error-free techniques for rational computations. The reader may find further material in the cited bibliography and, in particular, in [P92] and [BP94].

1 Introduction. Complexity of some basic computations with general matrices. BLAS 1, 2 and 3 levels.

We will review the computational complexity estimates for a real nonsingular linear system of n equations with n unknowns,

$$Ax = b \; . \qquad (1.1)$$

This will also enable us to review some effective algorithms supporting these estimates and some powerful techniques used in the design of these algorithms. We

*Supported by NSF Grant CCR 9020690 and PSC CUNY Award 666327.

G. Winter Althaus and E. Spedicato (eds.), Algorithms for Large Scale Linear Algebraic Systems, 1–24.

will focus on the cases where the $n \times n$ coefficient matrix A is sparse and/or structured, but we will start with the results in the case of general (possibly dense and unstructured) matrix A.

Gaussian elimination (with partial pivoting) enables us to compute the solution vector,

$$x = A^{-1}b\ , \tag{1.2}$$

by using $n^2+O(n)$ words of memory, $(2/3)n^3+O(n^2)$ arithmetic operations (hereafter, referred to as *ops*), and about $0.5n^2$ comparisons of the magnitudes of pairs of real numbers.

The latter memory space bound is nearly optimum (n^2+n words represent the input A and b), but the number of ops is not. In fact, $O(n^\beta)$ ops, for $\beta < 2.376$ (cf. [CW90], [P84]), suffice for solving (1.1), as well as for $n \times n$ matrix multiplication and inversion (hereafter, referred to as MI and MM, respectively). Due to (1.2), solving (1.1) can be reduced to MI (at least in principle), and it is instructive to recall how MI can be reduced to MM.

Let us write

$$A = \begin{pmatrix} B & C \\ D & E \end{pmatrix}, \qquad S = E - DB^{-1}C\ , \tag{1.3}$$

where B and S are assumed to be nonsingular matrices. Apply block Gauss-Jordan elimination and obtain block factorization of A and A^{-1}:

$$A = \begin{pmatrix} B & C \\ D & E \end{pmatrix} = \begin{pmatrix} 1 & O \\ DB^{-1} & 1 \end{pmatrix} \begin{pmatrix} B & O \\ O & S \end{pmatrix} \begin{pmatrix} 1 & B^{-1}C \\ O & 1 \end{pmatrix}\ , \tag{1.4}$$

$$A^{-1} = \begin{pmatrix} 1 & B^{-1}C \\ O & 1 \end{pmatrix} \begin{pmatrix} B^{-1} & O \\ O & S^{-1} \end{pmatrix} \begin{pmatrix} 1 & O \\ DB^{-1} & 1 \end{pmatrix}. \tag{1.5}$$

Here and hereafter, I and O denote the identity and null matrices of appropriate sizes, respectively. For simplicity, let $n = 2^h$ and let B be an $(n/2) \times (n/2)$ block of A. Let $mi(n)$ and $mm(n)$ denote the minimum numbers of ops required for $n \times n$ MI and MM, respectively. Then, (1.3) implies that

$$mi(n) \leq 2mi(n/2) + 6mm(n/2) + 4(n/2)^2$$

and, consequently,

$$\begin{aligned} mi(n) &\leq 4mi(n/4) + 12mm(n/4) + 8(n/4)^2 + 6mm(n/2) + 4(n/2)^2 \\ &\leq n \cdot mi(1) + 3\sum_{i=1}^{n} 2^i mm(n/2^i) + 2\sum_{i=1}^{n} 2^i (n/2^i)^2\ . \end{aligned}$$

It follows that

$$mi(n) < n + 3cn^\beta/(1-2^{1-\beta}) + 2n^2 = O(n^\beta) \tag{1.6}$$

provided that

$$mm(n) \leq cn^\beta\ , \quad \beta \geq 2\ . \tag{1.7}$$

This proves the claimed bound $O(n^\beta)$ on $mi(n)$ and on the complexity of solving (1.1) assuming (1.7) and nonsingularity of B, S and other auxiliary diagonal blocks. The nonsingularity assumption, however, is fulfilled for all the auxiliary matrices in the above recursive factorization applied to the symmetric positive definite matrix A^TA (note that A is assumed to be nonsingular). An alternative approach ensures nonsingularity based on pivoting (see [AHU76]).

The above reduction of MI to MM is an example of recursive divide-and-conquer algorithms, which reduce a given computational problem to two or several problems of smaller sizes. It is also an example of block matrix algorithms.

Another important example is given by recursive divide-and-conquer MM. Recall the following algorithm for 2×2 block MM, due to S. Winograd (see e.g. [P91]), which computes the product,

$$S = AB \ ,$$

of a pair of 2×2 block matrices

$$A = \begin{pmatrix} a_{11} & a_{12} \\ a_{21} & a_{22} \end{pmatrix} \quad \text{and} \quad B = \begin{pmatrix} b_{11} & b_{12} \\ b_{21} & b_{22} \end{pmatrix} ,$$

by using 7 block multiplications and 15 block additions/subtractions:

$$\begin{array}{llll} S_1 = a_{21} + a_{22}, & S_2 = S_1 - a_{11}, & S_3 = a_{11} - a_{21}, & S_4 = a_{12} - S_2, \\ S_5 = b_{12} - b_{11}, & S_6 = b_{22} - S_5, & S_7 = b_{22} - b_{12}, & S_8 = S_6 - b_{21}, \\ P_1 = S_2 S_6, & P_2 = a_{11} b_{11}, & P_3 = a_{12} b_{21}, & P_4 = S_3 S_7, \\ P_5 = S_1 S_5, & P_6 = S_4 b_{22}, & P_7 = a_{22} S_8, & \\ S_9 = P_2 + P_3, & S_{10} = P_1 + P_2, & S_{11} = S_{10} + P_4, & S_{12} = S_{10} + P_5, \\ S_{13} = S_{12} + P_6, & S_{14} = S_{11} - P_7, & S_{15} = S_{11} + P_6. & \end{array}$$

Then,

$$AB = \begin{pmatrix} S_9 & S_{13} \\ S_{14} & S_{15} \end{pmatrix} .$$

Let $n = 2^h$ again and let a_{ij} and b_{jk} be $(n/2)\times(n/2)$ blocks. By using s^2 *ops* per each addition/subtraction of $s \times s$ matrices and by recursively applying Winograd's algorithm for multiplication of blocks of recursively decreasing sizes, we finally arrive at a recursive algorithm that uses less than $3.92\ n^\alpha$ *ops* for $\alpha = \log_2 7 = 2.807\cdots$ (This algorithm of 1971 is sometimes called Strassen's for it has been preceded by a slightly inferior algorithm of [St69] that involved 7 multiplications and 18 block additions/subtractions for 2×2 block MM). Winograd's algorithm involves fewer ops than does the straightforward algorithm for MM (which uses $2n^3 - n^2$ *ops*). Let us point out an even more substantial advantage of operating with blocks. Compare the operations of basic linear algebra subroutines (BLAS) at level 1 of the form $\alpha x + y$ (called "saxpy" as an abbreviation for "scalar α (times vectors) x plus y") versus level 2 of the form $Ax + y$ (called "GAXPY" so as to abbreviate "general matrix A (times vector) x plus y"), and level 3 (block operation), of the form $AX + Y$, where A, X, and Y are matrices.

Let x and y be n-dimensional vectors, and let A, X and Y be $n \times n$ matrices. Then, we have $2n + 1$ data fetches, n writings of data, and $2n$ ops, at level 1, versus $n^2 + 2n$, n, and $2n^2$ at level 2, respectively, and $3n^2$, n^2, and $2n^3$ at level 3, respectively. Since data manipulation is much slower in practice than ops, the proportion $3n^2$, n^2, $2n^3$ achieved at BLAS 3 level is most desirable, the one achieved at BLAS 2 is the next desirable, and the one of BLAS 1 is least desirable. For this reason, performing computations at the BLAS 3 level is a desired goal, and this enhances the value of block matrix algorithms and, consequently, of methods for fast MM and MI.

This observation draws our attention to Newton's iteration,

$$X_0 = \alpha_0 A^T \quad , \quad X_{k+1} = \alpha_{k+1} X_k (2I - AX_k), \quad k = 0, 1, \ldots . \tag{1.8}$$

Already with $a_k = 1$, for all $k \geq 1$, the latter iteration converges to A^{-1} or, if A is singular, to A^+, the Moore-Penrose generalized (pseudo) inverse of A, thus defining the least-squares solution $x = A^+ b$ to the system (1.1). (Note that $A^+ = A^{-1}$ if A is nonsingular and see [GL89], [BP94] on some other properties of A^+.) For each k, the iteration step (1.8) is reduced to two matrix multiplications, and this step squares the residual matrix already for $\alpha_k = 1$:

$$I - AX_k = (I - AX_{k-1})^2 = (I - AX_0)^K \quad , \quad K = 2^k \; , \quad k = 1, 2, \ldots \tag{1.9}$$

The latter property implies local quadratic convergence and self-correcting of the approximations by X_k to A^{-1} (see [PS91] on the numerical stability of such an approximation for singular A). For an appropriate choice of α_k, $k = 1, 2, \ldots$, the convergence in roughly $\log_2 \kappa(A)$ steps [$\kappa(A)$ being the maximum ratio of two singular values of A] has been ensured in [PS91], with a further acceleration by twice for a Hermitian (real symmetric) A. [Furthermore, a modified version of (1.8), proposed in [PS91], converges to the matrix $A^+(e)$ obtained from the matrix A^+ by setting to zero all those singular values of A^+ that are less than an arbitrary fixed positive e; this modification can be further applied to some fundamental numerical computations of linear algebra.] The parallel complexity of such an approach is

$$O(\log \kappa(A)) mm^*(n) = O_A((\log n) \log \kappa(A), M(n))$$

[cf. (1.8), (1.9)]. Generally, $\kappa(A)$ is not easy to estimate, but $\kappa(A)$ is not large for many classes of linear systems, for instance, for linear systems representing discretizations of integral equations.

As an example of practical application of asymptotically fast algorithms, we recall that Winograd's algorithm has been implemented as the basis of the MM subroutines on several large modern computers [GL89]. (According to W. Kahan, for some narrow but important classes of matrices, one should guard carefully against numerical instability of this algorithm; to control the quality of the output, one may apply Frievald's simple test for correctness of MM via comparison of $(AB)v$ with Aw for $w = Bv$ and for a few random vectors v [BLRu90].) The known asymptotically faster algorithms for $n \times n$ MM, using Cn^{β} ops, for $\beta < \log_2 7$, do not promise to become practical because the constants $C = C(\beta)$ are very large and/or a much greater memory space is required to support these algorithms. The direct reduction of the system (1.1) to MI is rarely used in practice so far, since MI is generally harder than the solution of (1.1), but, due to intensive use of block matrix computations, decompositions (1.3)–(1.5) are increasingly popular.

We will conclude this section with some comments on partitioning the work for $n \times n$ MM among $p = 2n^2/S^2$ processors in parallel implementation. In this case, one may keep the computations at the BLAS 3 level by partitioning the input matrices A and B into $s \times s$ blocks and assigning each processor to handle one block. Then, each multiplication of blocks would require $3s^2$ data communication steps (of fetching and writing data) and (assuming a straightforward algorithm) $2s^3 - s^2$ ops. For large s, this is a desired proportion since data manipulation steps are practically slower than ops. More detailed treatment of various aspects of organization of parallel computing (including both BLAS 3 and BLAS 2 levels) can be found e.g in [Q94] and [GL89].

In this paper, we will present some estimate for parallel time and simultaneously for the number of arithmetic processors required by the algorithms that solve (1.1), so that $O(t, p)$ denotes the simultaneous bounds $O(t)$ on time and $O(p)$ on the number of processors. We will assume Brent's scheduling principle [Q94], according to which $O(s)$ time-steps of a single processor may simulate a single step of s processors, for any $s \geq 1$, so that $O_A(t, ps)$ implies $O_A(ts, p)$. We will rely on the following parallel complexity estimates for some basic linear algebra computations

	Parallel Computation for	Parallel time $(t_p M)$	Processors (p)	Source
1	FFT at n points (n is a power of 2)	$3(1 + \log_2 n)$	$2n$	[Pease68]
2	Summation of n numbers	$2\lceil \log_2 n \rceil$	$\lceil n/\log_2 n \rceil$	[Q94]
3	Inner product	$2\lceil \log_2 n \rceil + 1$	$\lceil n/\log_2 n \rceil$	(reduce to summation)
4	$n \times n$ matrix times a vector	$2\lceil \log_2 n \rceil 1$	$n\lceil n/\log_2 n \rceil$	(n inner products)
5	Product of $n \times n$ matrices	$2\lceil \log_2 n \rceil + 1$	$n^2\lceil n/\log_2 n \rceil$	(n^2 inner products)
6	$n \times n$ Toeplitz matrix by vector (n is a power of 2)	$1 + 9(2 + \log_2 n)$	$4n$	(reduce to 3FFT's at $2n$ points [AHU76], [BP94])

We also recall from [BP94] that, based on fast MM, one may (at least, theoretically) achieve the parallel cost bound

$$mm^*(n) = O_A(\log n, M^*(n)) \tag{1.10}$$

for $n \times n$ MM, where

$$M^*(n) \leq \text{minimum}[n^3/\log n, n^d/\log n, n^\beta], \quad \beta < 2.376. \quad \alpha = \log_2 7\,, \quad (1.11)$$

2 Computations with Complex, Integer, and Rational Matrices

The presented analysis can be easily extended to complex matrices $A = A_0 + A_1\sqrt{-1}$, where A_0 and A_1 are real matrices. A^T should be replaced by the Hermitian transpose $A^H = A_0^T + A_1^T\sqrt{-1}$. Note that a complex MM,

$$AB = (A_0 + A_1\sqrt{-1})(B_0 + B_1\sqrt{-1}) = A_0B_0 - A_1B_1 + (A_0B_1 + A_1B_0)\sqrt{-1}\,,$$

can be reduced to three (rather than four) real MMs, for instance, by using the matrix equation

$$AB = C_0 - C_1 + (C_2 - C_1 - C_0)\sqrt{-1}\,,$$

where $C_0 = A_0B_0$, $C_1 = A_1B_1$, $C_2 = (A_0 + A_1)(B_0 + B_1)$.

In some practical computations, particularly, in those coming from discrete mathematics and combinatorial analysis, one may need to operate with matrices filled with zeros and ones or with small (lower precision) integers. Let us show some advantages that can be potentially obtained in this case by using the *binary segmentation technique* (cf. [P91], P92] or [BP94], chapter 3, section 9), though the latter technique has not been implemented in the existent computer subroutines yet. Usually, the coefficients of a polynomial $p(x) = \sum_{i=0}^{n} p_i x^i$ can be recovered from its values at $n+1$ points, but the single value $V_h = p(2^h)$ would suffice provided that p_i are integers and $0 \leq p_i < 2^h$ for all i. Such a binary interpolation gives us the coefficients of $p(x) = u(x)w(x)$ if we are given the values $u(2^h)$ and $w(2^h)$ and if we know that $u(x)$ and $w(x)$ are polynomials with "short" integer coefficients. If these coefficients are known to range from 0 to M, then, it suffices to choose any $h > \log_2(nM^2)$. If the range is from $-M$ to M, we may choose $h > \log_2(2nM^2)$ and recover the coefficients of $p(x)$ from the value

$$v_h = p(2^h) + \sum_{i=0}^{n-1} 2^{hi+h-1}\,.$$

Then, a single multiplication and a single addition, with the computer precision b, suffices in order to obtain the $2n$ coefficients of the product $u(x)w(x)$, provided that $\lceil \log_2 v_h \rceil \leq b$. Without using binary segmentation, we would have need at least $9n\log_2 n + 2n$ or even $2n^2$ ops for the same computation (cf. [BP94]). Furthermore, if $\lceil \log_2 v_h \rceil / s \leq b < \lceil \log_2 v_h \rceil$ for some integer s, then we may partition each of the representations of $u(2^h)$ and $v(2^h)$ into s pieces of lengths at most b and reduce the computation to the evaluation of a bilinear expression in these s pairs of pieces.

We may also apply the binary segmentation techniques to computations of the inner and outer products of vectors, of matrix-by-vector products, and to MM (cf. [P84], [P91] and [P92]). For demonstration, let $\{u_i, w_j, \ i,j = 0,1,\cdots,n-1\}$ denote the n^2 entries of the outer product of two vectors $u = \{u_i\}$ and $w = \{w_j\}$, with the entries ranging from $-M$ to $M = 2^m$. In this case, we may first compute

$$p(2^h) = \left(\sum_{i=0}^{n-1} 2^{ih} u_i\right)\left(\sum_{j=0}^{n-1} 2^{jhn} w_j\right) = \sum_{i,j=0}^{n-1} u_i w_j 2^{(i+jn)h},$$

then $p(2^h) + \sum_{i,j=0}^{n-1} 2^{(i+jn+m)h}$, and finally recover the values $u_i w_j$ for all i and j, by means of binary segmentation.

The well known techniques and subroutines of computer algebra can be applied in order to perform matrix computations with no errors when the input values are integers or rationals (cf. e.g. [BP94]).

3 Direct and Iterative Algorithms for Linear Systems

The most celebrated practical algorithms for solving linear systems (1.1) are Gaussian elimination with pivoting, block Gaussian elimination, and their modifications. These are *direct algorithms*; in finite number of ops performed exactly, with no errors, these algorithms output exact solution to any given nonsingular system (1.1). There are several customary classes of *iterative algorithms* for linear systems (1.1), such as Jacobi's, Gauss-Seidel's, SOR, SSOR, and various methods of residual minimization in the Krylov space formed by the vectors $v, Av, A^2v, \cdots, A^mv$, for a fixed vector v and a fixed natural m, in particular, the conjugate gradient and Lanczos algorithms. These methods are supposed to compute only approximate solution to (1.1) in finite number of ops, even if these ops are performed exactly, with no errors. Typically, the complexity of each iteration step is dominated by the cost of performing one or two multiplications of the input matrix A, its transpose and/or their submatrices by vectors; this requires small storage space and relatively few ops for linear systems (1.1) with *special matrices* A (sparse and/or well structured). Such iterative algorithms are effective and are customarily recommended for the latter linear systems, provided that sufficiently fast convergence can be ensured. The convergence rate depends on the eigenvalues or singular values of A or CA. The choice of a preconditioner matrix C is crucial for the convergence acceleration, but it usually relies on some empirical rules and experimental tests. Newton's iteration (1.8) represents somewhat different class of iterative methods, which do not preserve sparsity but, after some modification, may preserve structure of some dense matrices [P93a].

We will further comment on direct and iterative methods in the next sections.

4 Sparse Linear Systems. General Review.

The solution of a linear system (1.1) with a sparse $n \times n$ matrix A having $\gamma(A)$ nonzero entries, where $\gamma(A) = o(n^2)$, usually involves fewer computational resources of time and storage space provided that the sparsity patterns are sufficiently regular, so that we may use appropriate data structures and avoid storing zero entries of A and operating with them. In particular, the Lanczos and conjugate gradient algorithms only require few (say, 1 or 2) multiplications of A or A^T by vectors at each iteration and converge to the solution in at most n iterations (if performed with no errors) and in $O(n)$ iterations [and, actually, in $o(n)$ iterations] in practice of finite precision computations, with round-off. (Symmetrization needed in these algorithms may be included implicitly and may rely on the matrix equations: $A^{-1} = (A^TA)^{-1}A^T$ or

$$\begin{pmatrix} O & A \\ A^T & O \end{pmatrix}^{-1} = \begin{pmatrix} O & A^{-T} \\ A^{-1} & O \end{pmatrix} \qquad [\text{ with } A^{-T} = (A^{-1})^T \,];$$

in the latter case, the matrix has dimension twice as large and is indefinite but better conditioned than A^TA, compare [GL89]). Thus, $O(n\gamma(A))$ ops and about $\gamma(A)$ words of memory shall always suffice.

Various modifications of the Lanczos and conjugate gradient algorithms, using, in particular, various preconditioners, have been devised in order to accelerate the convergence and to improve nnumerical performance of these algorithms with rounding off, for various more special classes of linear systems (cf. e.g. [FGN92], [A94]).

For many such classes, the linear system (1.1) can be solved in $o(n\gamma(A))$ ops. In particular, this is the case for the important class, where A is a block tridiagonal matrix with $w \times w$ blocks, $w = o(n)$. In this case, application of block Gaussian elimination yields the solution in $O(n/w)$ operations of $w \times w$ MI and MM; this costs $O((n/w)w^\beta)$ ops, where we have $\beta = 3$ in practice and $\beta < 2.376$ in theory (see section 1).

Furthermore, if we seek a faster parallel algorithm, it can be most effective (unless numerical stability problems arise) to treat A as an $(n/w) \times (n/w)$ tridiagonal block matrix with (generally dense) $w \times w$ blocks and to apply the *block cyclic reduction* (*BCR*) algorithm, which solves the system (1.1) in $\lceil \log_2(n/w) \rceil$ cyclic reduction steps, each step essentially amounting to $O(n/w)$ concurrent inversions and multiplications of the $w \times w$ blocks ([GL89], section 5.5). The BCR algorithm is also effectively applied for sequential solution of banded (block tridiagonal) linear systems.

Two recent alternative parallel algorithms for such a system have been presented in [PSA95]. One of them ([PSA95], sections 3-6) relies on the factorization (1.3)-(1.5) but exploits the band structure of the input matrix A in order to yield about the same parallel complexity estimates as by using the BCR algorithm. Another algorithm ([PSA95], sections 7-9) a little improves the parallel time bound (without using any extra ops) under the additional assumption of the nonsingularity of all subdiagonal or all superdiagonal blocks. (Numerically, the latter requirement takes a stronger form of well-conditioning of all such blocks.). Both of these algorithms, as well as (generalized) nested dissection and block cyclic reduction, consist of the harder stage of preprocessing the coefficient matrix and of the substantially simpler stage of solving (1.1) by using the results of the preprocessing stage. Consequently, the solution of several (say, s) systems (1.1) with the same matrix A requires a single application of the (harder) preprocessing stage and s applications of the (simpler) second stage.

Furthermore, $O(n)$ ops suffice in the case where the system (1.1) is obtained via discretization of a partial differential equation (PDE) and is solved by means of a multigrid algorithm [HT82], [McC87], [PR92], [PR93] (compare our section 6), whereas $O(n^{1.5})$ ops suffice in practice and $O(n^{1.184})$ in theory, in the case where the matrix A is associated with a graph having $O(\sqrt{n})$-separator family and is given with its $O(\sqrt{n})$-separator tree [GeLi81], [LRT79]. The latter approach is applied to symmetric and positive definite matrices A, and the algorithm, called ***(generalized) nested dissection algorithm*** (see the next section), computes Choleski's factorization $PAP^T = L^T L$, L being a lower triangular matrix and P being a permutation matrix, chosen so as to decrease the size of the ***fill in***, that is, of the set of the nonzeros of L and L^T that appear at the places of the zeros of A.

The most widely used and simple policy for choosing the permutation matrix P is to proceed with the symmetric version of Gaussian elimination, where in each step the pivotal entry (i, j) is chosen so as to minimize the value $(r(i)-1)(c(i)-1)$, $r(i)$ and $c(j)$ denoting the numbers of nonzeros in the i-th row and in the j-th column of the current coefficient matrix, respectively. To see the power of this policy (called ***the Markowitz rule***, or, in the symmetric case, ***the minimum degree rule***), consider a matrix filled with zeros except for its diagonal and its first row and column; then, without pivoting, the first step of Gaussian elimination turns the original sparse matrix into fully dense one, but no fill-in appears at all if the Markowitz rule is applied.

Another important case is where (1.1) is a triangular system. Then, Gaussian elimination supports the solution at the parallel cost $O_A(n, n)$. We will conclude this section by recalling the algorithm of [P94], which accelerates the parallel solution of a triangular system (1.1) by factor $\sqrt{n}/\log n$ (with no increase of the bound $O(n^2)$ on the total number of ops included). The algorithm relies on simple ***subtract-and-conquer*** techniques. Namely, one first represents the triangular matrix A as $w \times w$ block matrix, with $q \times q$ blocks, for an appropriate q and for $w = n/q$. Then, one scales the system (1.1) by premultiplying A and b by the inverse of the diagonal matrix formed by the blocks of A (which are triangular matrices). Finally, one applies block substitution algorithm to compute the solution. [P94] supplies a simple algorithm for the inversion of each of the $q \times q$ triangular blocks, at the cost $O_A(\log^2 q, q^3/\log^2 q)$. For concurrent inversion of all the n/q blocks, this gives $O_A(\log^2 q, nq^2/\log^2 q)$. The block substitution stage uses n/q steps, each of at most $n-1$ concurrent multiplications and substractions of $q \times q$ matrices, which bounds the overall cost of the substitution stage by $O((n/q)\log q, q^3/\log q)$. For $q = \sqrt{n}$, the overall cost of the solution of a triangular system (1.1) is bounded by $O_A(\sqrt{n}\log n, n^{3/2}/\log n)$. [PP95] shows some further extensions of such a subtract-and-conquer techniques to parallel solution of Toepltz and general linear systems.

5 (Generalized) Nested Dissection

For general sparse matrix A, computing a permutation matrix P that minimizes the fill-in is an NP-*complete* problem [AHU76], that is, we cannot hope to solve it even for moderate n (say, for $n = 100$). For a large and important class of linear systems (1.1), however, an effective method, called *(generalized) nested dissection* ([GeLi81], [GT87], [LRT79]) enables us to bound the fill-in and, consequently, the computer time and the memory space involved in the computation. Typically, the bounds are $O(n^{3/2})$ ops, $O(n)$ words of memory space, and $O(n)$ elements of the fill-in, for the (generalized) nested dissection algorithm that computes the Choleski factorization of a sparse and appropriately structured matrix A, having an $O(\sqrt{n})$-separator family and given with its $O(\sqrt{n})$-separator trees.

With this factorization available, $O(n)$ ops and $O(n)$ words of memory suffice to solve the system (1.1) for any vector b on the right-hand side, which is particularly attractive when we need to solve several such systems for a fixed A and variable b.

It has been observed that the block cyclic reduction (BCR) algorithm is a special case of the (generalized) nested dissection applied to a banded linear system (1.1).

As the first major step in the study of the nested dissection, we replace the nonzero entries of A by ones and thus arrive at the adjacency matrix of the associated graph $G = (V, E)$. The edges of E correspond to the nonzero entries of A, and the n vertices of V are identified with n unknown variables of the system (1.1). For simplicity, we will assume that the matrix A is symmetric, so that the graph G is undirected.

The elimination of a variable v corresponds to the deletion of a vertex v, together with all the edges adjacent to it, followed by adding the edge (u, w) to the graph as soon as its two edges (v, u) and (v, w) have been deleted. For a large and important class of linear systems (1.1), particularly, for many linear systems arising in the solution of ordinary and partial differential equations, the associated graphs $G = (V, E)$ have small separator families, that is, G can be recursively partitioned, so that removing a set S of $O(\sqrt{n})$ vertices from G (together with all edges adjacent to S) partitions G into two subgraphs $G_1 = (V_1, E_1)$ and $G_2 = (V_2, E_2)$ that have no edges remaining in G and connecting them and such that each of G_1 and G_2 has at most a fixed fraction (say, 2/3) of the vertices of G. Moreover, similar partitions are recursively defined for the subgraphs G_1 and G_2 [GT87] or, in an alternative definition of [LRT74], [PR93], for the graphs $\widetilde{G}_1$ and $\widetilde{G}_2$ obtained from G_1 and G_2, respectively, by adjoining to each of them the separator set S, together with all the edges of G adjacent to S. The recursive process ends where G_i (for $i = 1$ and/or for $i = 2$) has less than n_0 vertices, for a fixed (smaller) constant n_0.

To better understand this definition, the reader may examine the important special case where G is a grid graph (say, for the two-dimensional 7×7 grid), and to define the separators recursively, by removing the vertical and horizontal edges in the middle of the grid. In particular, by setting $n_0 = 1$ and by using the definition of [GT87] for the separators, we will finally arrive at the singletons. Computing the Choleski factorization [which we identify with the symmetric (Gauss-Jordan) elimination] proceeds in the order opposite to the partition (and again the reader may examine this process for the 7×7 grid graph G, as an exercise). In particular, at the last, d-th, step of the recursive partition, we arrive at the graph $G_{h,d}$, $h = 1, 2, \cdots, k(d)$, with the vertex sets of cardinalities less than n_0, whose vertices are eliminated first. This corresponds to the Choleski factorization of the block diagonal matrix of the northwestern part of the matrix $A_0 = PAP^T$. The remaining graph corresponds to the matrix A_1 of the southeastern block of A. Then, the remaining vertices of the graphs $G_{j,d-1}$, $j = 1, 2, \cdots, k(d-1)$, obtained at the $(d-1)$-st partition step, are eliminated; they correspond to the blocks of the block diagonal submatrix of A_1 lying in its northwestern corner. Then, the process is recursively repeated for the southeastern block A_2 of A_1, and so on.

Due to the bound on the size of the vertex set involved in each step of the recursive partition of G, the block diagonal matrices eliminated in the elimination stage $g, g \leq d \leq \log_{(3/2)} n$, have blocks of the size $O(\sqrt{(2/3)^{d-g}n})$, and this implies a small upper bound on the complexity of the elimination, even though the (small) eliminated blocks are generally dense. We refer the reader to [GT87] and [LRT79] for many details involed in the most general case, where the graphs G are given with their families of small separators (the important special cases of grid graphs and of finite element graphs are relatively simple).

Finally, we would like to mention the option of replacing the Choleski factorization by computing the recursive factorization (1.4) [with $C = D$], by using which we may still solve (1.1) in $O(n^{3/2})$ ops (see [PR93]). This version (unlike the other version) can be performed over semirings (where divisions and subtractions are not allowed) and thus supports the extension of the generalized nested dissection algorithm to many highly important computations of paths in planar graphs [P93], [PR89]. Both versions allow parallel implementation with parallel time of the order of $\log^3 n$ and $O(n^{3/2})$ processors (see [PR93]) and [Haf88]).

6 Multigrid Methods

An important example of effective framework for various iterative methods is the multigrid/multilevel approach to the solution of PDEs discretized over some region. In particular, on a grid with n points, the associated linear system (1.1) with $n \times n$ matrix A can be solved by using $O(n)$ ops and $O(n)$ memory space (up to the level of discretization). Moreover, a version [RP92] of the *compact multigrid* method enables us to solve and to store the solution to the system (1.1) arising from a (piecewise) constant coefficient linear PDE by using $O(n)$ bit-operations and $O(n)$ bits of memory, which was the order of $\log n$ time improvement of the previous approaches.

According to the multigrid scheme, a given partial differential equation (PDE) is discretized over a sequence of grids of points, starting with the coarsest grid G_0 of fewer (namely, $N_0 = G_0$) points, and successively refined to finer grids $G_1, G_2, \cdots, G_k$, containing, say, about

$$N_1 = G_1 = 2^d N_0,$$

$$N_2 = G_2 = 2^d N_1 = 2^{2d} N_0,$$

$$\vdots$$

$$N_k = G_k = 2^d N_{k-1} = 2^{kd} N_0 = N$$

points, respectively, where d is the dimension of the grids (the number of the variables of the PDE). An approximation to the solution to the (piecewise) linear PDE on the superimposed grids $G_0, G_1, \cdots, G_k$ is defined as the solutions of the associated systems of N_j linear equations with N_j unknowns,

$$D_j u_j = b_j, \quad j = 0, 1, \cdots, k \;, \tag{6.1}$$

where u_j and b_j, for each j, are N_j-dimensional vectors, b_j is given, u_j is unknown, and D_j is a given matrix, defined by the PDE, sparse and usually having a special structure. In particular, the number of nonzeros per row of D_j is bounded by a constant (independent of j), and for a large class of PDEs, iterative algorithms are available such that, in each iteration step, the norm of the approximation error vector,

$$e_j^{(h)} = u_j^{(h)} - u_j \;,$$

decreases at least by a fixed constant factor $\Theta < 1$, $u_j^{(h)}$ being an approximation to u_j computed in h iteration steps [HT82], [McC87].

The function $u(x)$ satisfying the PDE is usually smooth, so that its approximation on the grid G_j, within a discretization error bound E_j, usually has order of $O(1/N_j^c)$ for a fixed positive constant c (say, $c = 2$ or $c = 3$). Surely, we only need to approximate u_j within the error of the order of E_j and shall end the iteration when this is done.

Therefore, in the transition from the approximation u_{j-1} to u_j, we only need to decrease the approximation error norm by the factor of 2^{cd}, which only requires a constant number of iteration steps [of the order of $\log(2^{cd})/\log(1/\Theta) = cd/\log(1/\Theta)$]. Each step amounts to $O(1)$ multiplications of D_j (or D_j^T) and/or their submatrices by vectors, and this involves $O(N_j)$ ops, since D_j only has $O(1)$ nonzeros per row. Summarizing, we only need $\sum_j O(N_j) = O(\sum_j N_j) = O(\sum_{j=0}^k 2^{-d_j} N) = O(N)$ ops.

Furthermore, for a large class of linear PDEs with piecewise constant coefficients, every arithmetic operation in this computation only involves operands represented with a small constant number of binary bits: shift from (6.1) to the linear system

$$D_j e_j = r_j$$

(where $r = b_j - D_j u_j^*$, u_j^* is the current approximation to u_j^*) and observe that the entries of D_j and of e_j are $O(1)$-bit binary numbers.

There are various ways of taking advantage of operating only with lower precision numbers. Most importantly, we may decrease the storage space if, instead of storing the order of $\log_2 N_j$ bits of each component of the vector u_j, we will only store $O(1)$ bits of each component of u_0 and e_j for $j = 0, 1, \cdots, k$, $k = O(\log N)$. Thus, we decrease by the factor of $\log_2 N$ the overall storage used. Similarly, by the factor of $\log_2 N$ or more, we decrease the upper bound on the number of ops used if we may perform ks ops with $O(1)$-bit operands as fast as or faster than s ops with k-bit operands, and this can be ensured either by using a specialized computer (such as MASPAR) or special algorithms (such as binary segmentation, recalled in section 1).

The above observation (due to [PR92]) about a possible decrease (by the factor $\log_2 N$) of the bit-time and the bit-space complexity of multigrid computations can be extended to various other types of the discretization sets (in particular, this includes extensions to grids with step sizes variable over the points of the grids, to nonrectangular, say, triangular, grids, and so on), to linear PDEs with variable coefficients (then, we may approximate the PDE by using the matrix D_j whose coefficients are defined with the precision increasing with j), and to nonlinear PDEs still reduced to solving linear systems of equations (see [PR93]). Note the practical significance of reduction by the factor $\log_2 N$. Indeed, $\log_2 N > 10$ if $N > 1024$, and customary multigrid computations frequently go with much larger N.

7 Dense Structured Linear Systems, General Comments

In numerous applications to scientific and engineering computing, communication and statistics, the coefficient matrix A of (1.1) is dense but well structured. Then, the solution of (1.1) can be much simpler than in the general case. Typically, for the solution of a linear system (1.1) with a structured $n \times n$ matrix A, $O(n)$ words of memory and from $O(n \log n)$ to $O(n \log^2 n)$ ops suffice [versus the orders of n^2 words and of n^2 to n^3 ops required in the case of a general linear system (1.1)]. Numerical stability of the fast algorithms is easily ensured in the cases of symmetric positive definite or diagonally dominant input matrices A, as well as of input matrices A of smaller sizes [Bun85]. In the general case, we may still yield some effective solution algorithms by using symmetrization, that is, the transition from (1.1) to the equivalent linear systems $A^T A x = A^T b$ or $A A^T y = b$, where $x = A^T y$, or by shifting from numerical to symbolic computations. Computations with dense structured matrices are closely related to polynomial computations (cf. [BP94]).

8 Toeplitz and Hankel Linear Systems

Particularly important in applications are the classes of linear systems with Toeplitz and Hankel matrices, $T = [t_{i,j}]$ and $H = [h_{i,j}]$, such that $t_{i,j} = t_{i-j}$ and $h_{i,j} = h_{i+j}$, for all entries (i, j), respectively. Each such an $n \times n$ matrix has at most $2n - 1$ distinct entries, and is completely defined by its two columns or by its two rows (the first and the last); this reduces to n entries of the first column in the symmetric Toeplitz case. For any Hankel matrix H and for the reflection matrix J (zero everywhere except for its antidiagonal, filled with ones), JH and HJ are Toeplitz matrices, and $J^2 = I$, so that we will just study Toeplitz linear systems (1.1), and this will actually cover the Hankel case too.

One may facilitate operations with $n \times n$ Toeplitz matrices by exploiting their embedding in $N \times N$ circulant matrices $C = [c_{i,j}]$, for $N > n$. The latter matrices constitute a subclass of the class of $N \times N$ Toeplitz matrices, $c_{i,j} = c_{i-j \bmod N}$, $N = O(n)$. Computations with circulant matrices rely on the following major result (cf. [BP94], page 134):

Theorem 8.1 *Let c^T denote the first row of $(n+1) \times (n+1)$ circulant matrix C. Let $\omega = \exp\left(\frac{2\pi\sqrt{-1}}{n+1}\right)$, let D_1 and D be the two diagonal matrices with diagonal entries $1, \omega, \cdots, \omega^n$ and $d_0, d_1, \cdots, d_n$, respectively, where*

$$\Omega = (\omega_{i,j},\ i, j = 0, 1, \cdots, n),\ \omega_{i,j} = \omega^{ij}/\sqrt{n+1},\ \Omega^H = (\omega^{-ij}/\sqrt{n+1}),$$

so that $\Omega^H \Omega = I$, and where $d = (d_i) = \sqrt{n+1}\Omega D_1 c$. Then,

$$\Omega D_1 C D D_1^{-1} \Omega^H = D\,.$$

Based on this diagonalization result, one may immediately reduce the computation of Cb (for a given pair of a vector b and an $N \times N$ circulant matrix C) to performing [at $O(N \log N)$ ops] three FFTs, each at $O(N)$ points. Similarly, within this cost bounds, one may compute both $C^{-1}b$, that is, solve (1.1) for $A = C$, if C is nonsingular (note that C^{-1} is a circulant matrix too) and Tb, where T is any Toeplitz matrix of size $n \times n$, $2n \leq N$, for, surely, we may embed T in an $N \times N$ circulant matrix. (The complexity of the solution of the system (1.1) decreases almost to the level $O_A(\log n, n)$ if A is a triangular Toeplitz matrix and decreases even further if A is a banded Toeplitz matrix [BP94].)

The latter observation suggests using preconditioned conjugate gradient algorithms for Toeplitz systems (1.1), since every iteration step reduces to a few multiplications of Toeplitz matrices by vectors, that is, to a few FFTs, and thus costs $O(\log n, n)$. The power of such algorithms and their fast convergence have been proven for a large and important subclass of all Toeplitz systems (1.1) [Strang86], [BDB90].

For other Toeplitz systems (1.1), for which the convergence of the latter algorithms is slow, the most customary solution algorithms are Levinson's, Schur's (both using the order of n^2 ops), and modified Schur's [using $O(n \log^2 n)$ ops], (see [CB90], [AG89], [Kai89]; for some earlier solutions in $O(n \log^2 n)$ ops, see

[BGY80], [BA80], [Mus81], [dH87]). In section 10, we will recall another (more recent) approach. All these are direct algorithms (they compute the solution exactly if performed with infinite precision), and in their parallel implementation, at least n parallel arithmetic steps are required.

In some applications, a greater parallel acceleration is needed. For instance, in some signal processing applications, the solution to a Toeplitz linear system must be updated in real time, together with the (slowly) updated input Toeplitz matrix of the coefficients, so that the solution computed at the previous time-step can be used as a good initial approximation to the solution at the current time-step.

In such cases, Newton's iteration (1.8) of section 1 is a better choice (a good alternative is also given by various methods of the steepest descent). Specifically, each Newton's iteration step essentially amounts to a pair of matrix multiplications. In the case of a Toeplitz input matrix A, we may modify such a step and perform it at the low cost of $O_A(\log n, n)$ [P92a].

In particular, this approach may employ the Gohberg-Semencul formulae (theorem 8.2 below) or their further extensions [Io82], [T90].

Definition 8.1 Z is the matrix of lower shift (displacement), filled with zeros, except for its first subdiagonal, filled with ones; $L(v)$ is the lower triangular Toeplitz matrix defined by its first column vector v. J is the reflection matrix (see the beginning of this section).

Theorem 8.2 *[GS72] Let the matrices Z, $L(v)$ and J be defined according to definition 8.1. Let T be a nonsingular Toeplitz matrix, let x and y be the two columns of T^{-1}, the first and the last, respectively, and let x_0 be the first component of the vector x. Then,*

$$x_0 T^{-1} = L(x)L^T(Jy) - L(Zy)L^T(Z^T Jx) \ .$$

Now, for an arbitrary matrix W, let u, v and x_0 denote its first column, its last column and the first component of v, respectively, and let $x_0 \neq 0$. Then, we write

$$X(W) = (1/x_0)(L(u)L^T(Jv) - L(Zv)L^T(Z^T Ju)) \ . \tag{8.1}$$

Hereafter, let x_0 always be large enough so as not to let the division by x_0 cause any numerical stability problem in (8.1); otherwise, we could have just shifted to some modified versions of (8.1) and of theorem 8.2 (see [Io82] or [T90]).

Now, we define the following modification of Newton's iteration (1.8):

$$X_{k+1} = X(Y_{k+1}), \quad Y_{k+1} = \alpha_{k+1} X_k (2I - AX_k), \quad k = 0, 1, \cdots, \tag{8.2}$$

where $X(Y_{k+1})$ is defined by (8.1), for $W = Y_{k+1}$, and actually, we only need to compute the two columns (the first and the last), for each of the two matrices Y_{k+1} and X_{k+1} in the k-th iteration step (8.2), whose cost is thus bounded by $O_A(\log n, n)$.

On the other hand, due to (1.9), rapid convergence of the matrices X_k to A^{-1} is ensured if some (vector induced) matrix norm $I - AX_0$, induced by a vector norm, is substantially less than 1 (say, if it is less than 0.8). (The latter assumption is satisfied, for instance, in the cited real time computations.) Then, we may effectively apply the modification (8.2) of Newton's iteration, so as to ensure that $I - AX_k < 2^{-s}$ already for $k = O(\log s)$, that is, at the cost $O_A(\log s \log n, n)$. Similar comments apply if , say, the entries of A rapidly decrease their magnitude as they move away from the diagonal; then we have a good approximation to A and A^{-1} by a band Toeplitz matrix X_0 and by its inverse, respectively, where X_0 is obtained from A by zeroing its smaller entries that lie farther from the diagonal of A.

By using the subtract-and-conquer techniques of section 4, one may extend the above complexity bound so as to yield the bound $O_A(n^{1-a}\log^2 n, n^{1+a}\log n)$ for any a, $1/2 \leq a \leq 1$ (cf. [PP95]). The reader is referred to [P92], [P93a] and [BP94] on these algorithms, underlying techniques, and various extensions and related computations, in particular, to parallel algorithms for computing the Krylov sequence $v, Av, A^2v, \cdots, A^m v$ for general and Toeplitz matrices A and for the solution of a general linear system (1.1) at the cost $O_A(\log^2 n, M^*(n)/\log n)$ [cf. (1.10), (1.11)].

Now, suppose that no good initial approximation to A^{-1} is available; we may still extend the above approach [P92a]. The estimated parallel complexity grows to $O_A(\log^2 n, n^2/\log n)$. Moreover, the algorithms involve computation of the coefficients of the characteristic polynomial of A and, consequently, generally lead to some problems of numerical stability but, on the other hand, can be immediately extended to computing the Moore-Penrose generalized (pseudo) inverse A^+ and, consequently, to computing $x = A^+b$, a least-squares solution to (1.1) having the minimum norm, even in the case where A is a rank deficient matrix. The algorithms support the record parallel cost bounds $O_A(\log^2 n, n^2/\log n)$ and the record sequential time bound $O_A(n^2\log n, 1)$ for the latter problem.

Generally, the precision of these computations is prohibitively high for numerical computing (this, however, applies to any method for computing A^+b, if, say, A is a rank deficient matrix), but we may avoid such problems by using symbolic computation with no round-off and with modular arithmetic (see our section 11, [P92] and section 3 of chapter 3 of [BP94]).

9 Toeplitz-like and Hankel-like Linear Systems

Symmetrization of a Toeplitz or Hankel linear system (1.1) (required for the numerical stabilization of its most popular solution algorithms) involves matrices A^TA or AA^T. Computations with Toeplitz or Hankel matrices involve their products and inverses. All these are neither Toeplitz nor Hankel matrices but belong to a more general class of dense structured matrices characterized by a lower rank of their images in the application of certain operators of displacement and/or scaling ([KKM79], [P90], [BP94]). In particular, for the displacement matrix Z of definition 8.1, the next four displacement operators,

$$F_-(A) = A - Z^TAZ \ , \tag{9.1}$$

$$F_+(A) = A - ZAZ^T , \tag{9.2}$$

$$F^-(A) = AZ^T - Z^T A , \tag{9.3}$$

$$F^+(A) = AZ - ZA , \tag{9.4}$$

are naturally associated with Toeplitz and Toeplitz-like matrices, for which the images of such operators have lower ranks, in particular, ranks at most 2, for every Toeplitz matrix, and at most $m+n$, for every $m \times n$ block matrix with Toeplitz blocks. Representing an $n \times n$ matrix Y of a lower rank r as the product $Y = GH^T$, where G and H are $r \times n$ matrices, we will call the pair (G, H) *a generator of a length* r for Y. If $Y = F(A) = GH^T$ (for a displacement operator $F(A)$ of (9.1)-(9.4), that is, if $F(A) = GH^T$ is $F_-(A)$, $F_+(A)$, $F^-(A)$, $F^+(A)$, then, this pair of matrices (G, H) will also be called a *displacement generator* (or an *F-generator*, for a specific operator F) *of length r for A.*

Let us recall some fundamental properties of displacement generators (see [KKM79], [P92a], [BP94]).

Theorem 9.1 ([*KKM79*]). *Let*

$$G = [g_1, \cdots, g_r] , \quad H = [h_1, \cdots, h_r]$$

Then

$$A = \sum_{i=1}^{r} L(g_i)L^T(h_i) \text{ if } A - ZAZ^T = GH^T , \tag{9.5}$$

$$A = \sum_{i=1}^{r} L^T(Jg_i)L(Jh_i) \text{ if } A - Z^T AZ = GH^T . \tag{9.6}$$

It follows that an $n \times n$ matrix A can be defined by the $2dn$ entries of its displacement generator of length r, rather than by the n^2 entries of A. For the same matrix A, its $L^T L$- and LL^T-representations (9.5) and (9.6) are closely related to each other (see [BP94]), so that an F_+ (respectively, F_-)-generator of length r for a matrix A immediately defines its F_-(respectively, F_+)-generator of a length at most $r+2$.

Furthermore, the operators F_-, F_+, F^- and F^+ are closely related to each other:

Proposition 9.1 *Let* $i_1 = [1, 0, \cdots, 0]^T$, $i_n = [0, \cdots, 0, 1]^T$. *Then, for any matrix* A,

$$F^-(A)Z = F_-(A) - Ai_n i_n^T , \quad F_-(A)Z^T = F^-(A) + Z^T Ai_1 i_1^T ,$$

$$F^+(A)Z^T = F_+(A) - Ai_1 i_1^T , \quad F_+(A)Z = F^+(A) + ZAi_n i_n^T ,$$

$$ZF^-(A) = i_1 i_1^T A - F_+(A) , \quad ZF_-(A) = i_1 i_1^T AZ - F^+(A),$$

$$Z^T F^+(A) = i_n i_n^T A - F_-(A) , \quad Z^T F_+(A) = i_n i_n^T AZ^T - F^-(A).$$

Proof: Pre- and postmultiply the matrix equations (9.1)-(9.4) by Z and Z^T and combine the resulting equations with (9.1)-(9.4) and with the matrix equations $Z^T Z = I - i_n i_n^T$, $ZZ^T = I - i_1 i_1^T$. □

We will call A a *Toeplitz-like matrix* if $r = rank(F(A))$ is small, for a displacement operator F of (9.1)-(9.4) (say, if $r = O(1)$ as $n \to \infty$). We may define a *Hankel-like* matrix A as such that JA is a Toeplitz-like matrix, or we may define Hankel-like matrices by introducing associated displacement operators such as $F(A) = A - ZAZ$ or $F(A) = Z^T A - AZ$. We may operate with the displacement generators of Toeplitz-like matrices rather than with the matrices themselves (and similarly for Hankel-like matrices) by relying on the following results (cf. [BP94]):

Theorem 9.2 *Given a displacement operator F of (9.1)-(9.4) and a pair of F-generators of lengths a and b for a pair of $n \times n$ matrices A and B, respectively, we may immediately obtain an F-generator of a length at most $a+b$ for $A + \alpha B$ (for any fixed scalar α); furthermore, at the cost of $O_A(\log abn)$, we may also compute F-generators of lengths at most $a+b+\delta(F)$ for AB and $A^H A$, where $\delta(F) = 0$ if $F = F^-$ or $F = F^+$ and $\delta(F) = 1$ if $F = F_-$ or $F = F_+$.*

Proof: The result for $A + \alpha B$ is obvious. For AB (and similarly for $A^H A$), it suffices to consider the cases of $F = F^-$ and $F = F_-$. Let $F = F^-$. Then, the result follows since

$$\begin{aligned} F^-(AB) &= ABZ^T - Z^T AB = A(BZ^T - Z^T B) + (AZ^T - Z^T A)B \\ &= AF^-(B) + F^-(A)B = AG_B^-(H_B^-)^T + G_A^-(H_A^-)^T B \\ &= G_{AB}^-(H_B^-)^T + G_A^-(H_{AB}^-)^T \end{aligned}$$

provided that

$$F^-(A) = G_A^-(H_A^-)^T, \quad F^-(B) = G_B^-(H_B^-)^T, \quad G_{AB}^- = AG_B^-, \quad (H_{AB}^-)^T = (H_A^-)^T B .$$

For $F = F_-$, theorem 9.2 follows since

$$\begin{aligned} F_-(AB) &= AB - Z^T AIBZ = AB - (Z^T AZ)(Z^T BZ) + Z^T A i_1 i_1^T BZ \\ &= (A - Z^T AZ)B + Z^T AZ(B - Z^T BZ) + uv^T \\ &= F_-(A)B + Z^T AZF_-(B) + uv^T \\ &= G_-(A)(H_-^T(A)B) + (Z^T AZG_-(B))H_-^T(B) + uv^T \end{aligned}$$

provided that

$$F_-(A) = G_-(A)H_-^T(A), \quad F_-(B) = G_-(B)H_-^T(B), \quad u = Z^T A i_1, \quad v^T = i_1^T BZ .$$

The proof is similar for $F = F^+$ and $F = F_+$. □

The above machinery enable us to extend our algorithm of section 8 to solving Toeplitz-like linear systems (see the details in [P92a] or in [BP94]).

10 Some Other Classes of Dense Structured Linear Systems.

Two other important classes of dense structured matrices are represented by Vandermonde matrices,

$$V = [v_{i,k}], \quad v_{i,k} = v_i^k, \quad i,k = 0,1,\cdots,n-1, \tag{10.1}$$

and Cauchy (generalized Hilbert) matrices,

$$B = [b_{i,j}], \quad b_{i,j} = \frac{1}{s_i - t_j}, \quad i,j = 0,1,\cdots,n-1, \tag{10.2}$$

for some scalars v_i, s_i, t_j, such that $s_i \neq t_j$ for all i and j. Clearly, we may well define every matrix V of (10.1) by its n entries, and every matrix B of (10.2) is well defined by its $2n-1$ entries. Multiplication of such a matrix by a vector, as well as the solution of each linear system, $Vx = b$ or $Bx = b$, amounts [for (10.1)] or can be reduced [for (10.2)] to the multipoint polynomial evaluation and interpolation and thus can be performed at the cost $O_A(\log^2 n, n)$ (see, for instance, [BP94]). This computation may lead to numerical stability problems, but they can be avoided by using some alternate computations, with a certain increase of the computational cost bound in the Vandermonde case ([GL89]) and with shifting from the exact to an approximate solution in the Cauchy (generalized Hilbert) case ([BP94], chapter 3).

On the other hand, for symbolic computations (with infinite precision), the complexity estimates of $O_A(\log^2 n, n^2)$ and of $O(n \log^2 n)$ ops, for Hankel-like and Toeplitz-like linear systems (1.1), can be extended to the case of Vandermonde-like and Hilbert-like linear systems (1.1).

For demonstration, let $A = V^T = [a_{i,j}]$, $a_{i,j} = v_j^i$, be the transpose of a Vandermonde matrix. Then, $H = AA^T = V^T V = [h_{i,j}]$, $h_{i,j} = \sum_k v_k^{i+j}$, is a Hankel matrix, and we may solve the system (1.1) as follows [BP94]:

1. Compute the coefficients of the polynomial $p(x) = \sum_k (x - v_k)$.
2. Compute the power sums of the zeros of $p(x)$, $\sum_k v_k^s$, $s = 1, 2, \cdots, 2n$.
3. Solve the Hankel linear system $Hy = b$, where $H = V^T V$.
4. Compute the solution $x = Vy = V^{-T} b$ to the original linear system $V^T x = b$ [where $V^{-T} = (V^T)^{-1}$].

The well-known techniques of polynomial computation [BP94] enable us to perform each of stages 1.-4., and thus the entire algorithm, at the cost $O_A(\log^2 n, n)$.

The same complexity estimates for solving the system $V^T x = b$ can be similarly obtained by means of an alternate reduction [P90] of the original linear system to the linear system $V^T V^- y = b$, where $V^- = [v_{j,k}^-]$, $v_{j,k}^- = v_j^{-k}$, is a Vandermonde matrix, and $V^T V^{-1} = [t_{i,j}]$, $t_{ij} = \sum_k v_k^{i-j}$, is a Toeplitz matrix, so that the desired solution $x = V^{-T} b$ to (1.1) is obtained as $x = V^- y$.

These two ad hoc approaches to the solution of (1.1), for $A = V^T$, have been extended in [P90] to the general reduction of solving Vandermonde-like and Cauchy-like linear systems to Toeplitz-like and/or Hankel-like linear systems and vice versa.

The reduction relies on generalizing the concepts of the displacement ranks and the displacement generators (by allowing the operators of scaling instead of or in addition to the displacement operators and on a respective extension of theorem 9.1) and on the respective extention of theorem 9.2. The extension of theorem 9.2 ensures a low cost evaluation of an F-generator of a length at most $r_1 + r_2$ or $r_1 + r_2 + 1$ for $A_1 A_2$ given F_i-generators of lengths r_i for A_i, $i = 1, 2$, where the operator F is defined by the operators F_1 and F_2, based on displacement and scaling, and where r_1 and r_2 are small (cf. [P90] or [BP94]).

Later on, it was observed [GKO94] that the reduction from a Toeplitz-like linear system to a Cauchy-like linear system can be very simply achieved by using FFT. A substantial advantage of such a reduction is that pivoting (by means of row and column permutation) for numerical stabilization of Cauchy-like linear system does not destroy its structure. (The reader may observe that this property holds for the Cauchy matrix (10.2) but does not generally hold for Toeplitz and Hankel matrices.) Such observations enabled I. Gohberg, T. Kailath, and V. Olshevsky in [GKO94] to design effective and numerically stable practical algorithm for solving Toeplitz and Toeplitz-like linear systems. Later on, Ming Gu obtained a further (though relatively very minor) amelioration.

11 P-adic (Newton-Hensel's) Lifting.

In this section, we will recall an effective method from the area of symbolic computation that outputs the solution x to (1.1) reduced modulo a prime power p^k, that is, the vector $A^{-1}b \bmod p^k$, given a matrix A and a vector b with integer entries. The vector x can be easily recovered from $x \bmod p^k$ (with k of the order of $n \log A_2 / \log p$) by means of the continued fraction approximation algorithm [HW79].

This iterative method (called p-adic, Hensel's and Newton-Hensel's) reduces modulo p or p^2 most of the order of n^2 ops involved in each iteration, so that only $O(n)$ ops per iteration are reduced mod p^i for large i. Since p is relatively small, the computations modulo p and p^2 can be performed with a lower precision.

Here is a desired effective solution from [MC79], where p-adic lifting is also proposed for matrix inversion modulo p^k.

Input: Prime p, natural k, vector b, matrix A, such that $\det A \neq 0 \bmod p$.

Output: $A^{-1}b \bmod p^k$.

Step 0 (initialization). Compute

$$S(0) = A^{-1} \bmod p, \quad x_1 = S(0)b \bmod p, \quad v_1 = Ax_1 \bmod p^2.$$

Step $j, j = 1, \cdots, k$. Successively compute the vectors:

$$\begin{aligned} r_i &= b - v_i \bmod p^{i+1} \ , \\ z_i &= (S(0)w_i / p^i) \bmod p^2 \ , \\ v_{i+1} &= v_i + p^i A y_i \bmod p^{i+2} \ , \\ x_{i+1} &= x_i + p^i y_1 \bmod p^{i+1} \ . \end{aligned}$$

Note that only two matrix-by-matrix multiplications are needed in each step i, that is, of $S(0)$ by $(w_i/p^i) \bmod p^2$ and of A by y_i, and both of these vectors, w_i/p^i and y_i, have to be reduced modulo p^2, whereas all other computations of i involve $O(n)$ ops.

References

[A94] O. Axelsson, *Iterative Solution Methods*, Cambridge University Press, Cambridge, 1994.

[AG88] G.S. Ammar and W.G. Gragg, Superfast Solution of Real Positive Definite Toeplitz Systems, *SIAM J. Matrix Anal. Appl.*, 9, 1, 61-67, 1988.

[AHU76] A.V. Aho, J.E. Hopcroft, and J.D. Ullman, *The Design and Analysis of Computer Algorithms*, Addison-Wesley, Reading, Massachusetts, 1976.

[BA80] R.R. Bitmead and B.D.O. Anderson, Asymptotically Fast Solution of Toeplitz and Related Systems of Linear Equations, *Linear Algebra and Its Applics.*, 34, 103-116, 1980.

[BDB90] D. Bini and F. Di Benedetto, A New Preconditioner for the Parallel Solution of Positive Definite Toeplitz Systems, *Proc. 2nd Ann. ACM Symp. on Parallel Algorithms and Architectures*, 220-223, ACM Press, New York, 1990.

[BGY80] R.P. Brent, F.G. Gustavson, and D.Y.Y. Yun, Fast Solution of Toeplitz Systems of Equations and Computation of Pade Approximations, *J. of Algorithms*, 1, 259-295, 1980.

[BLRu90] M. Blum, M. Luby, and R. Rubinfeld, Self-Testing/Correcting with Applications to Numerical Problems, *Proc. 22nd Ann. ACM Symp. on Theory of Computing*, 73-83, ACM Press, New York, 1990.

[BP94] D. Bini and V.Y. Pan, *Polynomial and Matrix Computations, Volume 1: Fundamental Algorithms*, Birkhäuser, Boston, 1994.

[Bun85] J.R. Bunch, Stability of Methods for Solving Toeplitz Systems of Equations, *SIAM J. on Scientific and Statistical Computing*, 6, 2, 349-364, 1985.

[CB90] G. Cybenko and M. Berry, Hyperbolic Householder Algorithms for Factoring Structured Matrices, *SIAM J. Matrix Anal. Appl.*, 11, 4, 499-520, 1990.

[CW90] D. Coppersmith and S. Winograd, Matrix Multiplication via Arithmetic Progressions, *J. of Symbolic Computations*, 9, 3, 251-280, 1990 (short version in *Proc. 19th Ann. ACM Symp. on Theory of Computing*, 1-6, ACM Press, New Yok, 1987).

[dH87] F.R. deHoog, On the Solution of Toeplitz Systems, *Linear Algebra and Its Applics.*, 88/89, 123-138, 1987.

[FGN92] R.W. Freund, G.H. Golub, and N.M. Nachtigal, Iterative Solution of Linear Systems, *Acta Numerica*, 1, 57-100, 1992.

[GeLi81] J.A. George and J.W. Liu, *Computer Solution of Large Sparse Positive Definite Systems*, Prentice-Hall, New Jersey, 1981.

[GKO94] I. Gohberg, T. Kailath, and V. Olshevsky, Fast Gaussian Elimination with Partial Pivoting for Matrices with Displacement Structure, *submitted for publication*, 1994.

[GL89] G.H. Golub and C.F. van Loan, *Matrix Computations*, Johns Hopkins Univ. Press, Baltimore, Maryland, 1989 (new edition to appear).

[GS72] I.C. Gohberg and A.A. Semencul, On the Inversion of Finite Toeplitz Matrices and Their Continuous Analogs, *Mat. Issled.*, 2, 201-233 (in Russian), 1972.

[GT87] J.R. Gilbert and R.E. Tarjan, The Analysis of a Nested Dissection Algorithm, *Numer. Math.*, 50, 377-404, 1987.

[Ha88] H. Hafsteinsson, Parallel Sparse Choleski Factorization, Ph.D. Thesis and Tech. Report, TR 88-940, *Computer Science Dept.*, *Cornell Univ.*, 1988.

[HT82] W. Hackbusch and U. Trottenberg (eds.), *Multigrid Methods*, Springer's Lecture Notes in Math., vol. 960, 1982.

[HW79] G.H. Hardy and E.M. Wright *An Introduction to the Theory of Numbers*, Oxford Univ. Press, Oxford, 1979.

[Io82] I.S. Iohvidov, *Hankel and Toeplitz Matrices and Forms, Algebraic Theory*, Birkhaeuser, Boston, 1982.

[Kai87] T. Kailath, Signal Processing Applications of Some Moment Problems, *Proc. AMS Symp. in Applied Math.*, 37, 71-100, 1987.

[KKM79] T. Kailath, S.-Y. Kung, and M. Morf, Displacement Ranks of Matrices and Linear Equations, *J. Math. Anal. Appl.*, 68, 2, 395-407, 1979.

[LRT79] R.J. Lipton, D. Rose, and R.E. Tarjan, Generalized Nested Dissection, *SIAM J. on Numerical Analysis*, 16, 2, 346-358, 1979.

[MC79] R.T. Moenck and J.H. Carter, Approximate Algorithms to Derive Exact Solutions to Systems of Linear Equations, *Proc. EUROSAM, Lecture Notes in Computer Science*, 72, 63-73, Springer, Berlin, 1979.

[McC87] S. McCormick, editor, *Multigrid Methods*, SIAM, Philadelphia, 1987.

[Morf80] M. Morf, Doubling Algorithms for Toeplitz and Related Equations, *Proc. IEEE Internat. Conf. on ASSP*, 954-959, IEEE Computer Society Press, 1980.

[Mus81] B.R. Musicus, Levinson and Fast Choleski Algorithms for Toeplitz and Almost Toeplitz Matrices, Internal Report, *Lab. of Electronics, M.I.T.*, Cambridge, Massachusetts, 1981.

[P84] V.Y. Pan, *How to Multiply Matrices Faster*, Lecture Notes in Computer Science, 179, Springer Verlag, Berlin, 1984.

[P90] V.Y. Pan, On Computations with Dense Structured Matrices, *Math. of Computation*, 55, 191, 179-190, 1990.

[P91] V.Y. Pan, Complexity of Algorithms for Linear Systems of Equations, in *Computer Algorithms for Solving Linear Algebraic Equations (The State of the Art)*, edited by E. Spedicato, NATO ASI Series, Series F: Computer and Systems Science, 77, 27-56, Springer, Berlin, 1991.

[P87/92] V.Y. Pan, Linear Systems of Equations, in *Encyclopedia of Physical Sciences and Technology* (Marvin Yelles edit.), 7, 304-329, 1987 (first edition) and 8, 779-804, 1992 (second edition), Academic Press, San Diego, California.

[P92] V.Y. Pan, Complexity of Computations with Matrices and Polynomials, *SIAM Rev.*, 34, 9, 225-262, 1992.

[P92a] V.Y. Pan, Parametrization of Newton's Iteration for Computations with Structured Matrices and Applications, *Computers and Mathematics (with Applications)*, 24, 3, 61-75, 1992.

[P93] V.Y. Pan, Parallel Solution of Sparse Linear and Path Systems, in Synthesis of Parallel Algorithms (J.H. Reif editor), Chapter 14, pp.621-678. Morgan Kaufmann Publishers, San Mateo, California, 1993.

[P93a] V.Y. Pan, Concurrent Iterative Algorithm for Toeplitz-like Linear Systems, *IEEE Trans. on Parallel and Distributive Systems* 4, 5, 592-600, 1993.

[P94] V.Y. Pan, Improved Parallel Solution of a Triangular Linear System, *Computers and Math. (with Applications)*, 27, 11, 41-43, 1994.

[Pease68] M. Pease, An Adaptation of the Fast Fourier Transform for Parallel Processing, *J. of ACM*, 15, 252-264, 1968.

[PP95] V.Y. Pan and F.P. Preparata, Work-Preserving Speed-up of Parallel Matrix Computations, *SIAM J. on Computing*, 24, 4, 811-821, 1995.

[PR89] V.Y. Pan and J. Reif, Fast and Efficient Solution of Path Algebra Problems, *J. Computer and Systems Sciences*, 38, 3, 494-510, 1989.

[PR91] V.Y. Pan and J. Reif, The Parallel Computation of the Minimum Cost Paths in Graphs by Stream Contraction, *Information Processing Letters*, 40, 79-83, 1991.

[PR92] V.Y. Pan and J. Reif, Compact Multigrid, *SIAM J. Sci. Stat. Comput.*, 13, 1, 119-127, 1992.

[PR93] V.Y. Pan and J. Reif, Generalized Compact Multigrid, *Computers and Mathematics (with Applications)*, 25, 9, 3-5, 1993.

[PS91] V.Y. Pan and R. Schreiber, An Improved Newton Iteration for the Generalized Inverse of a Matrix, with Applications, *SIAM J. Sci. Stat. Comput.*, 12, 5, 1109-1131, 1991.

[Q94] M.J. Qiunn, *Parallel Computing: Theory and Practice*, McGraw-Hill, New York, 1994.

[St69] V. Strassen, Gaussian Elimination is Not Optimal, *Numer. Math.*, 13, 354-356, 1969.

[Strang86] G. Strang, A Proposal for Toeplitz Matrix Calculations, *Stud. Appl. Math.*, 74, 171-176, 1986.

[T90] W.F. Trench, A Note on a Toeplitz Inversion Formula, *Linear Algebra Appl.*, 29, 55-61, 1990.

Block Iterative Methods for Reduced Systems of Linear Equations

D. J. EVANS
Parallel Algorithms Research Centre, Loughborough University
Loughborough LE11 3TU, U.K.

Abstract

We study the numerical solution of a steady convection-diffusion equation in which the convection term is dominant. A nine-point fourth order difference scheme is examined and the well known block successive overrelaxation (BSOR) and the block alternating group explicit (BLAGE) iterative methods are considered for solving the sparse unsymmetric linear system. The size of the linear system is reduced by the Strides of 3 algorithm and the BSOR and BLAGE methods are applied to the reduced system of linear equations. This leads to greater convergence for both methods. In addition, a great deal of parallelism can be achieved by using BLAGE.

Key Words: nine-point discretisation, convection-diffusion equation, block successive overrelaxation, block alternating group explicit, Strides of 3 reduction algorithm

C.R Categories: G 1.3 , G 1.7

1 Block Alternating Group Explicit Method

An iterative method for the solution of large block tridiagonal linear systems is the block alternating group explicit (BLAGE) method of Evans and Yousif [1]. We consider the linear system of equations of the form

$$Au \equiv (G_1 + G_2)\,u = f \;, \tag{1}$$

G. Winter Althaus and E. Spedicato (eds.), Algorithms for Large Scale Linear Algebraic Systems, 25–35.

where G_1 and G_2 are $n^2 \times n^2$ sparse matrices. We suppose that n is odd and the exact forms of G_1 and G_2 are given by

$$G_1 = \begin{bmatrix} \frac{1}{2}M & & & & & \\ & \frac{1}{2}M & N & & & \\ & N & \frac{1}{2}M & & & \\ & & & \ddots & & \\ & & & & \frac{1}{2}M & N \\ & & & & N & \frac{1}{2}M \end{bmatrix}$$

and

$$G_2 = \begin{bmatrix} \frac{1}{2}M & N & & & & \\ N & \frac{1}{2}M & & & & \\ & & \ddots & & & \\ & & & \frac{1}{2}M & N & \\ & & & N & \frac{1}{2}M & \\ & & & & & \frac{1}{2}M \end{bmatrix}$$

where M and N are $n \times n$ tridiagonal matrices. We write

$$A = G_1 + G_2 = blocktridiag[\, N \,,\, M \,, N\,]$$

The iteration of the well known successive block over relaxation (SBOR) method are given by:

$$\begin{aligned} u_1^{(k+1)} &= u_1^{(k)} + \omega M^{-1}(Nu_2^{(k)} + f_1 - Mu_1^{(k)}) \\ u_j^{(k+1)} &= u_j^{(k)} + \omega M^{-1}(Nu_{j-1}^{(k+1)} + Nu_{j+1}^{(k)} + f_j - Mu_j^{(k)}) \quad j = 2,3,\ldots n-1 \\ u_n^{(k+1)} &= u_n^{(k)} + \omega M^{-1}(Nu_{n-1}^{(k+1)} + f_n - Mu_n^{(k)}) \end{aligned} \tag{2}$$

whilst the iteration of the BLAGE method for solving the linear system (1) is given by

$$\begin{aligned} (G_1 + \omega I)\, u^{(k+\frac{1}{2})} &= f - (G_2 - \omega I)\, u^{(k)} \\ (G_2 + \omega I)\, u^{(k+1)} &= f - (G_1 - \omega I)\, u^{(k+\frac{1}{2})} \end{aligned} \tag{3}$$

We suppose that $(G_1 + \omega I)$ and $(G_2 + \omega I)$ are non-singular for any $\omega > 0$. We denote $M_1 = \frac{1}{2}M - \omega I$ and $M_2 = \frac{1}{2}M + \omega I$, then following Evans and

Yousif [1], we obtain the following algorithm:-

First sweep

First step

For i = 1 : n-2 step 2

$$z_i^{(k)} = f_i - M_1 u_i^{(k)} - N u_{i+1}^{(k)} \quad (4)$$

$$z_{i+1}^{(k)} = f_{i+1} - N u_i^{(k)} - M_1 u_{i+1}^{(k)} \quad (5)$$

endfor

$$z_n^{(k)} = f_n - M_1 u_n^{(k)} \quad (6)$$

Second step

$$M_2 u_1^{(k+\frac{1}{2})} = z_1^{(k)} \quad (7)$$

For i = 2 : n-1 step 2

$$M_2 u_i^{(k+\frac{1}{2})} + N u_{i+1}^{(k+\frac{1}{2})} = z_i^{(k)} \quad (8)$$

$$N u_i^{(k+\frac{1}{2})} + M_2 u_{i+1}^{(k+\frac{1}{2})} = z_{i+1}^{(k)} \quad (9)$$

endfor

Second sweep

First step

$$z_1^{(k+\frac{1}{2})} = f_1 - M_1 u_1^{(k+\frac{1}{2})} \tag{10}$$

For i = 2 : n-1 step 2

$$z_i^{(k+\frac{1}{2})} = f_i - M_1 u_i^{(k+\frac{1}{2})} - N u_{i+1}^{(k+\frac{1}{2})} \tag{11}$$

$$z_{i+1}^{(k+\frac{1}{2})} = f_{i+1} - N u_i^{(k+\frac{1}{2})} - M_1 u_{i+1}^{(k+\frac{1}{2})} \tag{12}$$

endfor

Second step

For i = 1 : n-2 step 2

$$M_2 u_i^{(k+1)} + N u_{i+1}^{(k+1)} = z_i^{(k+\frac{1}{2})} \tag{13}$$

$$N u_i^{(k+1)} + M_2 u_{i+1}^{(k+1)} = z_{i+1}^{(k+\frac{1}{2})} \tag{14}$$

endfor

$$M_2 u_n^{(k+1)} = z_n^{(k+\frac{1}{2})} \tag{15}$$

It can be said that BLAGE needs twice as many operations as the BSOR method. However there is some kind of natural parallelism in the BLAGE method. For example, we can compute the values of $z_i^{(k)}$ by Eqs.(4-6) at the same time using n processors. Similarly Eqs.(8-9) can be solved for i= 2, 4, 6, ..., n-1 at the same time using different processors. A similar argument holds for each of Eqs.(10-12) and Eqs.(13-14). Thus much saving in computational time can be achieved by parallellisation of the BLAGE iterations.

We now outline a method to solve Eqs.(7-9).
Eq.(7) can be easily solved as the matrix $M_2 = \frac{1}{2}M + \omega I$ is tridiagonal. We can write Eqs.(8) and (9) in the form

$$u_i^{(k+\frac{1}{2})} + Pu_{i+1}^{(k+\frac{1}{2})} = M_2^{-1} z_i^{(k)} \tag{16}$$

$$Pu_i^{(k+\frac{1}{2})} + u_{i+1}^{(k+\frac{1}{2})} = M_2^{-1} z_{i+1}^{(k)} \tag{17}$$

where $P = M_2^{-1}N$. We multiply Eq.(17) by P and subtract Eq.(16) from the product, obtaining

$$\begin{aligned} \left(P^2 - I\right) u_i^{(k+\frac{1}{2})} &= PM_2^{-1} z_{i+1}^{(k)} - M_2^{-1} z_i^{(k)} \qquad (18) \\ (P-I)(P+I)\, u_i^{(k+\frac{1}{2})} &= M_2^{-1} N M_2^{-1} z_{i+1}^{(k)} - M_2^{-1} z_i^{(k)} \\ \left(M_2^{-1}N - I\right)\left(M_2^{-1}N + I\right) u_i^{(k+\frac{1}{2})} &= M_2^{-1} N M_2^{-1} z_{i+1}^{(k)} - M_2^{-1} z_i^{(k)} \\ (N - M_2)\, M_2^{-1} (N + M_2)\, u_i^{(k+\frac{1}{2})} &= N M_2^{-1} z_{i+1}^{(k)} - z_i^{(k)} \qquad (19) \end{aligned}$$

To solve Eq.(19), we let $\psi_i^{(k+\frac{1}{2})} = M_2^{-1}(N + M_2)\, u_i^{(k+\frac{1}{2})}$, thus we have to solve

$$(N - M_2)\, \psi_i^{(k+\frac{1}{2})} = N M_2^{-1} z_{i+1}^{(k)} - z_i^{(k)}$$

Since $(N - M_2)$ is a tridiagonal matrix, we can easily obtain $\psi_i^{(k+\frac{1}{2})}$ and then we have

$$(N + M_2)\, u_i^{(k+\frac{1}{2})} = M_2 \psi_i^{(k+\frac{1}{2})} \tag{20}$$

Thus $u_i^{(k+\frac{1}{2})}$ can be obtained from Eq.(20) by applying another tridiagonal solver. Finally, $u_{i+1}^{(k+\frac{1}{2})}$ is obtained from

$$u_{i+1}^{(k+\frac{1}{2})} = M_2^{-1} z_{i+1}^{(k)} - Pu_i^{(k+\frac{1}{2})} \tag{21}$$

To solve Eqs.(13-14), we proceed in a similar way as described above.

2 Strides of 3 reduction algorithm

The strides of 3 (SR3) reduction algorithm introduced by Evans [2, 3] is an extension of the well known odd-even cyclic reduction algorithm for the solution of tridiagonal linear systems. In this section, we apply one step of the SR3 algorithm to the block tridiagonal system to obtain a reduced system of approximately one-third the size of the full system. The reduced system is then solved by the BLAGE and BSOR methods. The resulting iterative techniques are denoted by SR3-BLAGE and SR3-BSOR respectively.

The SR3 reduction

Consider the following block tridiagonal system:

$$\begin{bmatrix} K & L & & & \\ L & K & L & & \\ & \ddots & \ddots & \ddots & \\ & & L & K & L \\ & & & L & K \end{bmatrix} \begin{bmatrix} u_1 \\ u_2 \\ \vdots \\ u_{n-1} \\ u_n \end{bmatrix} = \begin{bmatrix} f_1 \\ f_2 \\ \vdots \\ f_{n-1} \\ f_n \end{bmatrix} \tag{22}$$

Let us suppose that n is either of the form 3^p+2 or $2(3^p)-1$. To minimise the number of matrix multiplications and thus avoid troublesome rounding-off errors, we express Eq.(22) in the form shown below with $C = K^{-1}L$ and $d_i = K^{-1}f_i$ (i= 1, 2, ..., n).

$$\begin{bmatrix} I & C & & & \\ C & I & C & & \\ & \ddots & \ddots & \ddots & \\ & & C & I & C \\ & & & C & I \end{bmatrix} \begin{bmatrix} u_1 \\ u_2 \\ \vdots \\ u_{n-1} \\ u_n \end{bmatrix} = \begin{bmatrix} d_1 \\ d_2 \\ \vdots \\ d_{n-1} \\ d_n \end{bmatrix} \tag{23}$$

We now present the SR3 algorithm. Consider the second, third and fourth equations of (22) which are given by

$$\begin{aligned} Cu_1 + u_2 + Cu_3 &= d_2 && (24) \\ Cu_2 + u_3 + Cu_4 &= d_3 && (25) \\ Cu_3 + u_4 + Cu_5 &= d_4 && (26) \end{aligned}$$

We multiply Eqs.(24) and (26) each by the matrix C and add. We then subtract Eq.(25) from this sum to give

$$C^2 u_1 + \left(2C^2 - I\right) u_3 + C^2 u_5 = C d_2 + C d_4 - d_3 \tag{27}$$

In a similar way, we also have

$$C^2 u_3 + \left(2C^2 - I\right) u_5 + C^2 u_7 = C d_4 + C d_6 - d_5 \tag{28}$$

We add Eq.(27) to Eq.(28), thus obtaining

$$C^2 u_1 + \left(3C^2 - I\right) u_3 + \left(3C^2 - I\right) u_5 + C^2 u_7 = C d_2 + 2C d_4 + C d_6 - d_3 - d_5 \tag{29}$$

In addition we have the equation

$$C u_3 + u_4 + C u_5 = d_4 \tag{30}$$

Multiplying Eq.(30) by $(3C^2 - I)$ and multiplying Eq.(29) by C, we have by subtraction the following equation involving only u_1, u_4 and u_7.

$$C^3 u_1 + \left(I - 3C^2\right) u_4 + C^3 u_7 = C^2 \left(d_2 + d_6 - d_4\right) + d_4 - C \left(d_3 + d_5\right) \tag{31}$$

It is thus seen that the representative equation for the SR3 reduction is

$$C^3 u_{i-3} + \left(I - 3C^2\right) u_i + C^3 u_{i+3} = C^2 \left(d_{i-2} + d_{i+2} - d_i\right) + d_i - C \left(d_{i-1} + d_{i+1}\right) \tag{32}$$

for i= 3, 6, ..., 3q where $q = 3^{p-1}$ when $n = 3^p + 2$ and $q = 2\left(3^{p-1}\right) - 1$ when $n = 2\left(3^p\right) - 1$.

The SR3-BLAGE algorithm

Now Eqs.(4-15) define the SR3-BLAGE iterations with n replaced by q, $M = I - 3C^2$ and $N = C^3$. Thus the matrices M_1 and M_2 are given by $M_1 = \left(\frac{1}{2} - \omega\right) I - \frac{3}{2}C^2$ and $M_2 = \left(\frac{1}{2} + \omega\right) I - \frac{3}{2}C^2$. In this case Eqs.(8-9) can be solved as follows. We have

$$M_2 u_i^{\left(k+\frac{1}{2}\right)} + N u_{i+1}^{\left(k+\frac{1}{2}\right)} = z_i^{(k)} \tag{33}$$

$$N u_i^{\left(k+\frac{1}{2}\right)} + M_2 u_{i+1}^{\left(k+\frac{1}{2}\right)} = z_{i+1}^{(k)} \tag{34}$$

We remark that in this case, matrices M_2 and N commute. Thus, multiplying Eq. (33) by M_2 , Eq. (34) by N and using subtraction gives

$$(M_2 - N)(M_2 + N)\, u_i^{(k+\frac{1}{2})} = M_2 z_i^{(k)} - N z_{i+1}^{(k)} \tag{35}$$

Then

$$\left(C^3 + \frac{3}{2}C^2 - \left(\frac{1}{2} + r\right) I\right) \psi_i^{(k+\frac{1}{2})} = \tau_i \ , \tag{36}$$

where $\psi_i^{(k+\frac{1}{2})} = (M_2 + N)\, u_i^{(k+\frac{1}{2})}$ and $\tau_i = N z_{i+1}^{(k)} - M_2 z_i^{(k)}$. We factorise $\left(C^3 + \frac{3}{2}C^2 - \left(\frac{1}{2} + r\right) I\right)$ in the form $(C - \alpha)(C - \beta)\left(C - \overline{\beta}\right)$ where β and $\overline{\beta}$ are complex conjuguates and α is real.
Therefore we have to solve

$$\begin{aligned} \left(K^{-1}L - \alpha\right)\left(K^{-1}L - \beta\right)\left(K^{-1}L - \overline{\beta}\right) \psi_i^{(k+\frac{1}{2})} &= \tau_i & (37) \\ (L - \alpha K)\, K^{-1} (L - \beta K)\, K^{-1} \left(L - \overline{\beta} K\right) \psi_i^{(k+\frac{1}{2})} &= K\tau_i & (38) \end{aligned}$$

Now $(L - \alpha K)$, $(L - \beta K)$ and $\left(L - \overline{\beta} K\right)$ are all tridiagonal matrices so that Eq.(38) can be solved for $\psi_i^{(k+\frac{1}{2})}$.
To obtain $u_i^{(k+\frac{1}{2})}$ from $(M_2 + N)\, u_i^{(k+\frac{1}{2})} = \psi_i^{(k+\frac{1}{2})}$, we factorise $M_2 + N = \left(C^3 - \frac{3}{2}C^2 + \left(\frac{1}{2} + r\right) I\right)$ in the form $(C - \gamma)(C - \delta)\left(C - \overline{\delta}\right)$ and we proceed in a similar way as discussed above.
If $u_i^{(k+\frac{1}{2})}$ is known, then $u_{i+1}^{(k+\frac{1}{2})}$ is obtained from Eq.(33).
Now suppose that the SR3-BLAGE method has converged to the solution $(u_3, u_6, u_9, \ldots, u_{3q})$. Then u_1 and u_2 are determined from the matrix equation

$$\begin{bmatrix} K & L \\ L & K \end{bmatrix} \begin{bmatrix} u_1 \\ u_2 \end{bmatrix} = \begin{bmatrix} f_1 \\ f_2 - L u_3 \end{bmatrix} \tag{39}$$

Eq.(39) can be solved in a procedure analogous to the solution of Eqs.(8-9). Then the remaining unknowns are determined from Eq.(22) i.e.,

$$\begin{aligned} u_{3i+1} &= L^{-1}\left(f_{3i} - K u_{3i} - L u_{3i-1}\right) \\ u_{3i+2} &= L^{-1}\left(f_{3i+1} - K u_{3i+1} - L u_{3i}\right) \end{aligned} \tag{40}$$

for i = 1, 2, 3, ..., q-1 by a forward recurrence.
Finally, u_{n-1} and u_n are obtained from an equation analogous to Eq. (39).

The SR3-BSOR iterations can be obtained easily. Numerical experiments indicate that the SR3-BSOR iterations do not suffer from matrix multiplications so that the BSOR method can be applied to the reduced block tridiagonal system in their original form as given in [2].

3 Computational Experiments

In this section we report on the numerical experiments carried out to solve the convection-diffusion equation:

$$-\epsilon\,\Delta u + u_x = 0 \tag{41}$$

($\epsilon > 0$) on the unit square $(0,1)\times(0,1)$ subject to the boundary conditions:

$$u(0,y) = \sin \pi y \quad , \; u(1,y) = 2\sin \pi y \quad 0 \le y \le 1$$

$$u(x,0) = u(x,1) = 0 \quad 0 \le x \le 1 \tag{42}$$

using a fourth order nine-point difference scheme due to Jain et. al [4]. The exact solution is given by

$$u(x,y) = e^{\frac{x}{2\epsilon}} \frac{\sin \pi y}{\sinh \sigma}\left(2e^{-\frac{1}{2\epsilon}} \sinh \sigma x + \sinh \sigma (1-x)\right)$$

where $\sigma^2 = \pi^2 + \frac{1}{4\epsilon^2}$.
Table 1 shows the number of iterations required for convergence to a tolerance of 10^{-6} by the different iterative techniques described above. We remark that considerable gain in convergence rates can be obtained when the iterative techniques are applied to the reduced system. The number of multiplicative operations done per iteration step are approximately equal to $14n^2$ (BSOR), $45n^2$ (BLAGE), $10n^2$ (SR3-BSOR) and $50n^2$ (SR3-BLAGE). The extra work needed for computing the right hand side of the reduced system and to recover all the unknowns amounts to approximately $23n^2$. For example, in the case $h^{-1} = 30$, BSOR requires approximately 506 thousands of operations where as SR3-BSOR requires only about 154 thousands of multiplicative operations. The gains in convergence rates by solving the reduced system are 60% for BSOR and 50% for BLAGE.

h^{-1}	BSOR		BLAGE		SR3-BSOR		SR3-BLAGE	
	niter	ω^*	niter	ω^*	niter	ω^*	niter	ω^*
12	18	1.33-1.35	13	3.7-3.8	7	1.02-1.06	7	0.22-0.26
18	26	1.45-1.48	19	2.5-2.6	10	1.1-1.2	9	0.16-0.18
30	43	1.64-1.65	27	1.795-1.796	16	1.25-1.3	13	0.1-0.105
54	74	1.78-1.80	49	0.8-0.815	26	1.476-1.48	21	0.55-0.6
84	111	1.85-1.855	66	0.644-0.6455	40	1.62-1.625	29	0.04-0.045

Table 1: Number of iterations ($\epsilon = 10^{-1}$)

4 Conclusion

The block alternating group explicit method requires fewer number of iterations for convergence than the block successive overrelaxation method with considerable gain in convergence rates for small mesh sizes. Though the cost for BLAGE is more, the BLAGE iterations have an inherent parallelism, thus much computational time can be saved on parallel architectures.

We have applied two iterative techniques to a reduced sytem obtained by applying one step of the Stride of 3 algorithm. We have considered a strongly unsymmetric linear system and our results show that we obtain efficient iterative algorithms. Block tridiagonal matrices typically arise in finite difference or finite element discretisation of second-order elliptic problems and the techniques described in this paper can be successfully implemented in other cases.

References

[1] **D.J. Evans** and **W.S. Yousif**, The block alternating group explicit method (BLAGE) for the solution of elliptic difference equations, *Int. J. Comp. Math*, **22**, (1987), 177-185.

[2] **D.J. Evans**, The Strides Reduction algorithms for solving tridiagonal linear systems, *Int. J. Comp. Math*, **41**, (1992), 237-250

[3] **D.J. Evans**, Parallel algorithms in Numerical Linear Algebra, Computer Studies **734**, (1992), Loughborough University of Technology.

[4] **M.K. Jain**, **R.K. Jain** and **R.K. Mohanty**, Fourth order difference method for elliptic equations with nonlinear first derivative terms, *Numerical Methods for Partial Differential Equations*, **5**, (1989), 87-95.

PARALLEL IMPLICIT SCHEMES FOR THE SOLUTION OF LINEAR SYSTEMS

D. J. EVANS
Parallel Algorithms Research Centre, Loughborough University
Loughborough LE11 3TU, U.K.

Abstract
In this report we discuss the results of the parallel implementation of two implicit schemes for the solution of large and dense linear systems: the Gaussian Implicit Elimination and a new implicit version of the well known Gauss-Jordan method.
We present a complete computational analysis of both methods and we explain the superiority of implicit schemes on the base of the number of accesses to the problem data.
A new pivoting strategy for the implicit methods is presented. We prove that this strategy gives better results than the classical partial pivoting used with Gaussian Elimination.

Keyword
Gaussian elimination, Gauss-Jordan elimination, implicit elimination, shared memory architectures, partial pivoting, implicit partial pivoting, linear systems, LU factorisation, WZ factorisation.

1. Introduction

The central problem of linear algebra is the solution of linear equations. This problem arises in many scientific, engineering and economics topics. It is usually represented by the following notation

$$Ax = b \tag{1}$$

where A is the coefficient matrix, x is the vector of the unknowns and b is the right hand side vector, i.e. the known vector. In this report we only deal with problems where A is a square, dense and without any special pattern. Furthermore we assume that the dimension n of A is even.
The direct methods used to solve (1) are based on a factorisation of the coefficient matrix A into factors that are easy to invert. The Gaussian Elimination (i.e. the LU factorisation) is optimal between the methods involving only linear combinations of rows and columns [5] and this explains why it is the most popular algorithm both in its sequential and parallel versions.
Evans [1] introduced a scheme that simultaneously eliminates two elements at a time

G. Winter Althaus and E. Spedicato (eds.), Algorithms for Large Scale Linear Algebraic Systems, 37–54.

instead of just one as in Gaussian Elimination and it is called Implicit Gaussian Elimination. Here we present a new scheme that is the implicit version of the Gauss-Jordan method.
In this paper we compare these two pairs of algorithms (Gauss and Implicit Gauss, Gauss-Jordan and Implicit Gauss-Jordan) and their parallel implementation on a shared memory multiprocessor computer.
The implicit methods also need, like the classical algorithm, a pivoting procedure, i.e. a permutation of rows and/or columns to avoid the elements of the factorised matrix becoming arbitrarily large. Here we propose a new efficient pivoting strategy.

2. Elimination scheme survey

In this section we present all the four methods referred to above. For each of them we give the details of the sequential algorithm and their computational complexities. Then we present the parallel version of each sequential algorithm in quite a straight way, splitting the main loops and adding some synchronisation points. Then the number of accesses to the shared memory is determined.
With regard to the data distribution, we assume that each processor can access the whole matrix A, but that at each step we limit its scope to a continuous group of rows obtained by subdividing the working area (i.e. the rows we need to process in a certain iteration) evenly among all the processors.

2.1 Gaussian Elimination

Gaussian Elimination is certainly the most popular numerical algorithm. It consists of two main phases: an "elimination phase" in which the matrix A is reduced to an upper triangular matrix and a "solution phase" in which the obtained triangular system is solved through a process called "back-substitution".

$$Ux = b' \qquad (2)$$

The system (2) is obtained by eliminating all the elements below the diagonal as described in the "elimination loop" of Figure 1. This procedure is equivalent to multiplying A by (n-1) elementary matrices as follows:

$$M^{(n-1)}M^{(n-2)}\dots M^{(1)}A$$

where

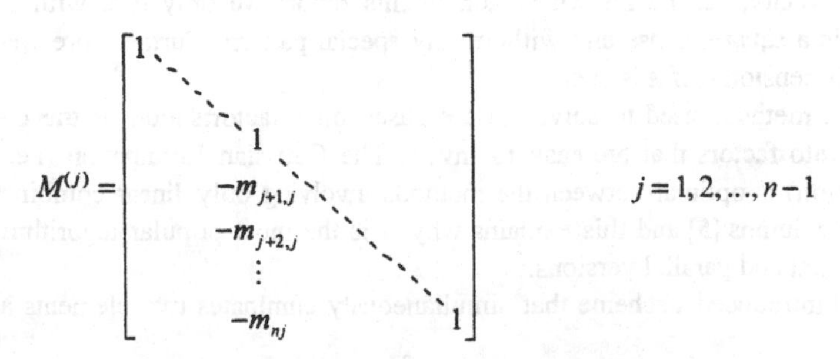

It can be easily observed that

$$L = \left[M^{(1)}\right]^{-1}\left[M^{(2)}\right]^{-1}\dots\left[M^{(n-1)}\right]^{-1}$$

is a unit lower triangular matrix, i.e. the Gaussian Elimination process is equivalent to the LU factorisation.

```
            (* elimination loop *)
            for j = 0, ..., n-1
PAR                 for i = j+1, ..., n-1

                            m = - a[ i,j] / a[ j,j]

                            (* update matrix A *)
                            for k = j+1, ..., n-1
                                    a[ i,k] = a[ i,k] + m* a[ j,k]

                            (* update right hand side *)
                            b[ i] = b[ i] + m* b[ j]

                    SYNC

            (* solution loop *)
            for j = n-1, ..., 0
                    x[ j] = b[ j] / a[ j,j]
                    (* update right hand side *)
PAR                 for i = j+1, ..., 0
                            b[ i] = b[ i] - a[ i,j] *x[ j]

                    SYNC
```

Figure 1 - Sequential Gaussian Elimination. No pivoting performed.

The computational complexity and the number of accesses to the shared memory, i.e. the number of times each processor needs to write or read data to/from the main memory, is shown in Table 1.

	Computational complexity(*)	Shared memory accesses
Elimination phase	$\left(\frac{n^3}{3}+\frac{n^2}{2}-\frac{5}{6}n\right)m+\left(\frac{n^3}{3}-\frac{n}{3}\right)a$	n^3+n^2-2n
Solution phase	$\left(\frac{n^2}{2}+\frac{n}{2}\right)m+\left(\frac{n^2}{2}-\frac{n}{2}\right)a$	$2n^2+n$
Total	$\left(\frac{n^3}{3}+n^2-\frac{n}{3}\right)m+\left(\frac{n^3}{3}+\frac{n^2}{2}-\frac{5}{6}n\right)a$	n^3+3n^2-n

Table 1 - Gaussian Elimination analysis.
(*) *m*: multiplications and divisions, *a*: additions and subtractions.

We can note that, while the number of floating point operation is $O(n^3/3)$, the number of accesses to the elements of the matrix A is $O(n^3)$. Most of the computations and accesses are located in the inner loop of the "elimination phase", where for each step j, i.e. for each of the n-1 transformations by the elementary matrices $M^{(j)}$, we need to update $(j-1)^2$ elements, see Figure 2.

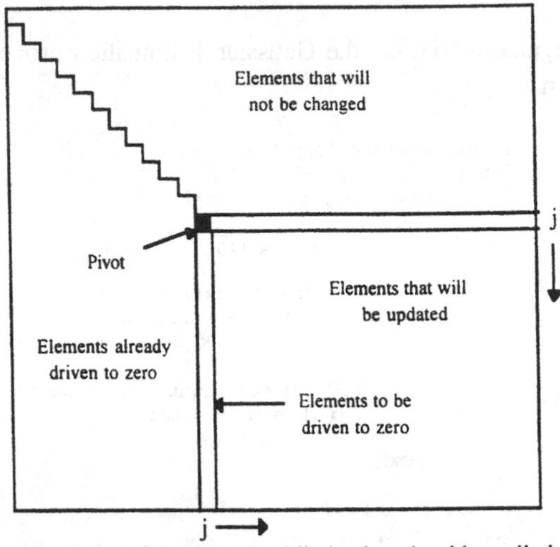

Figure 2 - During iteration j of the Gaussian Elimination algorithm. all elements in column j, for each row below row j, are driven to zero subtracting a multiple of row j from row i.

2.2 Implicit Gaussian Elimination

In the same way Gaussian Elimination is related to LU factorisation, Implicit Gaussian Elimination is related to WZ factorisation [1]. Also in this case the algorithm consists of two phases: an "elimination phase" in which the matrix A is transformed to a butterfly matrix Z as shown by (3), and a "solution phase" in which the system (4) is solved through a process called "bidirectional back substitution".

$$Z=\begin{bmatrix} a_{11}^{(0)} & a_{12}^{(0)} & a_{13}^{(0)} & \cdots & a_{1,n-2}^{(0)} & a_{1,n-1}^{(0)} & a_{1n}^{(0)} \\ & a_{22}^{(1)} & a_{23}^{(1)} & \cdots & a_{2,n-2}^{(1)} & a_{2,n-1}^{(1)} & \\ & & a_{33}^{(2)} & \cdots & a_{3,n-2}^{(2)} & & \\ & & & \cdots & & & \\ & & a_{n-2,3}^{(2)} & \cdots & a_{n-2,n-2}^{(2)} & & \\ & a_{n-1,2}^{(1)} & a_{n-1,3}^{(1)} & \cdots & a_{n-1,n-2}^{(1)} & a_{n-1,n-1}^{(1)} & \\ a_{n1}^{(0)} & a_{n2}^{(0)} & a_{n3}^{(0)} & \cdots & a_{n,n-2}^{(0)} & a_{n,n-1}^{(0)} & a_{nn}^{(0)} \end{bmatrix} \tag{3}$$

$$Zx = b' \tag{4}$$

The "elimination loop" of Figure 4 is equivalent to multiplying A by $n/2$-1 transformation matrices:

$$W^{(n/2-1)}\, W^{(n/2-2)} \ldots W^{(1)}\, A = Z$$

where

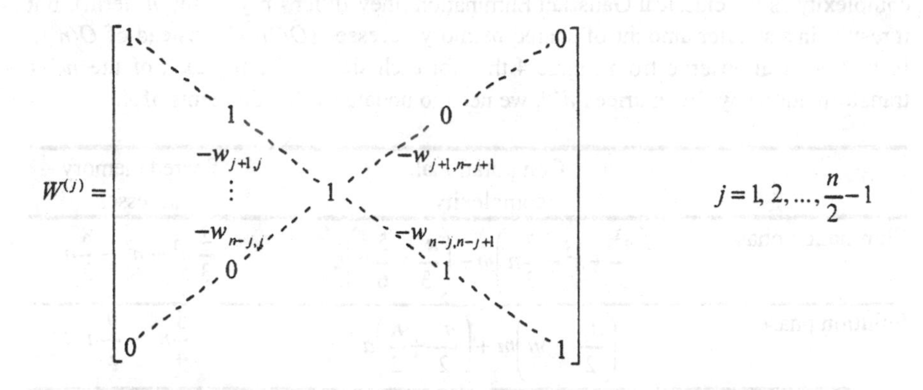

The w coefficients are computed two at a time solving n-$2j$ small 2×2 systems with w_{ij} and $w_{i,n-j+1}$ as unknowns.

```
                (* elimination loop *)
                for j = 0, 1, ..., n/2-2

                    j2 = n-j-1
                    det = a[ j,j]*a[ j2,j2] - a[ j2,j]*a[ j,j2]

PAR                 for i = j+1, ..., j2-1

                        (* compute the w coefficients *)
                        wj = (a[ j2,j]*a[ i,j2] - a[ j2,j2]*a[ i,j] )/det
                        wj2 = (a[ j,j2]*a[ i,j] - a[ j,j]*a[ i,j2] )/det

                        (* update matrix A *)
                        for k = j+1, ..., j2-1
                            a[ i,k] = a[ i,k] + wj*a[ j,k] + j2*a[ j2,k]

                        (* update right hand side *)
                        b[ i] = b[ i] + wj*b[ j] + wj2*b[ j2]

                        SYNC

                (* backward substitution loop *)
                for i = n/2-1, ..., 0
                    i2 = n-i-1

                    (* solve 2x2 system *)
                    det = a[ i,i]*a[ i2,i2] - a[ i,i2]*a[ i2,i]
                    x[ i] = (b[ i]*a[ i2,i2] - b[ i2]*a[ i,i2] ) / det
                    x[ i2] = (b[ i2]*a[ i,i] - b[ i]*a[ i2,i] ) / det

                    (* update upper right hand side *)
PAR                 for k = i+1, ..., 0
                        b[ k] = b[ k] - x[ i]*a[ k,i] - x[ i2]*a[ k,i2]

                    (* update lower right hand side *)
PAR                 for k = i2+1, ..., n-1
                        b[ k] = b[ k] - x[ i]*a[ k,i] - x[ i2]*a[ k,i2]

                    SYNC
```

Figure 3 - Sequential Implicit Gaussian Elimination. No pivoting performed.

The computational complexity and shared memory accesses are shown in Table 2. It can be seen immediately that Implicit Gaussian Elimination has the same asymptotic complexity as the classical Gaussian Elimination (they differ only in the n^2 term), but it results in a smaller amount of shared memory accesses ($O(2n^3/3)$ instead of $O(n^3)$). In fact, we can observe from Figure 4 that for each step j, i.e. for each of the $n/2$-1 transformations by the matrices $W^{(j)}$, we need to update $(n-2j)^2$ elements of A.

	Computational complexity	Shared memory accesses
Elimination phase	$\left(\frac{n^3}{3}+n^2-\frac{7}{3}n\right)m+\left(\frac{n^3}{3}-\frac{5}{6}n\right)a$	$\frac{2}{3}n^3+n^2-\frac{8}{3}n$
Solution phase	$\left(\frac{n^2}{2}+3n\right)m+\left(\frac{n^2}{2}+\frac{n}{2}\right)a$	$\frac{5}{4}n^2+\frac{9}{2}n$
Total	$\left(\frac{n^3}{3}+\frac{3}{2}n^2+\frac{2}{3}n\right)m+\left(\frac{n^3}{3}+\frac{n^2}{2}+\frac{n}{3}\right)a$	$\frac{2}{3}n^3+\frac{9}{4}n^2+\frac{11}{6}n$

Table 2 - Implicit Gaussian Elimination analysis

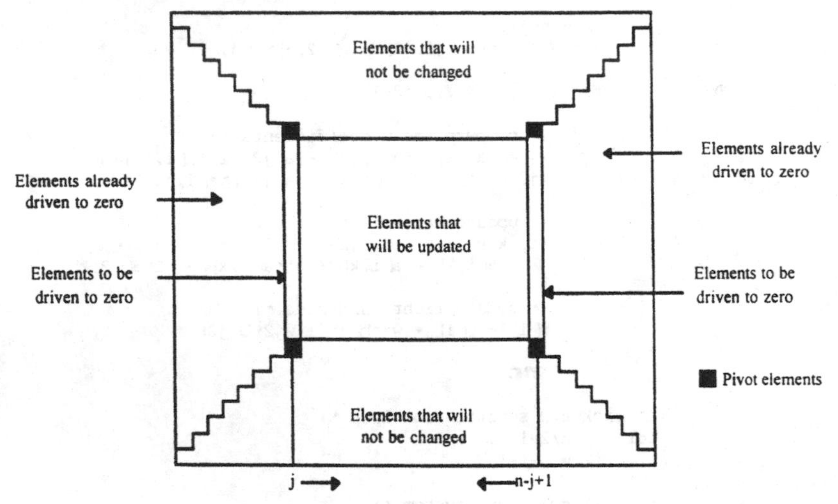

Figure 4 - During iteration j of the Implicit Gaussian Elimination algorithm, all elements in column j and n-j+1 between rows j and n-j+1 are driven to zero.

2.3 Gauss-Jordan Elimination

This also is a variant of Gaussian Elimination. The main difference is in the pattern of the elementary matrices it involves. At each step j, this algorithm annihilates all the elements of A in column j with row index different from j, i.e. above and below the diagonal. In this way, after n-1 steps, the coefficient matrix becomes a diagonal matrix D. See Figure 5 for further details. The "elimination loop" of Figure 5 is equivalent to the multiplication of A by n-1 elementary matrices:

$$M^{(n-1)}M^{(n-2)}\dots M^{(1)}A=D$$

where

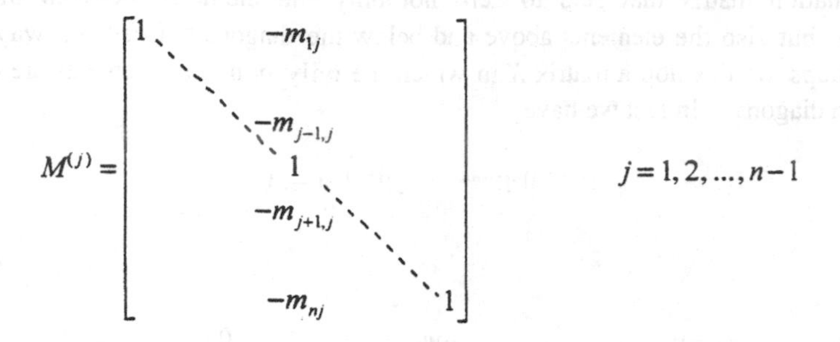

From Table 3 we can observe that the Gauss-Jordan "elimination phase" has a computational complexity of $O(n^3/2)$ that is mostly due to its inner loop. The extra computation to obtain the final diagonal matrix ($O(n^3/6)$) is not compensated by a lighter "solution phase" ($O(n)$) and this results in an algorithm which is more time consuming than Gaussian Elimination.

```
                (* elimination loop *)
                for j = 0, ..., n-1
PAR                  for i = 0, ..., j-1, j+1, ..., n-1

                          m = - a[ i,j] / a[ j,j]

                          (* update matrix A *)
                          for k = j+1, ..., n-1
                               a[ i,k] = a[ i,k] + m* a[ j,k]

                          (* update right hand side *)
                          b[ i] = b[ i] + m* b[ j]

          SYNC

                (* solution loop *)
PAR             for j = n-1, ..., 0
                     x[ j] = b[ j] / a[ j,j]
```

Figure 5 - Gauss-Jordan algorithm.

	Computational complexity	Shared memory accesses
Elimination phase	$\left(\frac{n^3}{2}+n^2-\frac{3}{2}n\right)m+\left(\frac{n^3}{2}-\frac{n}{2}\right)a$	$\frac{3}{2}n^3+2n^2-\frac{7}{2}n$
Solution phase	$(n)m$	$3n$
Total	$\left(\frac{n^3}{2}+n^2-\frac{n}{2}\right)m+\left(\frac{n^3}{2}-\frac{n}{2}\right)a$	$\frac{3}{2}n^3+2n^2-\frac{n}{2}$

Table 3 - Gauss-Jordan method analysis

2.4 Implicit Gauss Jordan Elimination

We can apply the same idea used to obtain the Gauss-Jordan method from the

Gaussian Elimination, to the Implicit Gaussian Elimination. This results in finding a transformation matrix that sets to zero not only the elements between the two diagonals, but also the elements above and below the diagonals. In such a way, after (n/2-1) steps we develop a matrix X in which the only non-zero elements are on the two main diagonals. In fact we have:

$$W^{(n/2-1)}\, W^{(n/2-2)} \ldots W^{(1)}\, A = X$$

where

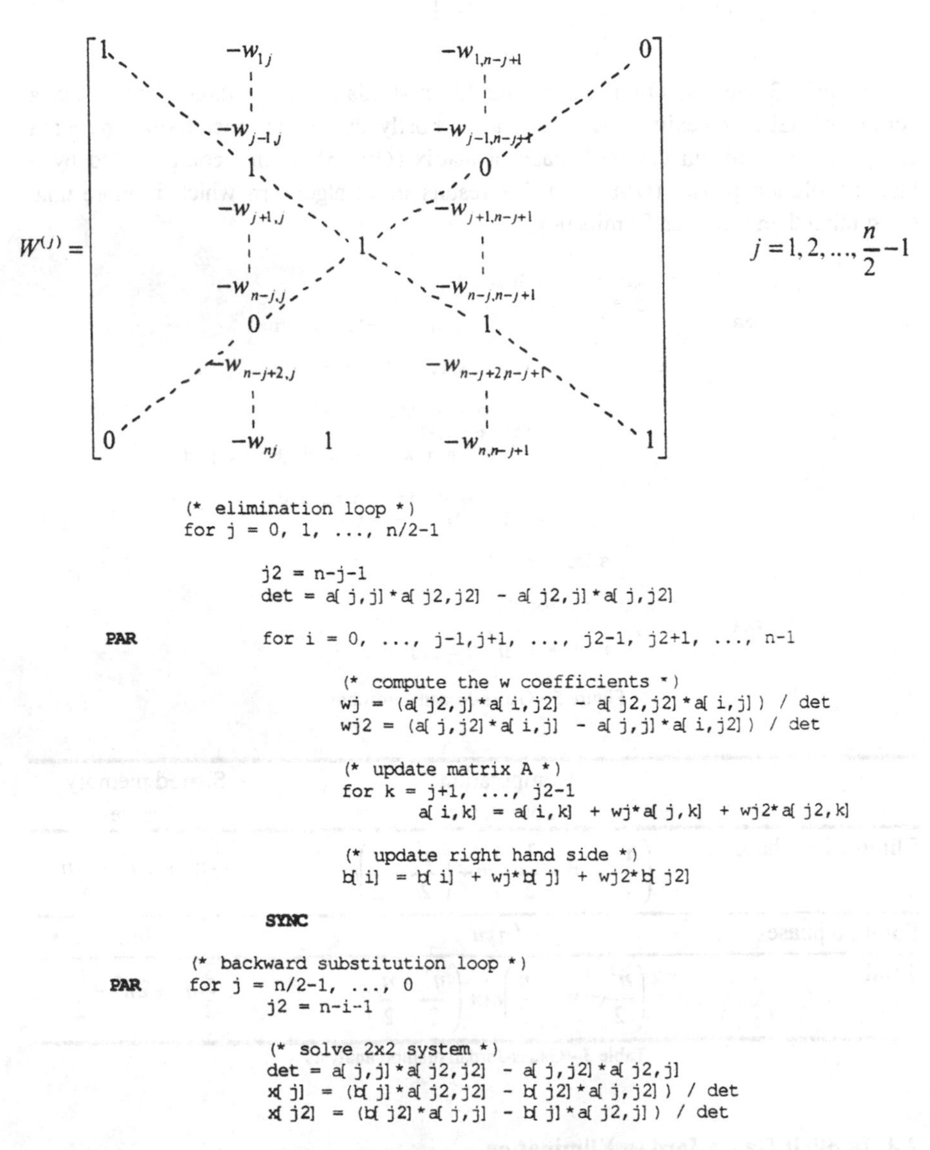


```
                (* elimination loop *)
                for j = 0, 1, ..., n/2-1

                        j2 = n-j-1
                        det = a[j,j]*a[j2,j2] - a[j2,j]*a[j,j2]

PAR                     for i = 0, ..., j-1,j+1, ..., j2-1, j2+1, ..., n-1

                                (* compute the w coefficients *)
                                wj = (a[j2,j]*a[i,j2] - a[j2,j2]*a[i,j]) / det
                                wj2 = (a[j,j2]*a[i,j] - a[j,j]*a[i,j2]) / det

                                (* update matrix A *)
                                for k = j+1, ..., j2-1
                                        a[i,k] = a[i,k] + wj*a[j,k] + wj2*a[j2,k]

                                (* update right hand side *)
                                b[i] = b[i] + wj*b[j] + wj2*b[j2]

                        SYNC

                (* backward substitution loop *)
PAR             for j = n/2-1, ..., 0
                        j2 = n-i-1

                        (* solve 2x2 system *)
                        det = a[j,j]*a[j2,j2] - a[j,j2]*a[j2,j]
                        x[j] = (b[j]*a[j2,j2] - b[j2]*a[j,j2]) / det
                        x[j2] = (b[j2]*a[j,j] - b[j]*a[j2,j]) / det
```

Figure 6 - Implicit Gauss-Jordan algorithm

The w coefficients are computed, as in the Implicit Gaussian Elimination, two at a time solving (n-2) 2×2 systems at each of the (n/2-1) steps. The "solution phase" consists in solving the resulting bi-diagonal system by means of (n/2) 2×2 systems. As in the classical Guass-Jordan this phase can be easily and efficiently executed in parallel. See Figure 6 for the algorithm details.
This method has the same computational complexity of the classical Gauss-Jordan method (they differ only in the n^2 term), but we have a remarkable reduction of the shared memory accesses. This is explained by the fact that using the Implicit Gauss-Jordan, at each step, we need to update $(n-1)\times(n-j-1)$ elements of A instead of the $(n-1)\times n$ elements of A using the classical Gauss-Jordan algorithm.

	Computational complexity	Shared memory accesses
Elimination phase	$\left(\frac{n^3}{2}+2n^2-5n\right)m+\left(\frac{n^3}{2}-\frac{3}{2}n\right)a$	n^3+2n^2-14n
Solution phase	$(4n)m+\left(\frac{3}{2}n\right)a$	$7n$
Total	$\left(\frac{n^3}{2}+2n^2-n\right)m+\left(\frac{n^3}{2}\right)a$	n^3+2n^2+7n

Table 4 - Implicit Gauss-Jordan method analysis

2.5 Implicit Partial Pivoting scheme

To ensure that these elimination schemes are numerically stable, we need to limit the growth of elements in the reduced coefficient matrices. One common way to achieve this is to avoid large m and w multipliers by means of row interchanges. This procedure is called pivoting.
For the Gaussian Elimination we can easily guarantee $|m_{ik}| \le 1$ using the Partial Pivoting procedure. Consider the kth elimination stage, we search over column k for a row r such that

$$|a_{rk}| = \max_{k \le i \le n} |a_{ik}|$$

Then we interchange equations r and k and the elimination proceeds as usual using the new pivot element a_{kk} to perform the eliminations.
For the Implicit Gaussian method the pivot is no longer a single element, in fact at the kth stage of the elimination we have the following multipliers:

$$w_{ik} = \frac{a_{ik}^{(k-1)} a_{n-k+1,n-k+1}^{(k-1)} - a_{i,n-k+1}^{(k-1)} a_{n-k+1,k}^{(k-1)}}{\Delta} \tag{5}$$

$$w_{i,n-k+1} = \frac{a_{k,n-k+1}^{(k-1)} a_{ik}^{(k-1)} - a_{ki}^{(k-1)} a_{i,n-k+1}^{(k-1)}}{\Delta} \qquad i = k+1, \ldots, n-k$$

where

$$\Delta = a_{kk}^{(k-1)} a_{n-k+1,n-k+1}^{(k-1)} - a_{k\,n-k+1}^{(k-1)} a_{n-k+1,k}^{(k-1)}$$

The safest strategy is to choose the largest Δ, i.e. the largest 2x2 pivot, but this involves a search over

$$\frac{n!}{2!(n-2)!} = \frac{n^2 - n}{2}$$

possible pivots at each step, with a consequent increase of the total computational complexity. Instead we consider the following Implicit Partial Pivoting scheme:

1) Find a row r such that

$$\left|a_{rk}^{(k-1)}\right| = \max_{k \le i \le n-k+1} \left|a_{ik}^{(k-1)}\right|$$

2) Exchange rows r and k.

3) Find a row s such that

$$\left|a_{s,n-k+1}^{(k-1)} - \frac{a_{k,n-k+1}^{(k-1)}}{a_{kk}^{(k-1)}} a_{sk}^{(k-1)}\right| = \max_{k+1 \le i \le n-k+1} \left|a_{i,n-k+1}^{(k-1)} - \frac{a_{k,n-k+1}^{(k-1)}}{a_{kk}^{(k-1)}} a_{ik}^{(k-1)}\right|$$

4) Exchange rows s and n-k+ 1.

We now prove that this implicit strategy ensures

$$|w_{ik}| \le 2 \qquad k = 1, \ldots, n \qquad i = k+1, \ldots, n-k \tag{6}$$

and

$$\left|w_{i,n-k+1}\right| \le 1 \qquad k = 1, \ldots, n \qquad i = k+1, \ldots, n-k \tag{7}$$

$$A^{(k-1)} = \begin{bmatrix} a_{11} & \cdots & & & a_{1n} \\ 0 & \ddots & & \iddots & 0 \\ & a_{kk}^{(k-1)} & \cdots & a_{k,n-k+1}^{(k-1)} & \\ & \vdots & & \vdots & \\ & a_{ik}^{(k-1)} & & a_{i,n-k+1}^{(k-1)} & \\ & \vdots & & \vdots & \\ & a_{n-k+1,k}^{(k-1)} & \cdots & a_{n-k+1,n-k+1}^{(k-1)} & \\ 0 & \iddots & & \ddots & 0 \\ a_{n1} & \cdots & & & a_{nn} \end{bmatrix}$$

Let $A^{(k-1)}$ be the reduced matrix after k-1 complete steps and the kth step pivoting procedure. We focus our attention on the kth and $(n\text{-}k+1)^{th}$ columns. We have, omitting the upper indexes,

$$|a_{k,k}| \geq |a_{i,k}| \qquad i = k+1, \ldots, n-k+1 \tag{8}$$

and

$$\left| a_{n-k+1,n-k+1} - \frac{a_{k,n-k+1}}{a_{k,k}} a_{n-k+1,k} \right| \geq \left| a_{i,n-k+1} - \frac{a_{k,n-k+1}}{a_{k,k}} a_{i,k} \right| \qquad i = k+1, \ldots, n-k+1 \tag{9}$$

It can be easily observed that (7) follows trivially from (9). To prove (6) is more complex. We proceed as follows.Considering each couple $\left(a_{i,k}^{(k-1)} ; a_{i,n-k+1}^{(k-1)}\right)$ as a point of the Cartesian plane, we have that $\left(a_{i,k}^{(k-1)} ; a_{i,n-k+1}^{(k-1)}\right)$ necessarily lies inside the shaded parallelogram of Figure 7.

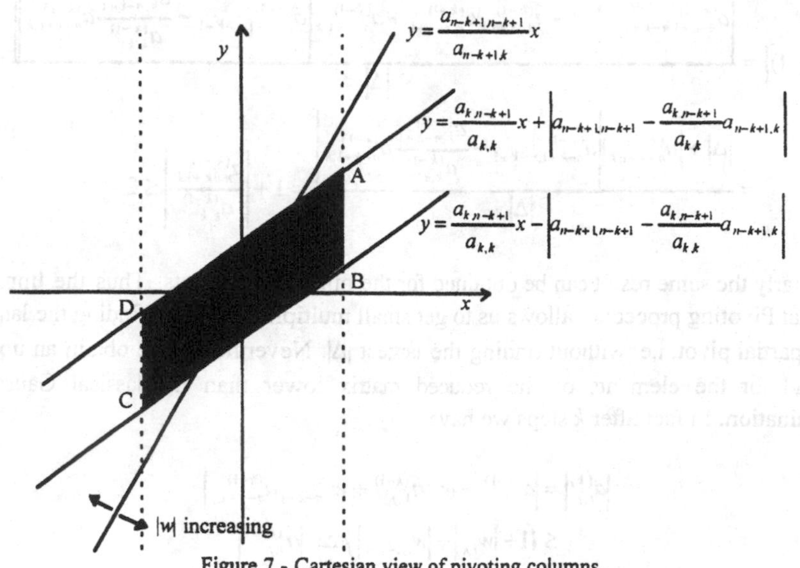

Figure 7 - Cartesian view of pivoting columns.

We can consider $|w_{i,k}|$ as a linear function of $\left(a_{i,k}^{(k-1)} ; a_{i,n-k+1}^{(k-1)}\right)$. This function is null along the line

$$y = \frac{a_{n-k+1,n-k+1}^{(k-1)}}{a_{n-k+1,k}^{(k-1)}} x$$

and it is increasing in both orthogonal directions. This allows us to restrict the search for its maximum, with respect to the constraints (8) and (9), to the following four points only

$$A\left(a_{k,k}^{(k-1)}\,;\,a_{k,n-k+1}^{(k-1)}+\left|a_{n-k+1,n-k+1}^{(k-1)}-\frac{a_{k,n-k+1}^{(k-1)}}{a_{k,k}^{(k-1)}}a_{n-k+1,k}^{(k-1)}\right|\right)$$

$$B\left(a_{k,k}^{(k-1)}\,;\,a_{k,n-k+1}^{(k-1)}-\left|a_{n-k+1,n-k+1}^{(k-1)}-\frac{a_{k,n-k+1}^{(k-1)}}{a_{k,k}^{(k-1)}}a_{n-k+1,k}^{(k-1)}\right|\right)$$

$$C\left(-a_{k,k}^{(k-1)}\,;\,-a_{k,n-k+1}^{(k-1)}-\left|a_{n-k+1,n-k+1}^{(k-1)}-\frac{a_{k,n-k+1}^{(k-1)}}{a_{k,k}^{(k-1)}}a_{n-k+1,k}^{(k-1)}\right|\right)$$

$$D\left(-a_{k,k}^{(k-1)}\,;\,-a_{k,n-k+1}^{(k-1)}+\left|a_{n-k+1,n-k+1}^{(k-1)}-\frac{a_{k,n-k+1}^{(k-1)}}{a_{k,k}^{(k-1)}}a_{n-k+1,k}^{(k-1)}\right|\right)$$

Consider the point A. We have

$$\left|w_{i,k}(A)\right|=\frac{\left|a_{n-k+1,n-k+1}^{(k-1)}a_{k,k}^{(k-1)}-a_{n-k+1,k}^{(k-1)}a_{k,n-k+1}^{(k-1)}+a_{n-k+1,k}^{(k-1)}\left|a_{n-k+1,n-k+1}^{(k-1)}-\frac{a_{k,n-k+1}^{(k-1)}}{a_{k,k}^{(k-1)}}a_{n-k+1,k}^{(k-1)}\right|\right|}{|\Delta|}=$$

$$=\frac{|\Delta|+\left|a_{n-k+1,k}^{(k-1)}\right|\left|a_{n-k+1,n-k+1}^{(k-1)}-\frac{a_{k,n-k+1}^{(k-1)}}{a_{k,k}^{(k-1)}}a_{n-k+1,k}^{(k-1)}\right|}{|\Delta|}=1+\left|\frac{a_{n-k+1,k}^{(k-1)}}{a_{k,k}^{(k-1)}}\right|\leq 2$$

Similarly the same result can be obtained for the other three points. Thus the Implicit Partial Pivoting procedure allows us to get small multipliers without finding the largest 2x2 partial pivot, i.e. without finding the largest $|\Delta|$. Nevertheless we obtain an upper bound for the elements of the reduced matrix lower than in classical Gaussian Elimination. In fact after k steps we have

$$\begin{aligned}\left|a_{i,j}^{(k)}\right|&=\left|a_{i,j}^{(k-1)}+w_{i,k}a_{k,j}^{(k-1)}+w_{i,n-k+1}a_{n-k+1,j}^{(k-1)}\right|\\&\leq\left(1+\left|w_{i,k}\right|+\left|w_{i,n-k+1}\right|\right)\max_{i,j}\left|a_{i,j}^{(k-1)}\right|\\&\leq 4\max_{i,j}\left|a_{i,j}^{(k-1)}\right|\leq 4^k\max_{i,j}\left|a_{i,j}^{(0)}\right|\end{aligned}$$

and after the complete reduction process, i.e. $k=n/2\text{-}1$

$$\left|a_{i,j}^{(n/2-1)}\right|\leq 2^{n-2}\max_{i,j}\left|a_{i,j}^{(0)}\right|$$

This upper bound is half of that for the elements of a matrix reduced using Gaussian Elimination [7]. This is more significant in view of the fact that the Gaussian Elimination upper bound is not oversized and that matrices exist for which it is attainable.

The Implicit Partial Pivoting procedure has a very low computational cost and requires only a slightly greater number of accesses to the shared memory than the classical Partial Pivoting scheme. For further details see Table 5.

	Computational Complexity	Shared Memory Accesses	Synchronisation Point
Partial Pivoting	0	$n^2 - n$	$n-1$
Implicit Partial Pivoting	$\left(\frac{n^2}{2} - \frac{3}{2}n\right)(m+a)$	$\frac{3}{2}n^2 - 3n$	$n-2$

Table 5 - Pivoting schemes analysis

We use the Implicit Partial Pivoting strategy for both implicit methods even if the w multipliers external to the diagonals could be arbitrarily large in the Implicit Gauss-Jordan algorithm. We do this in the same way as the classical Partial Pivoting is used with the Gauss-Jordan method. It was proved [6] that Gauss-Jordan with the Partial Pivoting is as stable as Gaussian Elimination and it is sensible to claim that this result holds for the Implicit Gauss-Jordan too.

3. Numerical results

The numerical tests were performed on Balance Sequent system with 10 processors. All processors plug into a single bus and share a single pool of memory. Each of them has 8 Kb of cache memory.

The algorithms were implemented in double precision using the Sequent C language and a static scheduling of tasks, i.e. distributing the computation load among processor *a priori*, before starting the execution of the parallel program. We divided the **PAR** marked loop iterations evenly between the processors, in other words we evenly subdivided the rows we need to process at each step among processors. This did not affect the load balancing because each iteration (that is between two synchronisation point) involves the same amount of computations for each processor.

We run only one process per processor to avoid the overhead of switching from one process to another. All the programs were written with the same accuracy in order to obtain meaningful results. For the same reason, even if the Dynix environment allows multiprogramming, all the tests were completed while no other user's task was running on the Balance.

Matrix A was chosen to be diagonally dominant when pivoting was not applied, and a random matrix when some kind of partial pivoting was performed.

Despite the computational analysis prediction, the implicit methods perform better than the classical methods for all the combination of problem size and number of processors.

The average gain is 20%-22% for both Implicit Gaussian Elimination and Implicit Gauss-Jordan Elimination. Furthermore the application of the Implicit Partial Pivoting procedure does not affect these gains. See Table 6, Table 7, Table 8 and Table 9.

		Processors			
		4	6	8	10
200	PIG	37620	25380	19240	15620
	PG	47260	31950	24250	19670
	%	20.4	20.6	20.7	20.6
400	PIG	302490	201690	151190	121670
	PG	380520	252620	190200	153300
	%	20.5	20.2	20.5	20.6
600	PIG	1030190	686500	515680	412050
	PG	1302670	864260	649130	516200
	%	20.9	20.6	20.6	20.2
800	PIG	2446500	1639320	1228650	983660
	PG	3100240	2073480	1557940	1246100
	%	21.1	20.9	21.1	21.1

Table 6 - Computation times in ms and relative gains of Parallel Implicit Gaussian Elimination (PIG) versus Parallel Gaussian Elimination (PG).

		Processors			
		4	6	8	10
200	PIJ	55730	38060	28180	22670
	PJ	70490	47950	35590	28590
	%	20.9	20.6	20.8	20.7
400	PIJ	451410	300260	224150	182950
	PJ	580150	385400	283490	227460
	%	22.2	22.1	20.9	19.6
600	PIJ	1531040	1021590	767900	614260
	PJ	1969060	1318450	980530	785140
	%	22.2	22.5	21.7	21.8
800	PIJ	3623670	2431250	1824820	1462760
	PJ	4684480	3136170	2361790	1897500
	%	22.6	22.5	22.7	22.9

Table 7 - Computation times in ms and relative gains of Parallel Implicit Gauss-Jordan (PIJ) method versus Parallel Gauss-Jordan (PJ) methods.

		Processors			
		4	6	8	10
200	PIG	45730	30920	23710	19420
	PG	58770	39840	30440	25010
	%	22.2	22.4	22.1	22.4
400	PIG	362530	243580	183770	148380
	PG	466010	313100	236290	191050
	%	22.2	22.2	22.2	22.3
600	PIG	1227190	820390	617210	497130
	PG	1574440	1051710	795710	639590
	%	22.1	22.0	22.4	22.3
800	PIG	2917500	1949050	1461200	1174790
	PG	3743480	2501340	1877400	1509060
	%	22.1	22.1	22.2	22.2

Table 8 - Computation times in ms and relative gains of Parallel Implicit Gaussian Elimination (PIG) versus Parallel Gaussian Elimination (PG). Partial pivoting.

		Processors			
		4	6	8	10
200	PIJ	68060	46340	34490	27900
	PJ	86650	59110	43780	35350
	%	21.5	21.6	21.2	21.1
400	PIJ	545020	364180	272660	218910
	PJ	698740	467380	348370	285410
	%	22.0	22.1	21.7	23.3
600	PIJ	1840590	1231800	923080	738180
	PJ	2372150	1587670	1187260	954010
	%	22.4	22.4	22.3	22.6
800	PIJ	4372040	2929730	2200650	1757020
	PJ	5650300	3783640	2832930	2266030
	%	22.6	22.6	22.3	22.5

Table 9 - Computation times in ms and relative gains of Parallel Implicit Gauss-Jordan method (PIJ) versus Parallel Gauss-Jordan method (PJ). Partial pivoting.

We now show that these better performances are due to a lesser number of accesses to the shared memory. Each implicit method involves an amount of accesses to the problem data, i.e. to elements of A and b, that is about 30% smaller than the corresponding classical method.

In view of the above considerations, we can propose the following simplified theoretical model for a parallel algorithm *Alg* running on a shared memory parallel computer

$$T_{Alg}(n,p) = CC_{Alg}(n)T_{lop} + SMA_{Alg}(n)T_{smt}(p) + SP_{Alg}(n)\,T_{sync}(p) \tag{10}$$

The total elapsed time T_{Alg} is obviously a function of the problem size n and of the number of available processors p. We can split this time into three parts:

- $CC(n)T_{lop}$
  The time to perform the floating point operations with the operands already stored in the local memory (that is, for the Sequent Balance, RAM cache, CPU cache and registers). $CC(n)$ is the computation complexity of the algorithm, that is the number of operations, and is function of n. T_{lop} is the time needed to complete a single operation, we suppose it constant for each kind of arithmetic operation and for each type of operand storage.
- $SMA(n)T_{smt}(p)$
  The time to transfer the operands from the global shared memory to the local memory of each processor. $SMA(n)$ is the number of accesses to the shared memory involved by the algorithm. $T_{smt}(p)$ is the time to transfer a single operand or result from/to the shared memory. It usually is a function of the number of available processors p.
- $SP(n)T_{sync}(p)$
  The time to synchronise the various processes. We can represent it as the product of the number of synchronisation points using the $SP(n)$ by the time we need to synchronise p processors, $T_{sync}(p)$. This is quite a big simplification because this term is function not only of n and p, but also of the computational load balancing, that is, if one processor has a greater number of operations to perform before a synchronisation point, then all the other processors have to wait for it.

	Computational complexity(*)	Shared memory accesses	Synch points
Gauss Elimination	$\frac{n^3}{3}+n^2-\frac{n}{3}$	n^3+3n^2-n	$2n$
Gauss-Jordan Method	$\frac{n^3}{2}+n^2-\frac{n}{2}$	$\frac{3}{2}n^3+2n^2-\frac{n}{2}$	n
Implicit Gauss Elimination	$\frac{n^3}{3}+\frac{3}{2}n^2+\frac{2}{3}n$	$\frac{2}{3}n^3+\frac{9}{4}n^2+\frac{11}{6}n$	n
Implicit Gauss-Jordan Elimination	$\frac{n^3}{2}+2n^2-n$	n^3+2n^2+7n	$\frac{n}{2}$

Table 10 - Gauss and Implicit method analysis.
(*)Additions omitted.

T_{lop}, $T_{smt}(p)$ and $T_{sync}(p)$ are a synthesis of the parallel computer architecture characteristcs. We computed these parameters solving the 3×3 systems built with equation (10) using the Gaussian Elimination algorithm timings and 3 different values of n to obtain satisfactory prediction of the results.

GAUSSIAN ELIMINATION

IMPLICIT MATRIX ELIMINATION

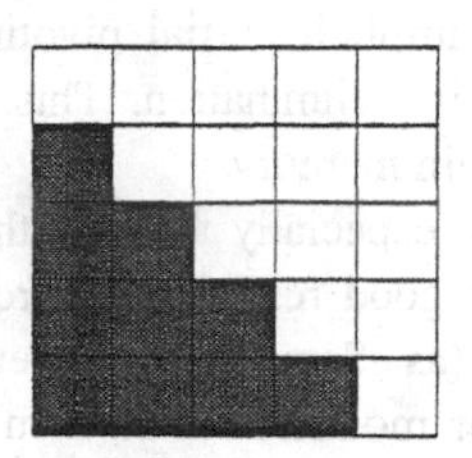

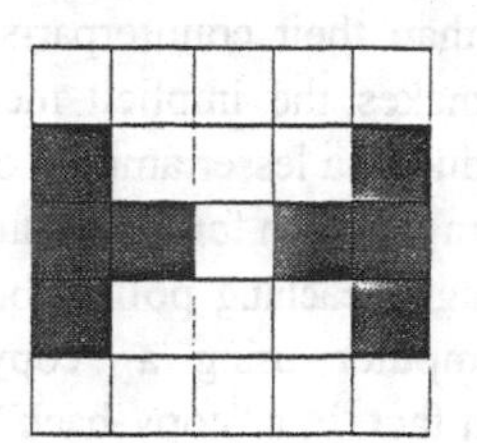

N=5

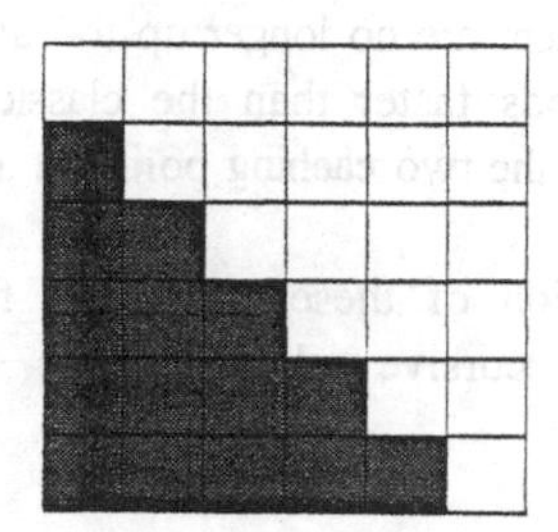

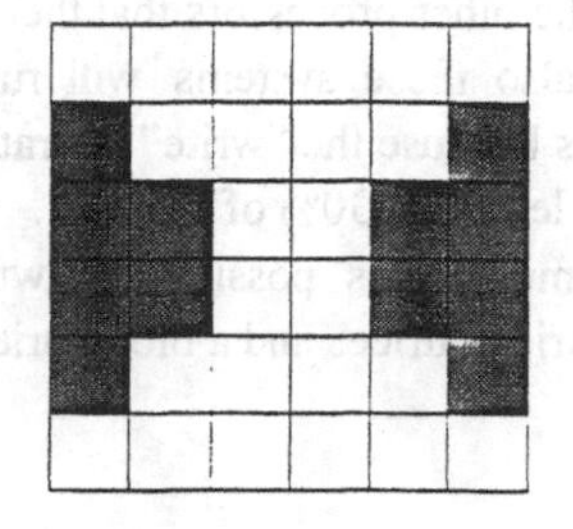

N=6

GAUSS-JORDAN ELIMINATION

IMPLICIT JORDAN ELIMINATION

N=5

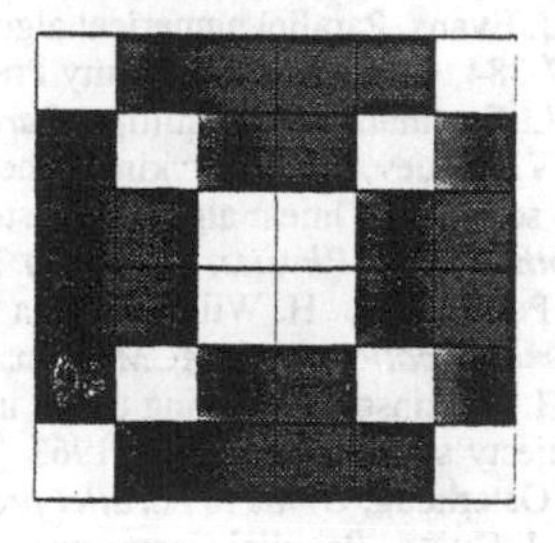

N=6

4. Conclusion and future work

On the Sequent Balance system the implicit algorithms proved to be superior (more than 20%) than their counterparts. The introduction of an implicit partial pivoting procedure makes the implicit methods as stable as Gaussian elimination. This is essentially due to a lesser amount of accesses to the ystem main memory.
We can claim this gain for all the shared memory architectures, especially those with a "write-through" caching policy, but it is sensible to expect good results even from parallel computer using a "copy-back" caching policy (as Sequent Symmetry systems). In fact, in a "copy-back" system, when a processor modifies a data item in its own cache, it does not write the update value to the main memory until it swaps out the cache block or until another processor needs that data item; however it has to signal the other processors that their copies of the data item are no longer up-to-date. Hence also these systems will run the implicit methods faster than the classical methods because the "write" operations (that distinguish the two caching policies) are usually less than 30% of the total.
Furthermore it is possible to write an $O(n^3/6)$ version of these algorithms for symmetric matrices and a block-oriented approach using recursive techniques.

References

[1]D. J. Evans, Implicit matrix elimination (IME) schemes, *Intern. J. Computer Math.*, Vol. 48, pp. 229-237, 1983
[2]D. J. Evans and R. Abdullah, The parallel implicit elimination (PIE) method for the solution of linear systems, *Parallel Algorithms and Applications*, Vol. 4, pp. 153-162, 1994
[3]D. J. Evans, Parallel numerical algorithms for linear systems, *Parallel Processing Systems*, pp. 357-384, Cambridge University Press, 1982
[4]T. L. Freeman and C. Phillips, *Parallel Numerical Algorithms*, Prentice Hall, 1992
[5]V. V. Klyuev, H. I. Kokvkin-Shcherbak, Minimisation of the number of arithmetic operations in the solution of linear algebraic systems of equations, *U.S.S.R. Computational Mathematics and Mathematical Physics*, Vol. 5, pp. 25-43, 1965
[6]G. Peters and J. H. Wilkinson, On the stability of Gauss-Jordan elimination with pivoting, *Communication of the ACM*, January 1975, Vol. 18, Number 1, pp. 20-24
[7]J. H. Wilkinson, Rounding errors in algebraic processes, *Notes on Applied Science N.32*, Her Majesty's Stationery Office, 1963
[8]A. Osterhaug, *Guide to parallel programming on Sequent computer systems*, Prentice Hall, 1989
[9]M. J. Quinn, *Parallel Computing - Theory and Practice*, McGraw-Hill, 1994

ADAPTIVE MULTIGRID METHODS FOR HYBRID FINITE ELEMENTS

L. FERRAGUT
Dpt. of Pure and Applied Mathematics
University of Salamanca, E37008, Salamanca, Spain.

Abstract. We consider in this paper the numerical approximation of second order elliptic problems, using adaptive multigrid finite element techniques focusing our attention in numerical methods for the dual problem. First a general result concerning the a-posteriori error estimation of the numerical solution of both primal and dual variables is stated. The relationship between hybrid formulation and non conforming approximation is used to propose and justify an adaptive multigrid algorithm in order that the computational cost be acceptable. Finally a representative numerical example is presented.

Key words: multigrid, hybrid finite elements, adaptive numerical methods.

1. Introduction

Reliable and efficient a posteriori error estimates are a fundamental tool for the development of adaptive finite element methods. At present efficient adaptive algorithms for a large class of problems have been constructed, specially for linear elliptic problems, for which many errors estimators have been developed with a solid theoretical foundation.

Between the main approaches in the development of a posteriori error estimations, there exist those based on residual tecniques, see among numerous contributions, the work by [1], [2], [3], and more recently [4]. Another technique is postprocessing, which consists basically on smoothing techniques applied to some derivatives of the numerical solutions in order to obtain superconvergence results, [8]. In what concerns to adaptive methods for nonlinear problems see for example [5], [6], and the references their in, see also the more recent work by [7] about adaptivity in elastoplas-

G. Winter Althaus and E. Spedicato (eds.), Algorithms for Large Scale Linear Algebraic Systems, 55–71.

tic solids where known error estimates for linear problems are appropriatly modified to account for non linearity.

We will consider the solution of the so called primal and dual problems for second order elliptic equations. Usually in practical problems, the solution of the dual problem (flux, stresses) is of primary interest, consequently we focus our attention in adaptive numerical methods for the dual problem. We note that most of the finite elements methods and a posteriori error estimators developed till now concern the numerical approximation of the solution of the primal problem. Furthermore, in some nonlinear problems as the case of perfect plasticity, the conventional approximation of the primal variable (displacement) is not possible due to lack of smothness, [9].

On the other hand one of the most effective iterative method for solving the systems obtained by the finite element discretization has become lately the multigrid method. Here we use non standard multigrid techniques which combine conforming and non conforming approximation in order to obtain an efficient adaptive multigrid algorithm.

The outline of this paper is the following: Section 2 presents the general mathematical framework and the statement of the problem. In section 3 we set two different variational approximations, the first one giving the standard conforming approximation and the second one based on the primal hybrid formulation. In section 4 we establish the basic a-posteriori error estimate. Section 5 and 6 explain how the primal and dual solution can be obtained using a multigrid method. Finally in section 7 we present a representative numerical example.

2. Mathematical preliminaries and statement of the problem

Throughout this paper Ω shall be a given domain with polygonal boundary Γ in the two dimensional, real Euclidean space.

We denote $\mathbf{H}^s(\Omega)$ the usual Sobolev space with norm $\|.\|_{s,\Omega}$. In particular on $\mathbf{H}^1(\Omega)$ we shall consider the trace operator γ: $\mathbf{H}^1(\Omega) \rightarrow \mathbf{L}^2(\Omega)$; we define $\mathbf{H}^{1/2}(\Gamma)$ as the image by γ of $\mathbf{H}^1(\Omega)$. Finally we introduce the space of functions of $\mathbf{H}^1(\Omega)$ with null trace on Γ, that is, $\mathbf{H}_0^1(\Omega)$= $\{v \in \mathbf{H}^1(\Omega); \gamma(v) = 0\}$. As it is well known on $\mathbf{H}_0^1(\Omega)$ the seminorm $|\cdot|_{1,\Omega}$, defined by $|v|_{1,\Omega} = (\int_\Omega |\nabla|^2)^{1/2}$ is a norm equivalent to the norm induced by the norm $\|.\|_{1,\Omega}$ of $\mathbf{H}^1(\Omega)$. As usual we call $\mathbf{H}^{-1}(\Omega)$ the dual space of $\mathbf{H}_0^1(\Omega)$.

Also we make use in the paper of the following vector Sobolev space $\mathbf{H}(\text{div};\Omega)$=$\{\mathbf{q}\in\mathbf{L}^2(\Omega)^2;\text{div}\mathbf{q}\in\mathbf{L}^2(\Omega)\}$ normed by

$$\|\mathbf{q}\|_{\mathrm{H}(\text{div};\Omega)} = (\|\mathbf{q}\|_{0,\Omega}^2 + \|\text{div}\mathbf{q}\|^2)^{1/2}$$

If $\mathbf{n}$ denotes the unit normal along Γ the mapping γ_ν: $\mathbf{q} \rightarrow \mathbf{q}.\mathbf{n}|_\Gamma$ is well defined on $\mathbf{H}(\mathrm{div};\Omega)$ and the image is $\mathbf{H}^{-1/2}(\Gamma)$ the dual space of $\mathbf{H}^{1/2}(\Gamma)$.

For given $f \in \mathbf{L}^2(\Omega)$, we search for u such that

$$\begin{aligned} -\mathrm{div}(\mathbf{A}\nabla u) &= f \text{ in } \Omega \\ u &= 0 \text{ on } \Gamma \end{aligned} \tag{1}$$

where $\mathbf{A}$ is a symmetric 2x2 matrix of $\mathbf{L}^\infty(\Omega)$ functions on Ω, verifying the ellipticity condition: $\exists \alpha > 0$ such that $\mathbf{A}(\mathrm{x})\xi.\xi \geq \alpha\xi.\xi$ $\forall \mathrm{x} \in \Omega$, $\forall \xi \in \mathbf{R}^2$.

The classical variational formulation of the problem (1) is:

$$\begin{aligned} (\nabla u, \nabla v)_{\mathrm{A}} &= (f, v) \; \forall v \in \mathrm{H}_0^1(\Omega) \\ u &\in \mathrm{H}_0^1(\Omega) \end{aligned} \tag{2}$$

where $(.,.)$ is the usual inner product in $\mathbf{L}^2(\Omega)$ and $(\mathbf{p},\mathbf{q})_{\mathbf{A}} = \int_\Omega \mathbf{A}\mathbf{p}.\mathbf{q}$ for all $\mathbf{p}$, $\mathbf{q} \in \mathbf{L}^2(\Omega)^2$.

To this problem we shall associate the energy norm

$$\|v\|_{\mathrm{A}} = (\nabla v, \nabla v)_{\mathbf{A}}^{1/2}$$

which is equivalent to the norms $\|\cdot\|_{1,\Omega}$ and $|\cdot|_{1,\Omega}$ defined on $\mathrm{H}_0^1(\Omega)$. Corresponding to this norm we consider in $\mathbf{H}^{-1}(\Omega)$ the following norm:

$$\|\varphi\|_{-1,\Omega} = \sup_{v \in \mathrm{H}_0^1(\Omega)} \frac{\langle \varphi, v \rangle}{\|v\|_{\mathrm{A}}}$$

where $\langle .,. \rangle$ denotes the duality product between $\mathbf{H}^{-1}(\Omega)$ and $\mathbf{H}_0^1(\Omega)$.

On the other hand for all vector functions $\mathbf{p}$, $\mathbf{q}$ belonging to the space $(\mathbf{L}^2(\Omega))^2$ we shall use the scalar product $(\mathbf{p},\mathbf{q})_{\mathrm{A}^{-1}} = \int_\Omega \mathbf{A}^{-1}\mathbf{p}.\mathbf{q}$ and the corresponding norm $\|\mathbf{q}\|_{\mathrm{A}^{-1}} = (\mathbf{q},\mathbf{q})_{\mathrm{A}^{-1}}^{1/2}$.

3. Finite Element approximation

In many cases, the vector function $\mathbf{p} = \mathbf{A}\nabla u$ has a practical physical meaning. Here we are concerned with the finite element approximation of both unknowns u and $\mathbf{p}$. In the following $\mathcal{T}$ will be an admisible triangulation of Ω, in the sense of [12] and for every $\tau \in \mathcal{T}$, $\mathbf{P}_{\mathrm{k}}(\tau)$ the space of polynomes of degree less or equal than k defined in τ.

Consider first the following approximation of (1):

$$\begin{aligned} -\mathrm{div}(\mathbf{A}\nabla \tilde{u}) &= \tilde{f} \text{ in } \Omega \\ \tilde{u} &= 0 \text{ on } \Gamma \end{aligned} \tag{3}$$

where $\tilde{f}$ is a piece-wise constant approximation of f, defined by:

$$\widetilde{f} : \Omega \rightarrow \mathbf{R} \;,\; \widetilde{f}|_\tau = \frac{1}{|\tau|}\int_\tau f \tag{4}$$

that is, $\widetilde{f}$ is the $\mathbf{L}^2(\Omega)$ projection of f on the space $\{v : \Omega \rightarrow \mathbf{R}; v|_\tau \in \mathbf{P}_0(\tau)\}$. We have the following error estimation between u and $\widetilde{u}$:

Theorem:

If $f \in \{f \in \mathbf{L}^2(\Omega); f|_\tau \in \mathbf{H}^1(\tau)\, \forall \tau \in \mathcal{T}\}$, then $\|u - \widetilde{u}\|_A \leq Ch^2(\sum |f|^2_{1,\tau})^{1/2}$ where h is the discretization parameter associated to the triangulation $\mathcal{T}$, and C is a constant independent of h.

Proof: We have

$$\|u\text{-}\widetilde{u}\|_{\mathrm{A}} = \left\|f\text{-}\widetilde{f}\right\|_{-1,\Omega} = \sup_{v \in \mathrm{H}^1_0(\Omega)} \frac{\langle f - \widetilde{f}, v\rangle}{\|v\|_{\mathrm{A}}} \tag{5}$$

But for all $v_0 \in \{v : \Omega \rightarrow \mathbf{R}; v|_\tau \in \mathbf{P}_0(\tau)\}$, we have for several constants C

$$\begin{aligned}\langle f\text{-}\widetilde{f},v\rangle \;&=\; \int_\Omega (f\text{-}\widetilde{f})\mathrm{v} = \sum_{\tau\in\mathcal{T}}\int_\tau (f\text{-}\widetilde{f})v = \sum_{\tau\in\mathcal{T}}\int_\tau (f\text{-}\widetilde{f})(v\text{-}v_0) \leq \\ &\sum_{\tau\in\mathcal{T}} \left\|f\text{-}\widetilde{f}\right\|_{0,\tau} \cdot \|v\text{-}v_0\|_{0,\tau} \leq \mathrm{C}h^2 \sum_{\tau\in\mathcal{T}} |f|_{1,\tau}\,|v|_{1,\tau} \leq \\ &\mathrm{C}h^2(\sum_{\tau\in\mathcal{T}} |f|^2_{1,\tau})^{1/2}\,\|v\|_{\mathrm{A}}\end{aligned}$$

and the conclusion follows.

In the following we consider first order finite element approximations of $\widetilde{u}$ and $\widetilde{\mathbf{p}} = \mathbf{A}\nabla\widetilde{u}$. Let u_h and $\mathbf{p}_h$ be such approximations, then by the above theorem

$$\|u - u_h\|_{\mathrm{A}} \leq \|u - \widetilde{u}\|_{\mathrm{A}} + \|\widetilde{u} - u_h\|_{\mathrm{A}} = \|\widetilde{u} - u_h\|_{\mathrm{A}} + \mathcal{O}(h^2) \tag{6}$$

Analogously,

$$\|\mathbf{p} - \mathbf{p}_h\|_{\mathrm{A}^{-1}} \leq \|\mathbf{p} - \widetilde{\mathbf{p}}\|_{\mathrm{A}^{-1}} + \|\widetilde{\mathbf{p}} - \mathbf{p}_h\|_{\mathrm{A}^{-1}} = \|\widetilde{\mathbf{p}} - \mathbf{p}_h\|_{\mathrm{A}^{-1}} + \mathcal{O}(h^2) \tag{7}$$

so u_h and $\mathbf{p}_h$ are also first order approximations of u and $\mathbf{p}$ respectively.

3.1. CONFORMING FINITE ELEMENT APPROXIMATION

We introduce the following conforming finite dimensional subspace $\mathbf{V}^c_h$ of $\mathrm{H}^1_0(\Omega)$,

$$\mathbf{V}^c_h = \{v \in \mathbf{H}^1_0(\Omega); v|_\tau \in \mathbf{P}_1(\tau)\;,\; \forall\tau \in \mathcal{T}\;\}$$

The approximate problem is, find u_h such that

$$\begin{aligned}(\nabla u_h, \nabla v)_A &= (\tilde{f}, v)\ \forall v \in \mathbf{V}_h^c \\ u &\in \mathbf{V}_h^c\end{aligned} \tag{8}$$

where $\tilde{f}$ has been defined in (4).

3.2. PRIMAL HYBRID FORMULATION AND NON CONFORMING FINITE ELEMENT APPROXIMATION

In order to get a better approximation of $\mathbf{p}=\mathbf{A}\nabla u$ following [11] we introduce the primal hybrid formulation of problem (3):

For an admissible triangulation $\mathcal{T}$ consider the functional spaces

$$\mathbf{H}=\{\ v \in \mathbf{L}^2(\Omega);\ v_\tau = v|_\tau \in \mathbf{H}^1(\tau)\ \forall \tau \in \mathcal{T}\}$$

$$\mathbf{M}=\{\mu = (\mu_\tau); \exists \mathbf{q} \in \mathbf{H}(\text{div};\Omega) \text{ such that } \mathbf{q.n}|_{\partial\tau} = \mu_\tau\ \forall \tau \in \mathcal{T}\ \}$$

With straightforward notation, problem (3) can be stated as: find a couple $(u,\lambda)\in$**H**x**M** such that

$$\begin{aligned}\sum_{\tau\in\mathcal{T}}(\nabla u_\tau, \nabla v_\tau)_{\mathbf{A},\tau} - \sum_{\tau\in\mathcal{T}}\langle\lambda_\tau, \gamma_\tau v_\tau\rangle_{\partial\tau} &= (\tilde{f},\mathrm{v})\ \forall v \in \mathbf{H} \\ \sum_{\tau\in\mathcal{T}}\langle\mu_\tau, \gamma_\tau u_\tau\rangle_{\partial\tau} &= 0\ \forall \mu \in \mathbf{M}\end{aligned} \tag{9}$$

where $\langle .,.\rangle_{\partial\tau}$ denotes the duality product between $\mathbf{H}^{-1/2}(\partial\tau)$ and $\mathbf{H}^{1/2}(\partial\tau)$ and γ_τ is the trace function defined on the boundary $\partial\tau$ of τ. It can be easily verified that the physical meaning of the function λ is the flow **p.n** through the edges of the triangulation $\mathcal{T}$.

The simplest approach of (9)can be constructed choosing finite dimensional subspaces $\mathbf{H}_h \subset \mathbf{H}$ and $\mathbf{M}_h \subset \mathbf{M}$ defined by

$$\mathbf{H}_h = \{v \in \mathbf{L}^2(\Omega);\ v|_\tau \in \mathbf{P}_1(\tau)\ \forall \tau \in \mathcal{T}\ \}$$

$$\mathbf{M}_h = \{\mu = (\mu_\tau) \in \Pi_{\tau\in\mathcal{T}}\mathbf{S}_\tau^0\ ;\exists \mathbf{q} \in \mathbf{H}(\text{div};\Omega) \text{ such that } \mathbf{q.n}|_{\partial\tau} = \mu_\tau\ ,\ \forall \tau \in \mathcal{T}\ \}$$

where $\mathbf{S}_\tau^0 = \Pi_{i=1,2,3}\mathbf{P}_0((\partial\tau)_i)$, $(\partial\tau)_i$ denoting the i-th edge of the triangle τ. The elements of $\mathbf{M}_h$ are functions $\mu = (\mu_\tau)$ with μ_τ constant on each edge of the triangulation $\mathcal{T}$ satisfying $\mu_{\tau_1} + \mu_{\tau_2} = 0$ on $\tau_1 \cap \tau_2$ for each pair of adjacent triangles τ_1 and τ_2. The primal hybrid finite element approximation can be written as: find a couple $(u_h^{nc},\lambda_h)\in\mathbf{H}_hx\mathbf{M}_h$ such that

$$\begin{aligned}\sum_{\tau\in\mathcal{T}}(\nabla u^{nc}_{h,\tau},\nabla v_\tau)_{\mathbf{A},\tau}-\sum_{\tau\in\mathcal{T}}\langle\lambda_{h,\tau},\gamma_\tau v_\tau\rangle_{\partial\tau} &= (\tilde{f},v)\ \forall v\in\mathbf{H}_h \\ \sum_{\tau\in\mathcal{T}}\langle\mu_\tau,\gamma_\tau u^{nc}_{h,\tau}\rangle_{\partial\tau} &= 0\ \forall\mu\in\mathbf{M}_h\end{aligned} \tag{10}$$

Usually, in practical computations, a reduced problem is solved: Consider the subspace $\mathbf{V}_h^{nc}\subset\mathbf{H}_h$ defined by

$$\mathbf{V}_h^{nc}=\{v\in\mathbf{H}_h;\ \sum_{\tau\in\mathcal{T}}\langle\mu_\tau,\gamma_\tau v_\tau\rangle_{\partial\tau}=0\ \forall\mu\in\mathbf{M}_h\ \}$$

then, the restriction of (10) to $\mathbf{V}_h^{nc}$ is : find $u_h^{nc}\in\mathbf{V}_h^{nc}$ such that

$$\sum_{\tau\in\mathcal{T}}(\nabla u^{nc}_{h,\tau},\nabla v_\tau)_{\mathbf{A},\tau}=(\tilde{f},v)\ \forall v\in\mathbf{V}_h^{nc} \tag{11}$$

which is a non conforming method to solve (3).

Now the recovering of λ_h is a simple postprocessing:

$$\lambda_{h,\tau}|_{(\partial\tau)_i}=\frac{1}{|\tau|}((\nabla u_h^{nc},\nabla v)_{\mathrm{A}}-(\tilde{f},v)) \tag{12}$$

where v is a function in $\mathbf{H}_h$ such that $v=1$ on the middle of the edge $(\partial\tau)_i$ and $v=0$ on the middle of $(\partial\tau)_j$ for $j\neq i$.

Finally, the flows values $\lambda_{h,\tau}$ allow us to construct a vector function

$$\mathbf{p}_h\in\{\mathbf{q}\in\mathbf{H}(\mathrm{div};\Omega);\mathbf{q}|_\tau=(\mathrm{a+cx}\ ,\ \mathrm{b+cy})^{\mathrm{t}},\forall\tau\in\mathcal{T}\ \} \tag{13}$$

such that $\mathbf{p}_h.\mathbf{n}|_{(\partial\tau)_i}=\lambda_{h,\tau}|_{(\partial\tau)_i}$ and it can be easily verified that $div\mathbf{p}_h=\tilde{f}$.

4. A posteriori error estimation

To build up an adaptive finite element method, we give an a-posteriori error estimation of the values u_h and $\mathbf{p}_h$ obtained in paragraph 3.1 and 3.2 respectively. This will be based on the following theorem:

Theorem:

Let u be the solution of problem (1), $\mathbf{p}=\mathbf{A}\nabla u$, defining

$$\mathbf{X}^f=\{\mathbf{q}\in\mathbf{H}(div;\Omega);div\mathbf{q}+f=0\}$$

Then we have for all $\mathbf{q}\in\mathbf{X}^f$ and for all $v\in\mathbf{H}_0^1(\Omega)$

$$\begin{aligned}&\tfrac{1}{2}(\nabla(u-v),\nabla(u-v))_{\mathbf{A}}+\tfrac{1}{2}(\mathbf{p}-\mathbf{q},\mathbf{p}-\mathbf{q})_{\mathbf{A}^{-1}}=\\ &J(u)+I(\mathbf{q})=\tfrac{1}{2}(\mathbf{q}-\mathbf{A}\nabla v,\mathbf{q}-\mathbf{A}\nabla v)_{\mathbf{A}^{-1}}\end{aligned} \tag{14}$$

where $J(.)$ and $I(.)$ denote the total and complementary energy functionals respectively, i.e.:

$$J(v) = \tfrac{1}{2}(\nabla v, \nabla v)_{\mathbf{A}} - (f, v)$$
$$I(\mathbf{q}) = \tfrac{1}{2}(\mathbf{q}, \mathbf{q})_{\mathbf{A}^{-1}} \tag{15}$$

Proof: Let u be the solution of problem (1), $\mathbf{p} = \mathbf{A}\nabla u$, then we have for all $\mathbf{q} \in \mathbf{X}^f$ and for all $v \in \mathbf{H}_0^1(\Omega)$

$$\begin{aligned}
\frac{1}{2}(\nabla(u-v), \nabla(u-v))_{\mathbf{A}} &= \\
\frac{1}{2}(\nabla u, \nabla u)_{\mathbf{A}} - (f, v) + \frac{1}{2}(\nabla v, \nabla v)_{\mathbf{A}} &= \\
-\frac{1}{2}(\nabla u, \nabla u)_{\mathbf{A}} + (f, u) - (f, v) + \frac{1}{2}(\nabla v, \nabla v)_{\mathbf{A}} &= J(v) - J(u)
\end{aligned}$$

Analogously, as $\mathbf{p} \in \mathbf{X}^f$, for all $\mathbf{q} \in \mathbf{X}^f$

$$\begin{aligned}
\frac{1}{2}(\mathbf{p}-\mathbf{q}, \mathbf{p}-\mathbf{q})_{\mathbf{A}^{-1}} &= \\
\frac{1}{2}(\mathbf{p}, \mathbf{p})_{\mathbf{A}^{-1}} - (\mathbf{p}, \mathbf{q})_{\mathbf{A}^{-1}} + \frac{1}{2}(\mathbf{q}, \mathbf{q})_{\mathbf{A}^{-1}} &= \\
-\frac{1}{2}(\mathbf{p}, \mathbf{p})_{\mathbf{A}^{-1}} + (\mathbf{p}, \mathbf{p}-\mathbf{q})_{\mathbf{A}^{-1}} + \frac{1}{2}(\mathbf{q}, \mathbf{q})_{\mathbf{A}^{-1}} &= I(\mathbf{q}) - I(\mathbf{p})
\end{aligned}$$

where we have taken into account that $(\mathbf{p}, \mathbf{z})_{\mathbf{A}^{-1}} = 0$, for all $\mathbf{z} \in \mathbf{X}^0$. It follows immediately that

$$\frac{1}{2}(\nabla(u-v), \nabla(u-v))_{\mathbf{A}} + \frac{1}{2}(\mathbf{p}-\mathbf{q}, \mathbf{p}-\mathbf{q})_{\mathbf{A}^{-1}} = J(v) + I(\mathbf{q})$$

Finally, it is easily seen that

$$J(v) + I(\mathbf{q}) = \frac{1}{2}(\mathbf{q} - \mathbf{A}\nabla v, \mathbf{q} - \mathbf{A}\nabla v)_{\mathbf{A}^{-1}}$$

and the theorem is proved.

Applying the last result to problem (3), taking u_h and $\mathbf{p}_h$ in the place of v and $\mathbf{q}$ we get the error expression:

$$\frac{1}{2}\|\tilde{u} - u_h\|_{\mathbf{A}}^2 + \frac{1}{2}\|\tilde{\mathbf{p}} - \mathbf{p}_h\|_{\mathbf{A}^{-1}}^2 = J(u_h) + I(\mathbf{p}_h) = \frac{1}{2}\|\mathbf{p}_h - \mathbf{A}\nabla u_h\|_{\mathbf{A}^{-1}}^2 \tag{16}$$

Remark: Assuming that errors $\tilde{u} - u_h$ and $\tilde{\mathbf{p}} - \mathbf{p}_h$ are of the same order, it could be concluded that

$$\|\tilde{u} - u_h\|^2_{\mathbf{A}} \approx \|\tilde{\mathbf{p}} - \mathbf{p}_h\|^2_{\mathbf{A}^{-1}} \approx J(u_h) + I(\mathbf{p}_h) \approx \frac{1}{2}\|\mathbf{p}_h - \mathbf{A}\nabla u_h\|^2_{\mathbf{A}^{-1}}$$

Remark: For the errors $u - u_h$ and $\mathbf{p} - \mathbf{p}_h$ we have

$$\|u - u_h\|^2_{\mathbf{A}} \approx \|\mathbf{p} - \mathbf{p}_h\|^2_{\mathbf{A}^{-1}} \approx \frac{1}{2}\|\mathbf{p}_h - \mathbf{A}\nabla u_h\|^2_{\mathbf{A}^{-1}} + \mathcal{O}(h^4)$$

5. Adaptive algorithm

Given a mesh $\mathcal{T}$ it is natural to refine the triangulation to increase the accuracy. For completness we very briefly describe some of the basic ideas underlying the adaptive method. Suppose $\delta > 0$ is a given tolerance and suppose we want to obtain finite element approximations $u_h, \mathbf{p}_h$ such that

$$E = (\|u - u_h\|^2_{\mathbf{A}} + \|\mathbf{p} - \mathbf{p}_h\|^2_{\mathbf{A}^{-1}})^{1/2} \leq \delta \tag{17}$$

From expressions (6) and (7) this condition will be asymptotically satisfied if

$$\widetilde{E} = (\|\tilde{u} - u_h\|^2_{\mathbf{A}} + \|\tilde{\mathbf{p}} - \mathbf{p}_h\|^2_{\mathbf{A}^{-1}})^{1/2} \leq \delta \tag{18}$$

Relying on the error estimate (16) we see that (18) holds if the corresponding finite element mesh $\mathcal{T}$ is chosen so that

$$\widetilde{E}^2 = \|\mathbf{p}_h - \mathbf{A}\nabla u_h\|^2_{\mathbf{A}^{-1}} = \sum_{\tau \in \mathcal{T}} \|\mathbf{p}_h - \mathbf{A}\nabla u_h\|^2_{\mathbf{A}^{-1},\tau} \leq \delta^2 \tag{19}$$

To determine a mesh satisfying (19) we may proceed as follows: Choose a mesh $\mathcal{T}$ and compute a corresponding finite element solution $(u_h, \mathbf{p}_h)$. Using u_h and $\mathbf{p}_h$, for each $\tau \in \mathcal{T}$ compute the errors indicators defined by

$$\eta^2_\tau = \|\mathbf{p}_h - \mathbf{A}\nabla u_h\|^2_{\mathbf{A}^{-1},\tau} \tag{20}$$

Now, if N is the actual number of triangles in the mesh, construct a new mesh subdividing into four triangles each triangle $\tau \in \mathcal{T}$ for which

$$\eta^2_\tau > \frac{\delta^2}{\mathrm{N}}$$

Remark: Additional refinement may be necessary in order to assure the admissibility of the new mesh. See for example [10].

Finally, the quality of the error estimator $\widetilde{E}$ can be assessed by means of the so called effectivity index, ϑ_{eff}, defined by

$$\vartheta_{eff} = \frac{\widetilde{E}}{E}$$

Clearly, $\widetilde{E} = E + \mathcal{O}(h^2)$, then as $E = \mathcal{O}(h)$, it results $\vartheta_{eff} = 1 + \mathcal{O}(h) \overset{h \to 0}{\to} 1$.

6. Multigrid algorithm

Let $\{\mathcal{T}_l\}_{l=0}^{k}$ be a family of nested triangulations. To each $\mathcal{T}_l$ we associate two finite element spaces $\mathbf{V}_{h_l}^{c}$ and $\mathbf{V}_{h_l}^{nc}$ corresponding to the conforming and non conforming finite element approximations (8) and (11) respectively. Then consider the sequence of finite element spaces

$$\mathcal{M}_0, \mathcal{M}_1, ..., \mathcal{M}_{2l}, \mathcal{M}_{2l+1}, ..., \mathcal{M}_{2k}, \mathcal{M}_{2k+1}$$

where $\mathcal{M}_{2l} = \mathbf{V}_{h_l}^{c}$ and $\mathcal{M}_{2l+1} = \mathbf{V}_{h_l}^{nc}$.

Let $(.,.)_m$ be a family of mesh dependent $\mathbf{L}^2$-inner products on $\mathcal{M}_m, \mathcal{M}_{m+1}$ such that

$$c^{-1} \|v\|_{0,\Omega}^2 \leq (v,v)_m \leq c \|v\|_{0,\Omega}^2 \;\; \forall v \in \mathcal{M}_m \oplus \mathcal{M}_{m+1}$$

If $(.,.)_m$ is adequately chosen, an easily computable "prolongation" operator $P_{m,m+1} : \mathcal{M}_m \to \mathcal{M}_{m+1}$ can be obtained by

$$(P_{m,m+1} v_m, w_{m+1})_{m+1} = (v_m, w_{m+1})_{m+1} \;\; \forall v_m \in \mathcal{M}_m \;\; \forall w_{m+1} \in \mathcal{M}_{m+1}$$

In practice, the calculation of $P_{m,m+1} v_m$ results in a suitable averaging of the nodal values of v_m. To see this consider the following examples:

a) Case $m = 2l$, $\mathcal{M}_{2l} \subset \mathcal{M}_{2l+1}$: Let $\{b_i\}$ be the set of midpoints of the edges of the triangulation $\mathcal{T}_l$

$$(u,v)_{2l+1} = (u,v) = \sum_{\tau \in \mathcal{T}_l} \frac{|\tau|}{3} \sum_{i=1,2,3} u(b_i) v(b_i)$$

and $P_{m,m+1}$ be the inclusion $v_m \in \mathcal{M}_m \to P_{m,m+1} v_m = v_m \in \mathcal{M}_{m+1}$.

b) Case $m = 2l+1$, $\mathcal{M}_{2l+1} \not\subset \mathcal{M}_{2l+2}$: Let $\{a_i\}$ be the set of vertices of the triangulation $\mathcal{T}_{l+1}$

$$(u,v)_{2l+2} = \sum_{\tau \in \mathcal{T}_{l+1}} |\tau| \sum_{i=1,2,3} u(a_i) v(a_i)$$

A function in $\mathcal{M}_{2l+1}$ is defined by its values on the midpoints of the edges of the triangulation $\mathcal{T}_l$. A function in $\mathcal{M}_{2l+2}$ is defined by its values on the vertex of the triangulation $\mathcal{T}_{l+1}$. For two representative points, a vertex A and a midpoint of an edge B, we get

$$P_{2l+1,2l+2}v_{2l+1}(A) = \frac{\sum\limits_{\tau\in\omega_a} |\tau|\, v_{2l+1}(A)}{\sum\limits_{\tau\in\omega_a} |\tau|}$$

$$P_{2l+1,2l+2}v_{2l+1}(B) = v_{2l+1}(B)$$

where ω_a is the set of triangles which have A as a common vertex. Note that in A the value $v_{2l+1}(A)$ could be different in each triangle $\tau \in \omega_a$.

Using the operators $P_{m,m+1}$ we are now in a position to formulate our multigrid algorithm. Let us introduce the notation

$$a_m(u_m, v_m) = \sum_{\tau\in\mathcal{T}_l} (\nabla u_m, \nabla v_m)_{\mathbf{A},\tau} \; ; \; m = 2l \text{ or } 2l+1$$

and consider the eigenvalue problem

$$\begin{aligned} a_m(\Psi_{m,i}, v_m) &= \lambda_{m,i}(\Psi_{m,i}, v_m)_m \; \forall v_m \in \mathcal{M}_m, \; \Psi_{m,i} \in \mathcal{M}_m \\ (\Psi_{m,i}, \Psi_{m,j})_m &= \delta_{i,j} \end{aligned}$$

Algorithm 1 (One iteration at level m with ν smoothing steps)
Set $\Lambda_m = \max\{\lambda_{m,i}\} \leq ch_l^{-2}$:
1.- Smoothing. Given $u_m^0 \in \mathcal{M}_m$ with $m = 2l$ or $2l+1$, for $i = 1, ..., \nu$ compute u_m^i from u_m^{i-1} by solving

$$(u_m^i, v_m)_m = (u_m^{i-1}, v_m)_m + \frac{1}{\Lambda_m}\{(f, v_m) - a_m(u_m^{i-1}, v_m)\} \; \forall v_m \in \mathcal{M}_m$$

2.- Coarse grid correction. Denote $u_{m-1}^* \in \mathcal{M}_{m-1}$ the solution of the coarse grid problem

$$a_{m-1}(u_{m-1}^*, v_{m-1}) = (f, P_{m-1,m}v_{m-1}) - a_m(u_m^\nu, P_{m-1,m}v_{m-1}) \forall v_{m-1} \in \mathcal{M}_{m-1} \tag{21}$$

If $m = 1$, set $\widetilde{u}_{m-1} = u_{m-1}^*$. If $m > 1$, compute an approximation $\widetilde{u}_{m-1}$ to u_{m-1}^* by applying $\gamma = 1$ or 2 iterations of the algorithm at level $m - 1$ to problem (21) with starting value zero. Put

$$u_m^{\nu,c} = u_m^\nu + P_{m-1,m}\widetilde{u}_{m-1}$$

The right-hand side of (21) in the coarse grid correction can easily be computed by a suitable averaging of the fine grid residuals. To see this, consider the two cases involved:

a) Case $m = 2l$, $\mathcal{M}_{2l} \subset \mathcal{M}_{2l+1}$, corresponding to the mesh $\mathcal{T}_l$

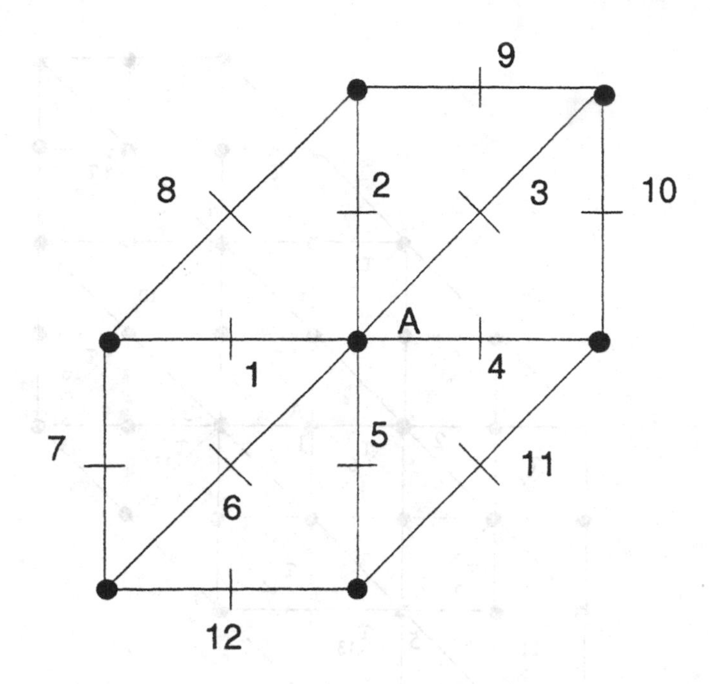

Denote $w_j \in \mathcal{M}_{2l+1}$ the basis functions corresponding to the midpoints of the edges of $\mathcal{T}_l$ and $v_A \in \mathcal{M}_{2l}$ the basis function corresponding to the point A. Put

$$r_j = (f, w_j) - a_{2l+1}(u^{\nu}_{2l+1}, w_j)$$

An easy calculation then yields

$$\begin{aligned} &(f, P_{2l,2l+1}v_A) - a_{2l+1}(u^{\nu}_{2l+1}, P_{2l,2l+1}v_A) \\ = \; &\frac{r_1 + r_2 + \ldots + r_6}{2} \end{aligned}$$

b) Case $m = 2l+1, \mathcal{M}_{2l+1} \not\subset \mathcal{M}_{2l+2}$, correspondig to the meshes $\mathcal{T}_l$ and $\mathcal{T}_{l+1}$.

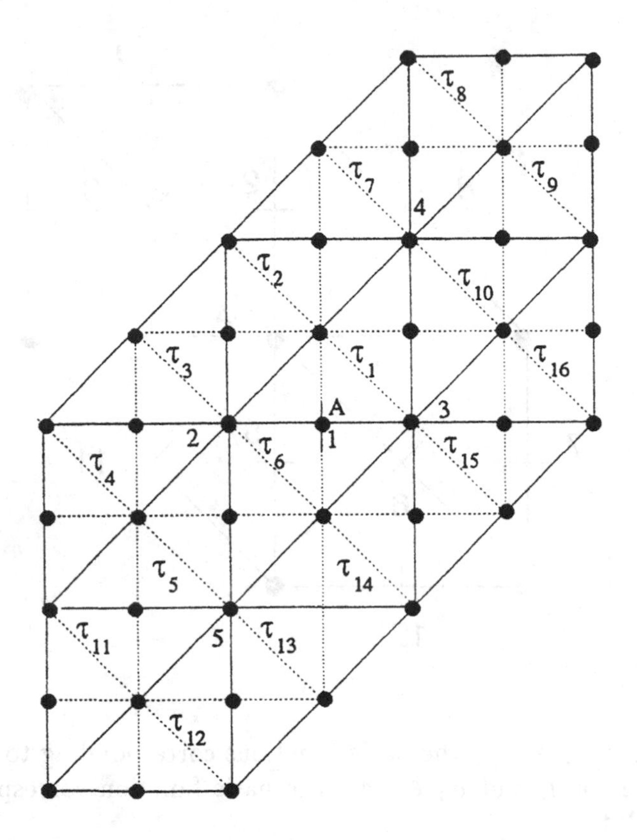

With the corresponding notation

$$(f, P_{2l+1,2l+2}v_A) - a_{2l+2}(u^{\nu}_{2l+2}, P_{2l+1,2l+2}v_A) = c_1 r_1 + \dots + c_5 r_5$$

where

$$c_1 = P_{2l+1,2l+2}v_A(x_1) = v_A(x_1) = 1$$

$$c_2 = P_{2l+1,2l+2}v_A(x_2) = \frac{|\tau_1| + 2|\tau_6|}{|\tau_1| + |\tau_2| + 2|\tau_3| + |\tau_4| + |\tau_5| + 2|\tau_6|}$$

$$c_3 = P_{2l+1,2l+2}v_A(x_3) = \frac{2|\tau_1| + |\tau_6|}{2|\tau_1| + |\tau_{10}| + |\tau_{16}| + 2|\tau_{15}| + |\tau_{14}| + 2|\tau_6|}$$

$$c_4 = P_{2l+1,2l+2}v_A(x_4) = \frac{-|\tau_1|}{|\tau_1| + |\tau_2| + 2|\tau_7| + |\tau_8| + |\tau_9| + 2|\tau_{10}|}$$

$$c_5 = P_{2l+1,2l+2}v_A(x_5) = \frac{-|\tau_6|}{|\tau_6| + 2|\tau_5| + |\tau_{11}| + |\tau_{12}| + 2|\tau_{13}|}$$

6.1. CONVERGENCE ANALYSIS OF THE MULTIGRID ALGORITHM

We will provide a convergence analysis for the two-level algorithm. The convergence of the multigrid algorithm can be obtained, at least for the W-cycle ($\gamma = 2$) following a recursive argument.

Theorem: Let u_m the solution of

$$a_m(u_m, v_m) = (f, v_m) \; \forall v_m \in \mathcal{M}_m, \; m = 2k \text{ or } 2k+1$$

Let $u_m^{\nu,ec}$ be the solution obtained after one iteration of the two grid algorithm (that is, algorithm 1 replacing step 2 by an exact coarse grid correction). Set $e_m^0 = u_m - u_m^0$ the initial error, then we have

$$\|u_m - u_m^{\nu,ec}\|_{0,\Omega} \leq \frac{c}{\nu+1} \left\| e_m^0 \right\|_{0,\Omega}$$

Proof: It is a simple adaptation of the analysis given in [14].

It should be noticed that the analysis of [14] uses that w, the solution of (24), verifies $w \in \mathbf{H}^2(\Omega)$. We now obtain a two level convergence rate with respect to the energy norm with a weak regularity assumption. With each space $\mathcal{M}_m$, we define discrete norms

$$\|v\|_{m,s}^2 = \sum c_i^2 \lambda_{m,i}^s \text{ for } v = \sum c_i \Psi_{m,i}$$

where $-2 \leq s \leq 2$. Note that $\|v\|_{m,1} = \|v\|_{\mathbf{A}}$ is the usual energy norm on level m. Further the following norm equivalence holds: There exist a constant C independent of the level m such that

$$C^{-1} \|v\|_{s,\Omega} \leq \|v\|_{m,s} \leq C \|v\|_{s,\Omega} \tag{22}$$

Theorem: Assume that there exists a constant $0 < \alpha \leq 1$, such that, for all $f \in \mathbf{H}^{\alpha-1}(\Omega)$ there exists a unique solution $u \in \mathbf{H}^{1+\alpha}(\Omega) \cap \mathbf{H}_0^1(\Omega)$ of problem (1) and that the spaces $\mathcal{M}_m$ satisfy the standard approximation property

$$\|u - u_h\|_{\mathbf{A}} \leq ch^\alpha \|u\|_{1+\alpha,\Omega}$$

Then we obtain the two level convergence rate:

$$\|e_m^{\nu,ec}\|_{\mathbf{A}} \leq c(2\nu + \alpha)^{-\alpha/2} \left\| e_m^0 \right\|_{\mathbf{A}}$$

Proof: The effect of the smoothing iteration can be easily computed. For every $0 < \alpha \leq 1$ we obtain, expanding the initial error e_m^0 in terms of the eigenfunctions

$$\begin{aligned} \|e_m^\nu\|_{m,1+\alpha}^2 &= \sum c_i^2 \lambda_{m,i}^{1+\alpha}(1-\frac{\lambda_{m,i}}{\Lambda_m})^{2\nu} \\ &= \Lambda_m^\alpha \sum c_i^2 \lambda_{m,i}(\frac{\lambda_{m,i}}{\Lambda_m})^\alpha (1-\frac{\lambda_{m,i}}{\Lambda_m})^{2\nu} \\ &\leq \Lambda_m^\alpha \max_{x\in[0,1]} |x^\alpha(1-x)^{2\nu}| \sum c_i^2 \lambda_{m,i} \\ &\leq c h_m^{-2\alpha}(2\nu+\alpha)^{-\alpha} \|e_m^0\|_{m,1}^2 \end{aligned} \tag{23}$$

In order to analyse the coarse grid correction we require the following energy norm consistency, (see [13] for a proof). For all $m \geq 1$ there exists a positive constant c, independent of m, such that

$$a_m(P_{m,m+1}v, P_{m,m+1}v) \leq c a_{m-1}(v,v) \; \forall v \in \mathcal{M}_{m-1}$$

Now, as in [14], define the function $r_m \in \mathcal{M}_m$ by

$$(r_m, v_m)_m = a_m(e_m^\nu, v_m) \; \forall v_m \in \mathcal{M}_m$$

then

$$a_{m-1}(u_{m-1}^*, v_{m-1}) = (r_m, v_{m-1})_m \; \forall v_{m-1} \in \mathcal{M}_{m-1}$$

Let $w \in \mathbf{H}_0^1(\Omega)$ be the solution of

$$\begin{aligned} -div(\mathbf{A}\nabla w) &= r_m \text{ in } \Omega \\ w &= 0 \text{ on } \Gamma \end{aligned} \tag{24}$$

we conclude taht e_m^ν and u_{m-1}^* are finite element approximations of w. Then by the regularity assumption

$$\|w\|_{1+\alpha,\Omega} \leq c \|r_m\|_{\alpha-1,\Omega}$$

Using the equivalence of norms (22), for different constants c

$$\begin{aligned} \|e_m^\nu - u_{m-1}^*\|_{m,1} &= \|e_m^\nu - u_{m-1}^*\|_{\mathbf{A}} \leq \\ \|e_m^\nu - w\|_{\mathbf{A}} + \|w - u_{m-1}^*\|_{\mathbf{A}} &\leq c h^\alpha \|w\|_{1+\alpha,\Omega} \leq c h^\alpha \|r_m\|_{\alpha-1,\Omega} \leq \\ c h^\alpha \|r_m\|_{m,\alpha-1} &\leq c h^\alpha \|e_m^\nu\|_{m,\alpha+1} \end{aligned} \tag{25}$$

Finally, using the energy consistency property we have

$$\|e_m^\nu - P_{m,m-1}u_{m-1}^*\|_{m,1} \leq c \|e_m^\nu - u_{m-1}^*\|_{m,1}$$

and by virtue of (23) and (25) the theorem is proved.

In this subsection we give an adaptive multigrid algorithm combining the refinement technique described in section 5 with the present multigrid algorithm using the nested iteration.

Let $MG(m, i, u_m^0, u_m^i)$ denote i iterations of algorithm 1 at level $m \neq 0$ where u_m^0 is the starting value and u_m^i the result after i iterations.

Algorithm 2
$Tol := \varepsilon < 1.$
$\widetilde{E} = 1.$
$k = k_{\max}$,
$l = 0$
While $\widetilde{E} > Tol$ and $l \leq k$ do
If $l = 0$
$\widetilde{u}_0 = u_0$ solution of $a_0(u_0, v) = (\widetilde{f}, v)\ \forall v \in \mathcal{M}_0$
else
$u_{2l}^0 = P_{2l-1,2l}\widetilde{u}_{2l-1}$
$MG(2l, i, u_{2l}^0, u_{2l}^i)$
$\widetilde{u}_{2l} = u_{2l}^i$
endif
$u_{2l+1}^0 = P_{2l,2l+1}\widetilde{u}_{2l}$
$MG(2l+1, i, u_{2l+1}^0, u_{2l+1}^i)$
$\widetilde{u}_{2l+1} = u_{2l+1}^i$
Compute $\widetilde{\mathbf{p}}_{2l+1}$ from $\widetilde{u}_{2l+1}$ using (12) and (13)
Compute η_{τ_l} for all $\tau_l \in \mathcal{T}_l$ using expression ((20))
Compute $\widetilde{E} = \widetilde{E}_l$ by expression (19)
If $\widetilde{E} > Tol$: Refine the mesh $\mathcal{T}_l$ to construct $\mathcal{T}_{l+1}$
$l = l + 1$
end while
end

7. Numerical example

To illustrate the foregoing theory, we describe the results for one representative example. See [15] for another example. Consider the problem:

$$\begin{aligned} -div(\mathbf{A}(x,y)\nabla u) &= f(x,y) \text{ in } [0,1]^2 \\ u(x,0) &= u(0,y) = u(x,1) = u(1,y) = 0 \end{aligned}$$

where

$$\mathbf{A}(x,y) = \frac{1}{\alpha} + \alpha(r-\beta)^2$$

$$
\begin{aligned}
f(x,y) &= 2[1+\alpha(r-\beta)\text{arg}\tan(\alpha(r-\beta)+\text{arg}\tan\alpha r)] \\
r &= \sqrt{x^2+y^2}, \alpha = 100., \beta = 0.36388
\end{aligned}
$$

Mesh	$(\mathbf{J}(\mathbf{u}_c) - \mathbf{I}(\sigma_h))/2.$	Error^2(*Estimation*).	Rel.Error(%).	Effect. index
1	-14.3112	9.5202	33.2618	1.9069
2	-13.0122	5.3750	20.6555	1.4174
3	-12.9054	4.1703	16.1576	1.4342
4	-11.9725	1.6559	6.9168	1.0393
5	-11.8222	0.4887	2.0768	1.0128

TABLE 1. Accuracy attained in solution.

The solution has an interior layer in the neighborhood of $r = \beta$. The values of the total and complementary energies are shown in figure (1). As can be seen in Table 1 the effectivity index tends to unity very fast confirming the quality of the error estimator. The number of stabilized iterations of the multigrid algorithm is 4.

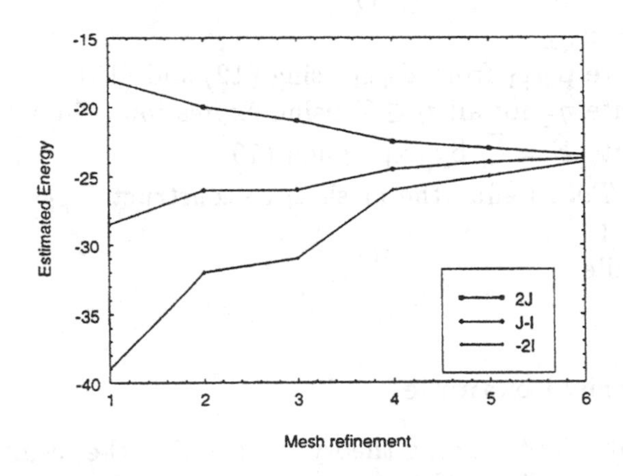

Figure 1. Estimated energy.

References

1. I. Babuska and W.C. Rheinboldt, *Error estimates for adaptive finite element computation.* SIAM J. Numer. Anal., vol. 15, pp. 736-754, 1978.
2. I. Babuska and W.C. Rheinboldt, *A-posteriori error estimates for finite element method.* Int. J. Numer. Methods Eng.,vol.12, pp. 1597-1615, 1978.

3. R.E. Bank and A. Weiser, *Some a posteriori error estimators for elliptic partial differential equations.* Math. Comput., vol. 44, pp. 283-301, 1985.
4. J.L. Liu and W.C. Rheinboldt, *A posteriori finite element error estimator for indefinite boundary value problems.* Numer. Funct. Anal. and Optimiz., vol. 15, pp. 335-356, 1994.
5. C. Johnson and P. Hansbo, *Adaptive finite element methods in computational mechanics.* Comput. Methods Appl. Mech. Eng., vol.101, pp. 143-181, 1992.
6. C. Johnson, *Adaptive finite element methods for the obstacle problem.* Mathematical Models and Methods in Applied Sciences, vol. 2, pp. 483-487, 1992.
7. D. Peric, J. Yu and D.R.J. Owen, *On error estimates and adaptivity in elastoplastic solids: Applications to the numerical solutions of strain localization in classical and Cosserat continua.* Int. J. Numer. Methods Eng., vol. 37, pp. 1351-1379, 1994.
8. O. Zienkiewicz and J.Z. Zhu, *A simple error estimator and adaptive procedure for practical engineering analysis.* Int. J. Numer. Methods Eng., vol. 24, pp. 337-357, 1987.
9. R. Teman, *Mathematical Methods in Plasticity Problems.* Gauthiers Villars, Paris, 1982.
10. M.C. Rivara, *Algorithms for refining triangular grids suitable for adaptive and multigrid techniques.* Int. J. Numer. Methods Eng., vol. 20, pp. 745-756, 1984.
11. P.A. Raviart and J.M. Thomas, *Primal hybrid finite element methods for second order elliptic equations.* Math. Comput., vol.31, pp. 391-413, 1977.
12. Ph. Ciarlet, *The Finite Element Method for Elliptic Problems.* North Holland, Amsterdam, 1978.
13. P. S. Vassilevski and J. Wang, *An application of the abstract multilevel theory to nonconforming finite element methods.* SIAM J. Numer. Anal., vol. 32, pp. 235-248, 1995.
14. D. Braess and R. Verfurth, *Multigrid methods for nonconforming finite element methods.* SIAM J. Numer. Anal., vol. 27, pp. 979-986, 1990.
15. L. Ferragut, R. Montenegro, *Accurate a-posteriori error estimation using duality.* Neural, Parallel and Scientific Computations, vol. 3, pp. 549-564, 1995.

ON FINDING AND ANALYZING THE STRUCTURE OF THE CHOLESKY FACTOR

ALAN GEORGE
Department of Computer Science
University of Waterloo
Waterloo, Ontario, Canada
EMAIL: jageorge@sparse1.uwaterloo.ca

Abstract. The first two steps in solving a sparse symmetric positive definite system of equations are to find an ordering which limits fill, followed by an analysis of the symmetrically permuted matrix to discover various structural properties of its Cholesky factor. This latter step is called symbolic factorization, and its purpose is to aid in implementing efficient data structures and computational algorithms. In the past, it was common to accept whatever low-fill ordering was supplied by the ordering algorithm, and proceed to the symbolic factorization step. However, much progress has been made in recent years in "massaging" the ordering so as to preserve its quality in terms of fill and to enhance other desirable properties in the structure of $\boldsymbol{L}$. Finding these reorderings involves the use of some fundamental algorithms and structures. The objective of this paper is to examine these basic tools and their use in some recent advances associated with discovering properties of $\boldsymbol{L}$.

1. Introduction and Background

This paper is tutorial in nature, and deals with some advances in an area of solving large sparse symmetric positive definite systems. It borrows heavily from a number of sources, particularly from recent articles by Gilbert et. al.[16] and Liu et. al.[26], as well as earlier work by numerous researchers [11, 12, 14, 15, 17, 24, 33, 34].

Consider the problem of solving the $n \times n$ sparse system of equations

$$\boldsymbol{Ax} = \boldsymbol{b},$$

G. Winter Althaus and E. Spedicato (eds.), Algorithms for Large Scale Linear Algebraic Systems, 73–105.

where the method to be used is Gaussian elimination. A triangular factorization of the matrix is computed, followed by the solution of two triangular systems in order to obtain the solution $\boldsymbol{x}$. For a general indefinite sparse matrix $\boldsymbol{A}$, some form of *pivoting* (row and/or column interchanges) is necessary to ensure numerical stability. Given $\boldsymbol{A}$, one normally obtains a factorization of $\boldsymbol{PA}$ or $\boldsymbol{PAQ}$, where $\boldsymbol{P}$ and $\boldsymbol{Q}$ are permutation matrices of the appropriate size. Thus, the process has two stages:

1. factor the matrix $\boldsymbol{PAQ}$ into the product of a lower triangular matrix $\boldsymbol{L}$ and an upper triangular matrix $\boldsymbol{U}$: $\boldsymbol{PAQ} = \boldsymbol{LU}$.
2. compute $\boldsymbol{x}$ using $\boldsymbol{L}, \boldsymbol{U}, \boldsymbol{P}, \boldsymbol{Q}$ and $\boldsymbol{b}$: solve $\boldsymbol{Ly} = \boldsymbol{Pb}$ and $\boldsymbol{Uz} = \boldsymbol{y}$, and then set $\boldsymbol{x} = \boldsymbol{Qz}$.

Sparse problems are interesting (or annoying, frustrating) because the coefficient matrix $\boldsymbol{A}$ normally suffers some *fill* when Gaussian elimination is applied to it. Its factors will generally have nonzeros in positions that are zero in $\boldsymbol{PA}$ or $\boldsymbol{PAQ}$. Thus, when $\boldsymbol{A}$ is sparse, the permutations above are determined *during the factorization* by a combination of (usually competing) numerical stability and sparsity requirements (Duff [8]). Different matrices, even though they may have the same nonzero pattern, will usually yield different $\boldsymbol{P}$ and $\boldsymbol{Q}$, and therefore have factors with different sparsity patterns. In other words, for general sparse matrices, it normally is not possible to predict where fill will occur before the computation begins. Thus, some form of *dynamic* storage scheme is required which allocates storage for fill as the computation proceeds.

However, symmetric and positive definite coefficient matrices have very important advantages. Gaussian elimination applied to such matrices does not require interchanges (pivoting) to maintain numerical stability. Since $\boldsymbol{PAP}^T$ is also symmetric and positive definite for any permutation matrix $\boldsymbol{P}$, one can choose to reorder $\boldsymbol{A}$ symmetrically

1. without regard to numerical stability and
2. before the numerical factorization begins.

This has major practical implications. Since the ordering can be determined before the factorization begins, the locations where the fill will occur during the factorization can also be determined before any numerical computation is performed, and the data structure used to store $\boldsymbol{L}$ can be constructed. The computation then proceeds with the storage structure remaining *static* (unaltered). Thus, the usual scenario for solving sparse positive definite systems is as follows:

1. ordering: determine a permutation $\boldsymbol{P}$ so that $\boldsymbol{PAP}^T$ has a sparse triangular factorization: $\boldsymbol{PAP}^T = \boldsymbol{LL}^T$.
2. symbolic factorization: determine the structure of $\boldsymbol{L}$ and set up a compact data structure for storing the nonzeros of $\boldsymbol{L}$.

3. numerical factorization: store the nonzeros of A in the fixed data structure determined at step 2 and compute L in place.
4. triangular solutions: solve $Ly = Pb$ and $L^T z = y$, and then set $x = P^T z$.

The first two steps depend on the structure of A only, and there are contexts in which it is desirable to perform these steps separate from the numerical computation. For example, one may need to solve many problems with the same structure but different numerical values, or many problems differing only in their right hand side. In such cases steps 1 and 2 need be done only once.

An objective of step 1 above is to find an ordering that yields low fill. In the past, it was common to accept whatever low-fill ordering was supplied by the ordering algorithm, and proceed to the symbolic factorization step. However, much progress has been made in recent years in adjusting the ordering so as to preserve its quality in terms of fill and to enhance other desirable properties in the structure of L. The objective of this article is to study some of the basic tools and techniques used in this process, and to show how they are used to discover some important structural properties of L.

An outline of the paper is as follows. In the remainder of this section some notation is introduced. Section 2 contains some basic facts about the structure of L which are direct consequences of the Gaussian elimination algorithm. These facts are then used to derive some additional standard results about the structure of L; the section concludes with a description of a basic but efficient symbolic factorization algorithm due to Sherman[34]. Section 3 introduces an important structure called the *elimination tree*, along with some standard results about it and an algorithm for obtaining it very efficiently. Important subtrees of the elimination tree called *row subtrees* are also introduced. Section 4 deals with reorderings of A that are induced by various topological orderings of the elimination tree, with special emphasis on postorderings of the tree. For completeness, a standard algorithm for finding a postordering of the tree is also included in this section[1, pp. 78-82]. Section 5 contains a study of row subtrees when the tree has been postordered. Following Liu et. al.[26], an efficient algorithm is developed to find the row subtrees, given A and the elimination tree. Section 6 introduces the notion of *supernodes* which are, loosely speaking, sequences of consecutive columns of L having the "same" structure. Identifying them is important in connection with setting up efficient data structures and computational algorithms for Cholesky factorization. Using techniques from sections 4 and 5, an efficient algorithm[26] for finding supernodes of L is described, given only the structure of A and the elimination tree. Finally, in section 7 algorithms from sections 4-6 are used as a foundation to develop an efficient algorithm due to Gilbert et. al.[16] to determine the number of nonzeros in the rows and columns of $\boldsymbol{L}$, given only the structure of $\boldsymbol{A}$ and the elimination tree. Section 8 contains concluding remarks.

1.1. NOTATION AND ASSUMPTIONS

The i-th row and i-th column of a matrix $\boldsymbol{A}$ are denoted by $\boldsymbol{A}_{i*}$ and $\boldsymbol{A}_{*i}$ respectively; $\Omega(\boldsymbol{v})$ denotes the set of subscripts corresponding to nonzeros in the vector $\boldsymbol{v}$ and $\Omega(\boldsymbol{A})$ denotes the set of subscripts pairs corresponding to nonzeros in the matrix $\boldsymbol{A}$. Finally, $\eta(\boldsymbol{v})$ denotes the number of nonzero components in $\boldsymbol{v}$ and $|S|$ denotes the number of members in a set S. Thus, $\eta(\boldsymbol{v}) = |\Omega(\boldsymbol{v})|$.

It is desirable to be able to refer to the upper or lower triangular parts of matrices, sometimes including the diagonal and at other times excluding it. To this end the notation $\boldsymbol{A}^{\triangle}$ is used to refer to the matrix obtained from $\boldsymbol{A}$ by setting all of its elements above the main diagonal to zero, and $\boldsymbol{A}^{\llcorner}$ denotes the matrix obtained by setting all the elements above the main diagonal as well as the diagonal itself to zero. Thus, for example, $\Omega(\boldsymbol{A}_{*i}^{\urcorner})$ is the set of row subscripts of nonzeros above the diagonal in column i of $\boldsymbol{A}$.

If $\boldsymbol{A}$ can be symmetrically permuted into block diagonal form, it is said to be *reducible*; otherwise, it is *irreducible*. It is assumed throughout this paper that $\boldsymbol{A}$ is irreducible.

The usual assumption that exact cancellation does not occur is also made. Addition or subtraction of nonzero quantities is always assumed to yield a nonzero element. Thus, with this assumption, the structure of $\boldsymbol{L}$ is completely determined by the structure of $\boldsymbol{A}$.

2. Basic structure of L and symbolic factorization

Given a (reordered) $\boldsymbol{A}$, the next step in the process of solving a sparse positive definite system usually is to set up a data structure to accommodate the corresponding Cholesky factor $\boldsymbol{L}$. In order to minimize notation, it is assumed that $\boldsymbol{A}$ has already been symmetrically permuted. That is, $\boldsymbol{A}$ rather that $\boldsymbol{PAP}^T$ will be referred to below.

As noted earlier, symbolic factorization refers to the process of simulating the factorization of $\boldsymbol{A}$ in some way in order to discover the structure of its factor $\boldsymbol{L}$. Of course, a naive way is to simulate the Cholesky factorization itself: $\boldsymbol{L}_{*i}$ is a linear combination of those columns $\boldsymbol{L}_{*k}$ for $k \in \Omega(\boldsymbol{L}_{i*}^{\llcorner})$, together with $\boldsymbol{A}_{*i}^{\triangle}$. Thus, $\Omega(\boldsymbol{L}_{*i})$ can be determined by examining $\Omega(\boldsymbol{A}_{*i}^{\triangle})$ along with $\Omega(\boldsymbol{L}_{*k})$ for $k \in \Omega(\boldsymbol{L}_{i*}^{\llcorner})$.

However, one can do much better than this by looking carefully at the structure of $\boldsymbol{L}$. One begins with a basic relationship between $\Omega(\boldsymbol{A})$ and $\Omega(\boldsymbol{L})$ due to Parter[31]:

Lemma 2.1 *Let $i > j$. Then $\ell_{ij} \neq 0$ if and only if at least one of the following conditions holds:*

a) $a_{ij} \neq 0$
b) *for some $k < j$, $\ell_{ik} \neq 0$ and $\ell_{jk} \neq 0$.*

■

In words, the lemma says that ℓ_{ij} will be nonzero if a_{ij} is already nonzero or, even if $a_{ij} = 0$, ℓ_{ij} will become nonzero if $\ell_{ik} \neq 0$ and $\ell_{jk} \neq 0$ for some k less than i and j.

This lemma may seem somewhat unnatural since the fill is characterized in terms of the lower triangle only. However, it is simply a statement about what happens when ordinary Gaussian elimination is applied to a matrix having symmetric structure. (The matrix does not have to be numerically symmetric.) Suppose ordinary Gaussian elimination is applied to such a matrix. It is easy to confirm that the structures of the sequence of successively smaller matrices that are generated by the algorithm will also have symmetric structure. Now consider the k-th step of the algorithm, which involves modifying the remaining $(n-k) \times (n-k)$ part of the matrix by the outer product of column k and row k. As depicted in Figure 1, a nonzero in position jk implies that there is also a nonzero in position kj, which will cause fill to occur in position ij when row k is used to eliminate the element in position ik. This is the essence of Lemma 2.1.

In the sequel, frequent reference will be made to a_{ij}, often in connection with whether it is zero or nonzero. Unless explicitly stated otherwise, it should be understood that $i > j$. To avoid tedious duplication, apart from in lemmas or theorems the qualifier will be omitted.

Figure 1. Illustration of Lemma 2.1.

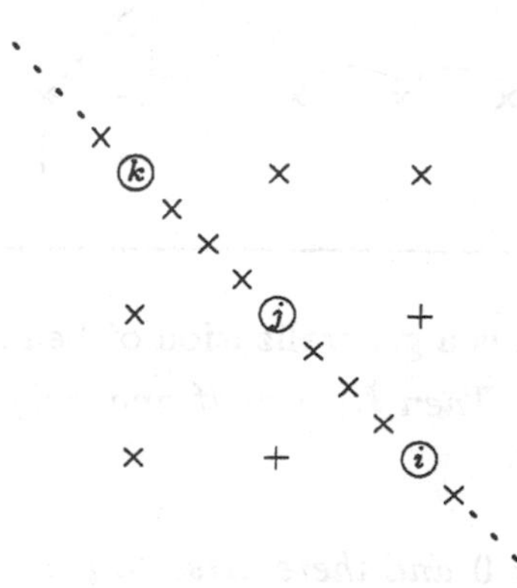

Let m_k denote the row index of the first off-diagonal nonzero in column k of $\boldsymbol{L}$. Recall that $\boldsymbol{A}$ is assumed to be irreducible; in this case it can be shown that $\eta(\boldsymbol{L}_{*k}) > 1$, which implies $m_k > k, 1 \leq k < n$. In what follows, sequences of the form $k, m_k, m_{m_k}, \cdots$ will be discussed; the notation $m^0(k)$ will mean k, $m(k) = m^1(k)$ will mean m_k, $m^2(k)$ will mean m_{m_k}, and so on. The next theorem is a simple consequence of Lemma 2.1.

Theorem 2.2 *Let $i > j$ and $a_{ij} \neq 0$. Then there exists a $p > 0$ such that $m^p(j) = i$ and $\ell_{is} \neq 0$ for $s = j, m(j), m^2(j), \cdots, m^p(j)$.* ■

In words, Theorem 2.2 says that if $i > j$ and $a_{ij} \neq 0$, then the sequence of column indices $j, m(j), m^2(j), \cdots$ will include i, and there will be nonzeros in the row corresponding to those column indices in the sequence that are less than or equal to i. Thus, each nonzero in the row of $\boldsymbol{A}$ can be viewed as being responsible for a sequence of (potential) fills in row i. Some may not be actual fills since the element may already be nonzero in $\boldsymbol{A}$. Each nonzero element a_{ij} in $\boldsymbol{A}^{\llcorner}_{i*}$ begins a sequence of column subscripts $j, m(j), m^2(j), \cdots$ which contains i and ends at n. Some of the sequences will partially overlap, while some will be subsequences of others, but they all eventually terminate at n. Figure 2 is a pictorial illustration of Theorem 2.2.

Figure 2. Illustration of Theorem 2.2.

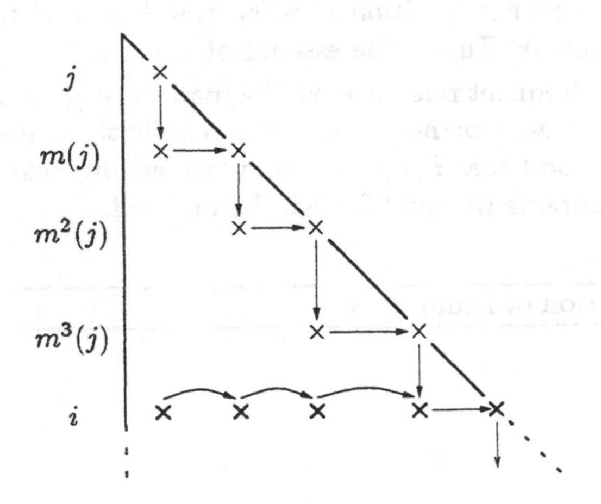

The following theorem is a generalization of Lemma 2.1.

Theorem 2.3 *Let $i > j$. Then $\ell_{ij} \neq 0$ if and only if at least one of the following conditions holds:*

a) $a_{ij} \neq 0$

b) *For some $k < j$, $a_{ik} \neq 0$ and there exists a $p > 0$ such that $m^p(k) = j$ and $\ell_{is} \neq 0$ for $s = k, m(k), m^2(k), \cdots, m^p(k)$.* ■

In words, Theorem 2.3 says that $\ell_{ij} \neq 0$ because $a_{ij} \neq 0$ or there is a nonzero a_{ik} to the left of it in the row that causes fill to occur in position ij because of the existence of the sequence $m_k, m^2(k), \cdots, m^p(k) = j$. Note that if both conditions hold, a_{ij} could be set to zero without changing the structure of $\boldsymbol{L}$, since fill will occur in position ij anyway as a result of the second condition being true. Thus, Theorem 2.3 could be used to determine a set of *redundant elements* of $\boldsymbol{A}$ that could be discarded without altering the structure of $\boldsymbol{L}$. On the other hand, if $a_{ij} \neq 0$ and condition b) does not hold, then a_{ij} is an *essential element* in the sense that setting it to zero would change the structure of $\boldsymbol{L}$. Thus, there are two classes of nonzeros in $\boldsymbol{A}$: those that are redundant in terms of generating fill, and those that are not. This notion will reappear numerous times later in this paper. Figure 3 is a pictorial illustration of Theorem 2.3.

Figure 3. Illustration of Theorem 2.3.

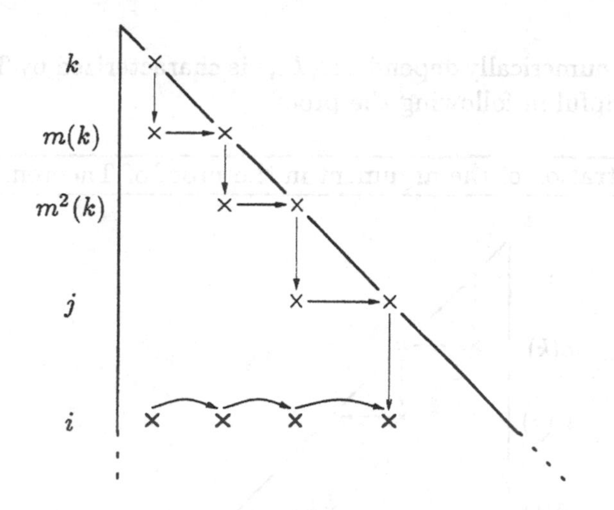

So far, Lemma 2.1 has been used to obtain some results about $\Omega(\boldsymbol{L}_{i*})$. It is also useful in obtaining results about column structure; Lemma 2.4 is an immediate consequence of Lemma 2.1 and is stated without proof. The result is illustrated in Figure 4.

Lemma 2.4

$$\Omega(\boldsymbol{L}_{*k}^{\triangleright}) \subseteq \Omega(\boldsymbol{L}_{*m_k}).$$

■

Thus, for $i \in \Omega(\boldsymbol{L}_{*k}^{\triangleright})$ and $i \geq m_k$, $\ell_{ik} \neq 0 \Rightarrow \ell_{i,m_k} \neq 0$. Using this allows one to determine $\Omega(\boldsymbol{L}_{*i})$ by examining a special subset of those columns

Figure 4. Illustration of Lemma 2.4.

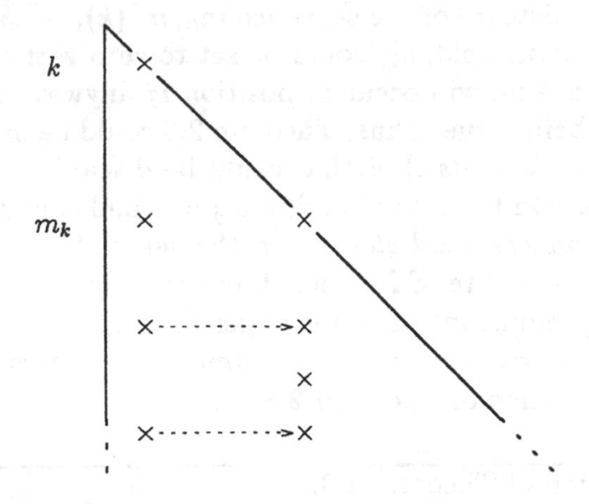

upon which it numerically depends; $\Omega(L_{*i})$ is characterized by Theorem 2.5. Figure 5 is helpful in following the proof.

Figure 5. Illustration of the argument in the proof of Theorem 2.5.

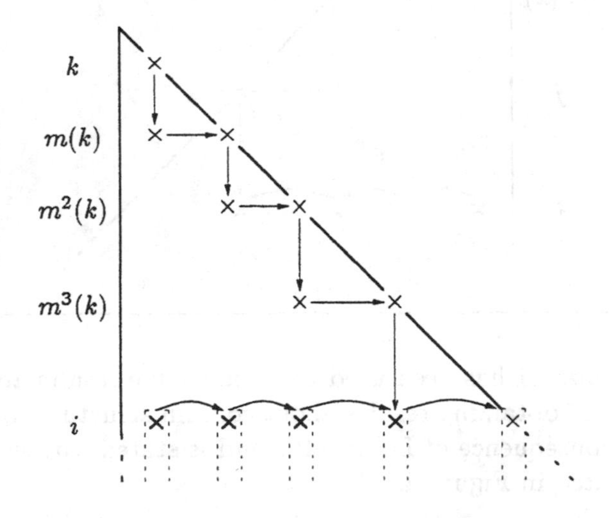

Theorem 2.5

$$\Omega(L_{*i}) = \Omega(A^{\triangleright}_{*i}) \bigcup \left(\bigcup_{k<i} \{\Omega(L^{\triangleright}_{*k}) \mid m_k = i\} \right).$$

Proof: Consider any k for which $i \in \Omega(L^{\llcorner}_{*k})$ and let $S = \{1, 2, \ldots, i-1\}$. If $m_k = i$ there is nothing to prove, so assume that $m_k < i$. Note that Theorem 2.2 holds with the condition $a_{ij} \neq 0$ replaced by the condition $\ell_{ij} \neq 0$. Thus, there exists a $p > 0$ such that $m^p(k) = i$. Using this together with Lemma 2.4 implies

$$\{\Omega(L_{*k}) - S\} \subseteq \{\Omega(L_{*,m(k)}) - S\} \subseteq \cdots \subseteq \{\Omega(L_{*,m^p(k)}) - S\}.$$

The result then follows. ■

In words, Theorem 2.5 says that the structure of column i of L is given by the structure of column i of A (excluding the portion above the diagonal), together with the structures of those columns of $L^{\llcorner}$ whose first off-diagonal nonzeros are in row i. An example is given in Figure 6 below. The structure of column 5 of L is given by the union of the structure of column 5 of $A^{\triangle}$ and the structures of columns 1 and 4 of $L^{\llcorner}$. Even though column 5 depends numerically on column 2, column 2 does not have to be considered when determining the structure of column 5 because the structure of column 2 is contained in column 4.

Figure 6. Symbolic factorization example.

$$A = \begin{pmatrix} \times & & & & \times & & & \times \\ & \times & & \times & \times & & & \\ & & \times & \times & & & & \times \\ & \times & \times & \times & & & \times & \\ \times & \times & & & \times & \times & & \\ & & & & \times & \times & & \times \\ & & & \times & & & \times & \\ \times & & \times & & & \times & & \times \end{pmatrix} \quad L = \begin{pmatrix} \times & & & & & & & \\ & \times & & & & & & \\ & & \times & & & & & \\ & \times & \times & \times & & & & \\ \times & \times & & + & \times & & & \\ & & & & \times & \times & & \\ & & & \times & + & + & \times & \\ \times & & \times & + & + & \times & + & \times \end{pmatrix}$$

Theorem 2.5 above leads to the following efficient algorithm due to Sherman[34] for performing symbolic factorization. In Algorithm 1 below, $\mathcal{S}_i$ is used to record the columns of L whose structures will affect that of column i of L. At the end of its execution, $\Omega(L_{*i})$ is given by elements of $\mathcal{L}_i$. Note that $\mathcal{S}_i \cap \mathcal{S}_l = \phi$ for $i \neq l$, and the total number of elements in all these sets together can never exceed n. Thus, all the $\mathcal{S}_i$ can be represented as needed using a linked list of length n. Since each column of L participates in the algorithm exactly once, it is not difficult to show that its execution time is $O(\eta(L))$. For details, see [12, Chapter 5].

In later sections the problem of detecting consecutive columns of L which have essentially identical structure will be considered. That is, the problem of determining, for some i, whether $\Omega(L_{*i}) = \Omega(L^{\llcorner}_{*,i-1})$. Theorem 2.5 provides information about what conditions are needed for such

Algorithm 1 Symbolic Factorization.

```
for i = 1,2,...,n do
    S_i ← φ
end for
for i = 1,2,...,n do
    L_i ← Ω(A^◭_{*i})
    for k ∈ S_i do
        L_i ← L_i ∪ L_k − {k}
    end for
    determine m_i
    if m_i > i then
        S_{m_i} ← S_{m_i} ∪ {i}
    end if
end for
```

pairs of columns to occur. Consideration is limited to pairs $\{i-1, i\}$ for which $|\{k \mid m_k = i\}| = 1$ and $m_{i-1} = i$. Using these conditions in Theorem 2.5 yields

$$\Omega(\boldsymbol{L}_{*i}) = \Omega(\boldsymbol{A}^{\triangle}_{*i}) \cup \Omega(\boldsymbol{L}^{\llcorner}_{*,i-1}).$$

That is, $\Omega(\boldsymbol{L}_{*,i}) = \Omega(\boldsymbol{L}^{\llcorner}_{*,i-1})$ implies $\Omega(\boldsymbol{A}^{\triangle}_{*i}) \subseteq \Omega(\boldsymbol{L}^{\llcorner}_{*,i-1})$: nonzeros of $\boldsymbol{A}^{\triangle}_{*i}$ (e.g., c in Figure 7) are necessarily redundant. They will be filled in (by b in Figure 7) in any case. The converse is also true: if c is an essential element of $\boldsymbol{A}$ in the sense described earlier, then b must be zero which means $\Omega(\boldsymbol{L}_{*,i}) \neq \Omega(\boldsymbol{L}^{\llcorner}_{*,i-1})$. The conclusion to this argument is encapsulated as Theorem 2.6 below.

Figure 7. Illustration of successive columns with "equal" structure.

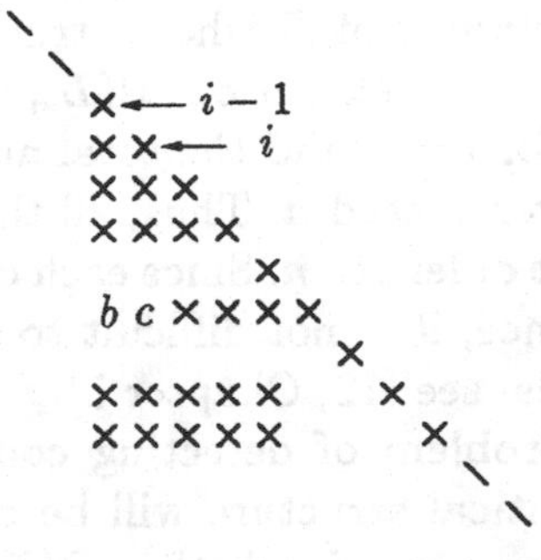

```
\
 \
  \
   ×← i−1
   ××← i
   ×××
   ××××
       ×
   bc××××
          ×
   ×××××   ×
   ×××××    ×
             \
              \
```

Theorem 2.6 *Let* $|\{k \mid m_k = i\}| = 1$ *and* $m_{i-1} = i$. *Then* $\Omega(L_{*,i}) = \Omega(L^{\llcorner}_{*,i-1})$ *if and only every nonzero in* $\Omega(A^{\llcorner}_{*i})$ *is redundant.* ■

Thus, if the conditions of Theorem 2.6 hold, the existence of *any* essential nonzero in $\Omega(A^{\llcorner}_{*i})$ implies that $\Omega(L_{*,i}) \neq \Omega(L^{\llcorner}_{*,i-1})$.

3. Elimination trees

3.1. DEFINITION AND RELATION TO L

The previous section demonstrated the pivotal role that sequences of the form $k, m(k), m^2(k), \cdots$ play in describing the structure of L. Indeed, the set $\{m_k, 1 \leq k < n\}$ together with $\Omega(A^{\llcorner})$ is all that is required to determine $\Omega(L)$. As seen below, the elimination tree is simply a special graph defined in terms of $m_k, 1 \leq k < n$.

Elimination trees play a key role in sparse factorization, as well as in other aspects of sparse matrix computation. Although the notion was used informally earlier by numerous authors, Schreiber[33] appears to have been the first to define elimination trees formally and draw attention to their fundamental role in sparse symmetric factorization; he provided many important results about them. Subsequently, Liu exploited the structure in various contexts and provided numerous additional uses of it and results about it[18, 20, 21, 23]. This section draws heavily from his excellent review article[24].

Let $A = LL^T$ as in previous sections. The *elimination tree* of A is an n-vertex graph, labelled from 1 to n according to the rows or columns of A, with an edge joining vertex i to vertex j if and only if the first nonzero below the diagonal in column i of L is in position j; that is, if and only if $j = m_i$, where m_i was defined in the previous section. The tree can be conveniently represented by the $PARENT[*]$ vector. For $j < n$, $PARENT[j] = m_j$, with $PARENT[n]$ arbitrarily set to zero. An example of a matrix, its elimination tree, and the $PARENT$ vector is given in Figure 8.

Recall Theorem 2.2 and the discussion that followed it. Each nonzero a_{ij} in $A^{\llcorner}$ begins a column sequence $j, m_j, m^2(j), \cdots$ which contains i and terminates at n. The elimination tree can be viewed as a convenient representation of the union of all those sequences, which correspond to paths in the tree.

The elimination tree of A will be denoted by $\mathcal{T}(A)$, and $\mathcal{T}[i]$ will denote the subtree rooted at vertex i; that is, the tree obtained by discarding everything in $\mathcal{T}$ except vertex i and those vertices below it. The notation $\mathcal{T}[i]$ to represent both the subtree rooted at vertex i and the set of vertices associated with that subtree, and similarly for row subtrees, defined later.

Figure 8. Example of an elimination tree.

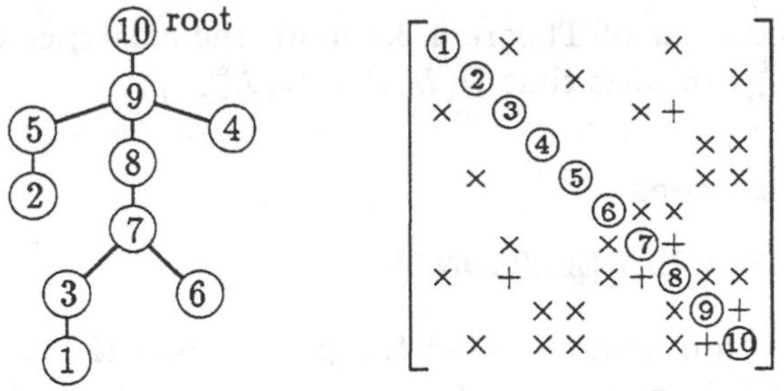

$L + L^T$

$PARENT[j]$	3	5	7	9	9	7	8	9	10	0
j	1	2	3	4	5	6	7	8	9	10

The explicit dependence on A will be dropped when the matrix being considered is clear from context.

By construction, the ordering of the tree is a *topological ordering*. That is, child vertices have lower numbers than their parents.

Theorem 3.1 below is a restatement of Theorem 2.2 in terms of $\mathcal{T}$. A pictorial illustration of it is contained in Figure 9.

Theorem 3.1 *Let $i > j$ and $a_{ij} \neq 0$. Then vertex i is an ancestor of vertex j in $\mathcal{T}$, and $\ell_{ik} \neq 0$ for $k = j, j_1, j_2, \ldots, j_r, i$, where $j_1, j_2, \ldots, j_r$ is the sequence of vertices in $\mathcal{T}$ on the path joining vertex j to vertex i.* ■

Paralleling the discussion after Theorem 2.2, this theorem says that each nonzero a_{ij} in A begins a sequence of (potential) fills in row i corresponding to the vertices on the path from vertex j to vertex i in $\mathcal{T}$. The theorem also says that if $a_{ij} \neq 0, i > j$, then vertex j is in $\mathcal{T}[i]$.

Suppose *all* such paths generated by nonzeros in row i of A are traced, with the objective of determining $\Omega(L_{i*})$. The following are key facts about these paths:

Fact #1 The union of the paths will contain exactly the set of vertices corresponding to nonzeros in row i of L. This is stated formally as Theorem 3.2 below, and is a restatement of Theorem 2.3 in terms of $\mathcal{T}$. The union of these paths determines a subtree of $\mathcal{T}$ called a *row subtree*. This subtree is denoted by $\mathcal{T}_r[i]$; its root is vertex i.

Fact #2 Some of the paths might be redundant because they completely overlap. If $a_{ik} \neq 0$ and $a_{i,k'} \neq 0$ and vertex k' lies on the path from vertex k to vertex i in $\mathcal{T}$, then tracing the path beginning at vertex

Figure 9. Pictorial illustration of Theorem 3.1.

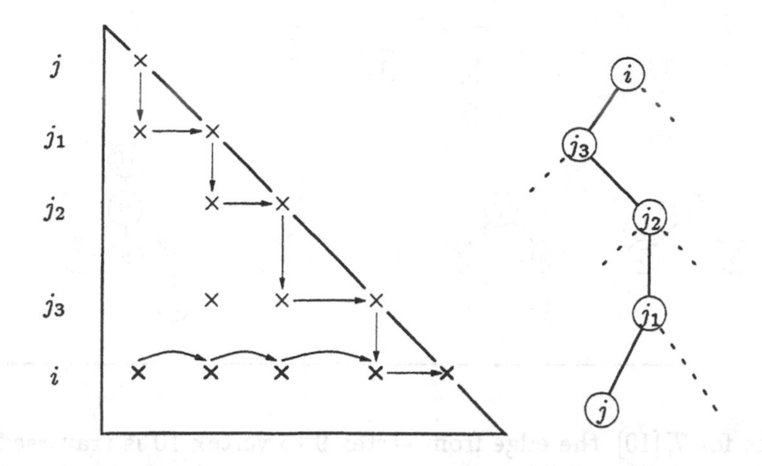

k' is unnecessary. Element $a_{i,k'}$ is *redundant* in the terminology of the previous section. These vertices k' correspond to *internal* vertices of $\mathcal{T}_r[i]$. Starting vertices in $\mathcal{T}_r[i]$ that correspond to non-redundant paths are *leaves* or *leaf vertices* of $\mathcal{T}_r[i]$, and correspond to *essential elements* of A in the terminology of the previous section. $\mathcal{T}_r[i]$ is obtained from $\mathcal{T}[i]$ by *pruning* it; that is, discarding those parts of it below the leaf vertices.

Fact #3 As noted above, nonzeros $a_{ik'}$ corresponding to internal vertices of $\mathcal{T}_r[i]$ are redundant. They do not contribute to fill in the sense that other nonzeros in the row will cause the corresponding fill anyway. If $a_{ik'}$ were set to zero, the structure of L would not change. Doing this for *all* rows would, in general, yield a matrix A^- having fewer nonzeros than A; it is the "sparsest" matrix whose Cholesky factor has the same structure as L[18].

Fact #4 Even though paths may not be redundant, they may still partly overlap. This means that it is possible to discover all of the structure of L_{i*} without completely tracing all paths in $\mathcal{T}_r[i]$.

Figure 10 displays the row subtrees corresponding to the matrix and elimination tree contained in Figure 8. Note that $a_{10,5} \neq 0$ and vertex 5 is an internal vertex of $\mathcal{T}_r[10]$ (Fact #2 above). It is unnecessary to trace the path beginning at vertex 5 in the elimination tree because it is a subpath of the path beginning at vertex 2. Also note that if $a_{10,5}$ were set to zero, the structure of L would not change, since $a_{10,2}$ would lead to fill in position $(10,5)$ anyway (Fact #3 above). Finally, note that in tracing the paths for $\mathcal{T}_r[8]$, the edge from vertex 7 to vertex 8 is traversed twice, and in tracing

Figure 10. Row subtrees corresponding to the example in Figure 8.

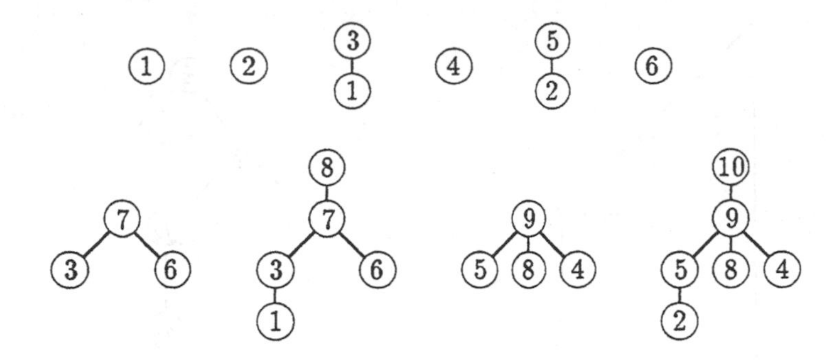

the paths for $\mathcal{T}_r[10]$, the edge from vertex 9 to vertex 10 is traversed three times (Fact #4 above.)

Theorem 3.2 *Let $i > j$. Then $\ell_{ij} \neq 0$ if and only if vertex j is an ancestor of some vertex k such that $a_{ik} \neq 0$.* ■

The elimination tree and its row subtrees convey important dependency information among computations in Cholesky factorization. For easy reference, these are included as additional "facts".

Fact #5 The structure of column i is the union of the structures of its children in the tree, together with $\Omega(A^{\llcorner}_{*i})$. This follows directly from Theorem 2.5

Fact #6 Column i of L depends numerically on exactly those columns in A corresponding to vertices below vertex i in $\mathcal{T}_r[i]$. This follows directly from Theorem 3.2.

Fact #7 Note that if $\mathcal{T}[s] \cap \mathcal{T}[t] = \phi$, then $\mathcal{T}_r[s'] \cap \mathcal{T}_r[t'] = \phi$ for all $s' \in \mathcal{T}[s]$ and $t' \in \mathcal{T}[t]$. Thus, the elimination tree exhibits the computational independence among the columns of L. This together with Fact #6 means that if $\mathcal{T}[s] \cap \mathcal{T}[t] = \phi$, then columns s and t can be computed completely independently. This explains why the elimination tree is such an important instrument in developing and analyzing parallel algorithms for the Cholesky factorization[6, 13].

Fact #8 As noted above, if $\mathcal{T}[s] \cap \mathcal{T}[t] = \phi$, then columns s and t can be computed completely independently since $\mathcal{T}_r[s] \cap \mathcal{T}_r[t] = \phi$. In particular, one could compute all of the columns in $\mathcal{T}[s]$ followed by all of those in $\mathcal{T}[t]$, or alternate among them. The only restriction is that within each subtree, child columns must be computed before their parents. The immediate implication of this is that *all topological orderings of the elimination tree are equivalent in terms of computation and fill.*

One way to find the elimination tree is to carry out the symbolic factorization algorithm derived earlier, since the m_k's of L are precisely the parent vertices of each vertex of $\mathcal{T}$.

However, one can do much better. The material in the previous subsection provides the key to a "bottom-up" algorithm to find the elimination tree. Suppose $\mathcal{T}[i-1]$, the elimination tree for the leading $(i-1) \times (i-1)$ principal submatrix of A, has been generated. In order to generate $\mathcal{T}[i]$, the tree for the leading $i \times i$ submatrix of A, one needs to discover which vertices are adjacent to vertex i in $\mathcal{T}[i]$. One way to do this is to determine $\mathcal{T}_r[i]$. The ℓ_{ij} which are nonzero can be discovered by traversing "up" the tree from each vertex k for which $a_{ik} \neq 0$. This is the basis for the algorithm due to Liu[18] which is presented as Algorithm 2 below. Figure 11 shows the steps of the algorithm for a small example. A more efficient one also due to Liu[18] is presented later.

Algorithm 2 A simple algorithm to find the elimination tree.

```
for i = 1, ..., n do
    PARENT(i) ← 0
    for {k | k < i and a_{i,k} ≠ 0} do
        r ← k
        while PARENT(r) ≠ 0 and PARENT(r) ≠ i do
            r ← PARENT(r)
        end while
        if PARENT(r) = 0 then PARENT(r) ← i
    end for
end for
```

How can the algorithm be made more efficient? Suppose in the process of determining $\mathcal{T}_r[i]$ a path from vertex k to vertex i is traced. During the process of determining $\mathcal{T}_r[s]$, $s > i$, one may need to trace a path from vertex k to vertex s. It is desirable to avoid tracing the path from vertex k to vertex i again and instead skip directly from vertex k to vertex i, and proceed up the tree from there to vertex s. This is achieved by employing a technique called *path compression*; a second *virtual tree* is maintained along with the elimination tree. Whenever a path from vertex k to vertex i is traced in this new tree, the parent of vertex k is set to vertex i. This modified algorithm is given as Algorithm 3, and its application to the example given in Figure 11 is provided in Figure 12. The vector $ANCESTOR$ is used to store the parent information for the virtual tree.

Figure 11. Illustration of the basic tree-finding algorithm.

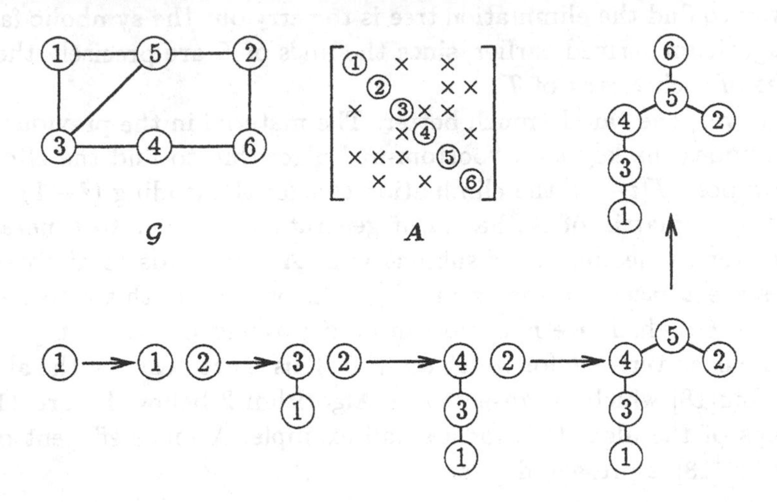

Algorithm 3 A fast algorithm to find the elimination tree.

for $i = 1, \ldots, n$ **do**
  $PARENT(i) \leftarrow 0$; $ANCESTOR(i) \leftarrow 0$
  for $\{k \mid k < i$ **and** $a_{i,k} \neq 0\}$ **do**
    $r \leftarrow k$
    while $ANCESTOR(r) \neq 0$ **and** $ANCESTOR(r) \neq i$ **do**
      $t \leftarrow ANCESTOR(r)$; $ANCESTOR(r) \leftarrow i$; $r \leftarrow t$
    end while
    if $ANCESTOR(r) = 0$ **then**
      $ANCESTOR(r) \leftarrow i$; $PARENT(r) \leftarrow i$
    end if
  end for
end for

This algorithm is indeed very efficient. It can be viewed as performing a sequence of disjoint set union operations. As shown by Liu, in that context the results of Tarjan[35] can be used to show that the execution time complexity of a slightly modified[1] version of this algorithm is

[1] The algorithm presented here is the version used in practice. There are contrived examples for which its execution time is $O(\eta(A)\log_2 n)$, but the more sophisticated version

Figure 12. Example of the fast tree-finding algorithm.

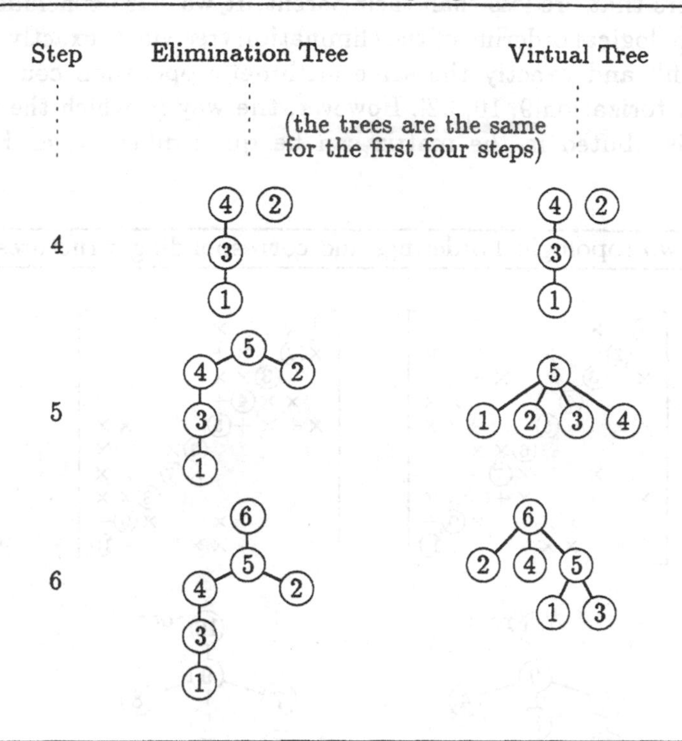

$O(\eta(A), \alpha(\eta(A), n))$, where $\alpha(m, n)$ is an astoundingly slowly growing inverse of Ackermann's function defined by Tarjan[35]. For all conceivable values of m and n, $\alpha(m, n) < 4$. Thus, for large n, algorithms with this complexity can be regarded for practical purposes as executing in time proportional to $\eta(A)$.

4. Topological orderings and postorderings

The elimination tree $\mathcal{T}(A)$ has n vertices corresponding to the rows and columns of A. A relabelling of the vertices corresponds to a symmetric permutation PAP^T of A. An arbitrary relabelling of $\mathcal{T}$ will obviously change where and how much fill occurs when PAP^T is factored. However, there is an important class of relabellings of $\mathcal{T}$ (choices of P) which preserve the amount of fill that occurs, and the arithmetic required, when PAP^T is factored. It was noted earlier that the definition of the elimination tree

for which the better bounds can be proved is more complicated; experience suggests that it is slower for real problems. Apparently, the structure of real problems is such that the log factor does not materialize.

implies that its labelling is a topological ordering. That is, child vertices have numbers that are less than their parent. It was also concluded earlier that *any* topological ordering of the elimination tree yields exactly the same amount of fill and exactly the same arithmetic operation count for the Cholesky factorization[9, 10, 32]. However, the way in which the nonzeros in $\boldsymbol{L}$ are distributed in the matrix can be quite different, as Figure 13 illustrates.

Figure 13. Two topological orderings and corresponding structures of $\boldsymbol{L}$.

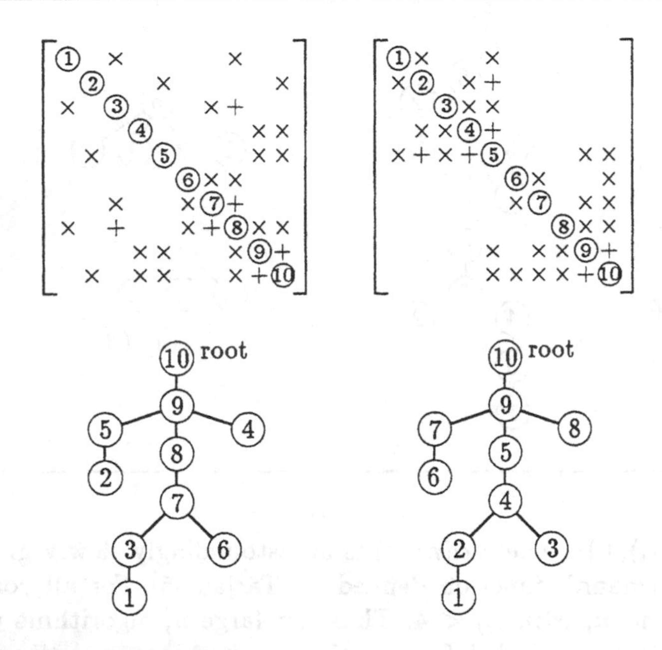

An important type of topological ordering of $\mathcal{T}$ is a *postordering* of it. A postordering of the tree is one in which the vertices in each subtree are numbered *consecutively* before the root of the subtree is numbered. Many algorithms, including some considered later, assume implicitly that the problem has been reordered so that it corresponds to a postordering of the tree. The topological ordering on the right in Figure 13 is a postordering. For future reference the row subtrees of $\mathcal{T}$ corresponding to that postordering are included in Figure 14.

Intuitively, postorderings arrange that columns which affect a succeeding column are as "close" as possible to it. This is important in the effective use of memory hierarchies, in paging environments, in multifrontal methods for implementing the numerical factorization, and in the use of special vector hardware features[2, 3, 5, 7, 22, 25, 27]. Note that there are many

Figure 14. Row subtrees corresponding to a postordering.

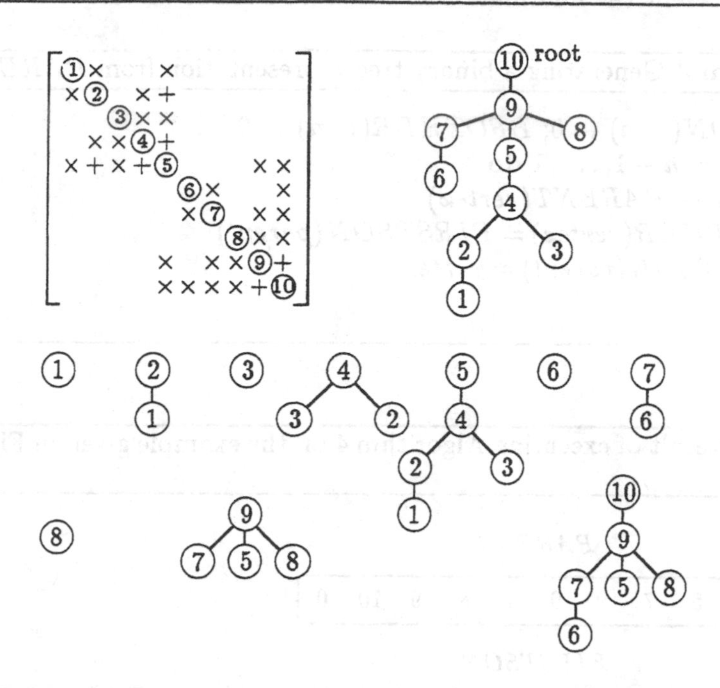

postorderings of the tree, and some have advantages over others in certain circumstances. Liu[19] provides a discussion of this issue and a method for choosing the best one in the context of multifrontal schemes.

How does one find such a reordering of the tree? The approach is standard[1], and is included here for completeness only. The method consists of two stages: the first involves finding a so-called binary tree representation of the elimination tree, which allows one to traverse the tree from its root "downward." Obviously, the parent representation used thus far is ideal for traversing "up" the tree, but is not efficient for going in the other direction.

The binary tree representation of $\mathcal{T}$ utilizes two vectors: $FIRSTSON$ and $BROTHER$. The vector $FIRSTSON(i)$ contains the vertex number of one of the child vertices of vertex i. If vertex i is a leaf of $\mathcal{T}$, then $FIRSTSON(i) = 0$. If there is more than one child vertex of vertex i, then $BROTHER(FIRSTSON(i))$ contains the vertex number of the second child, $BROTHER(BROTHER(FIRSTSON(i))$, contains that of the third, and so on. If vertex k has no more brothers, then $BROTHER(k) = 0$.

An algorithm for generating the vectors $FIRSTSON$ and $BROTHER$, given $PARENT$, is provided as Algorithm 4. The result of executing the algorithm using the $PARENT$ vector corresponding to the elimination tree and ordering from Figure 8 is displayed in Figure 15. The arrows pointing down represent information in the *FIRSTSON* vector, while the horizontal arrows represent the information in the *BROTHER* vector.

Algorithm 4 Generating a binary tree representation from *PARENT*.

```
FIRSTSON(1:n) ← 0; BROTHER(1:n) ← 0
for vertex = n − 1, ..., 1 do
    parent ← PARENT(vertex)
    BROTHER(vertex) = FIRSTSON(parent)
    FIRSTSON(parent) = vertex
end for
```

Figure 15. Result of executing Algorithm 4 on the example given in Figure 8.

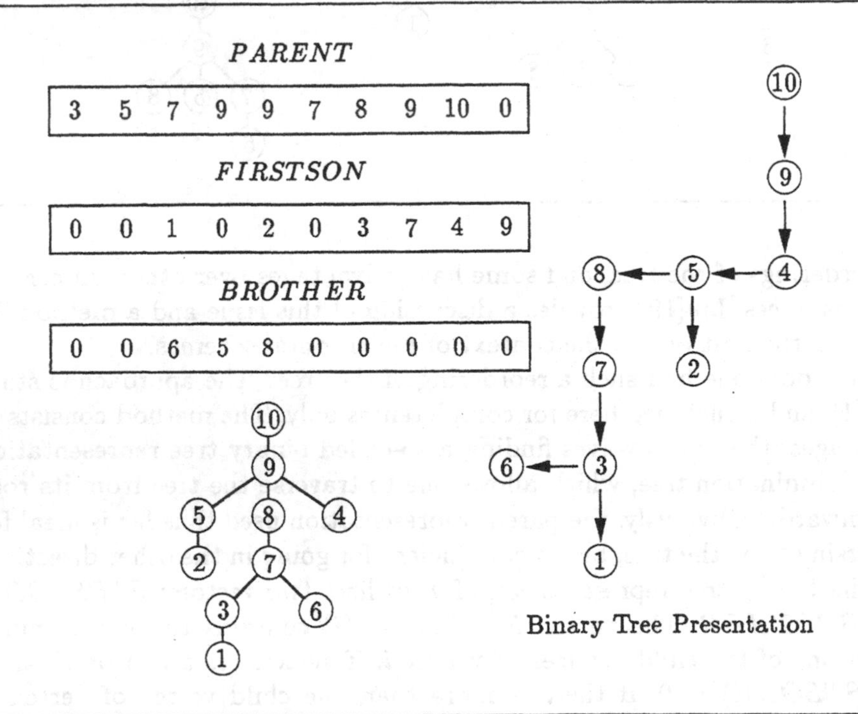

Once the binary tree representation of the elimination tree has been generated, a postorder traversal of the tree is performed, generating a new ordering. A standard depth-first traversal of the tree is executed, using a stack to record the vertices visited[1, pp. 215-218]. The algorithm is provided as Algorithm 5 below; its application to the example in Figure 15 is provided in Figure 16. The small numbers beside the vertices of the tree represent the new ordering.

Algorithm 5 Postorder traversal of the binary tree representation of $\mathcal{T}$.

$stacktop \leftarrow 0; vertex \leftarrow n; m \leftarrow 0$
while $m < n$ **do**
  while $vertex > 0$ **do**
    $stacktop \leftarrow stacktop + 1$
    $STACK(stacktop) \leftarrow vertex; vertex \leftarrow FIRSTSON(vertex)$
  end while
  while $vertex = 0$ **and** $stacktop > 0$ **do**
    $vertex = STACK(stacktop); stacktop \leftarrow stacktop - 1$
    $m \leftarrow m + 1; NEWNUMBER(vertex) \leftarrow m$
    $vertex \leftarrow BROTHER(vertex)$
  end while
end while

5. Row subtrees and postorderings

Here, and in subsequent parts of this paper, the problem is assumed to have been reordered corresponding to a postordering of the elimination tree. The following lemma is immediate. It says that the number of vertices "below" vertex i in $\mathcal{T}$ is one less than the total number of vertices in $\mathcal{T}[i]$, the subtree of $\mathcal{T}$ rooted at vertex i. This is true for any topological ordering, regardless of whether it is a postordering.

Lemma 5.1 *The number of proper descendents of vertex i is $|\mathcal{T}[i]| - 1$.* ■

Now suppose the ordering is a postordering, which means that the vertices in $\mathcal{T}[i]$ are numbered *consecutively,* with vertex i being numbered last. This provides a mechanism to test whether a vertex j is in $\mathcal{T}[i]$, since j must satisfy $j \leq i$ and $j > i - |\mathcal{T}[i]|$. This yields Lemma 5.2.

Lemma 5.2 *Let $\mathcal{T}$ be postordered and $j < i$. Then vertex j is a proper descendent of vertex i if and only if $j > i - |\mathcal{T}[i]|$.* ■

How can one test whether vertex j is a leaf vertex of $\mathcal{T}_r[i]$? Clearly, in order for vertex j to be a candidate, it must satisfy Lemma 5.2 and a_{ij} must be nonzero. If these conditions hold, then it is either a leaf vertex or an internal vertex of $\mathcal{T}_r[i]$. If it is an internal vertex, then (recall Fact #2 above) there must be some other vertex j' in $\mathcal{T}[i]$ "below" vertex j in the tree for which $a_{ij'} \neq 0$. Otherwise, vertex j must be a leaf vertex of $\mathcal{T}_r[i]$. The conclusion to this argument is encapsulated as Theorem 5.3.

Figure 16. **Result of executing a postorder traversal of the example given in Figure 15.**

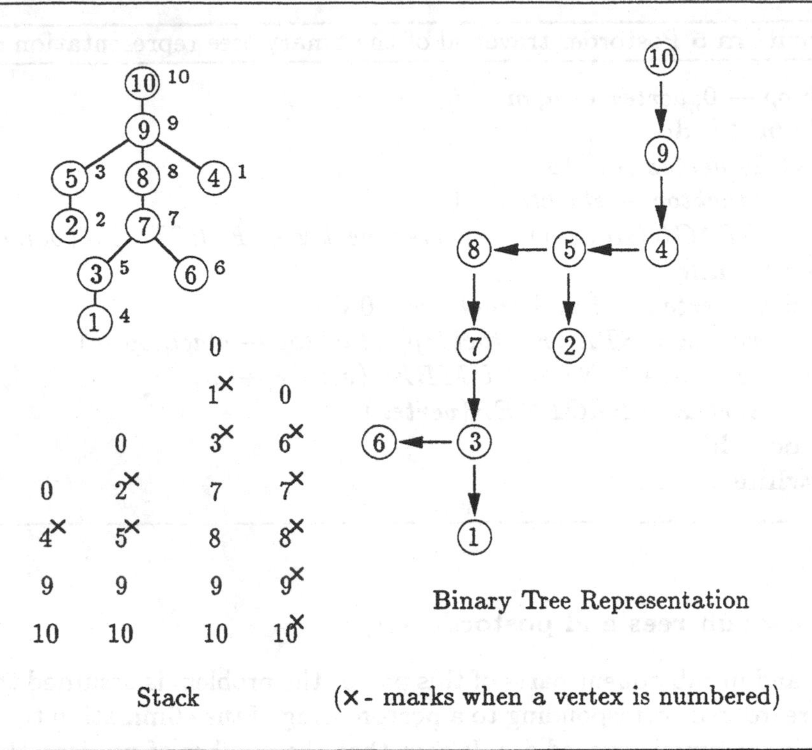

Theorem 5.3 *Vertex $j \in \mathcal{T}[i]$ is a leaf vertex of the i-th row subtree of* $\boldsymbol{L}$ *if and only if $a_{ij} \neq 0$ and $a_{ik} = 0$ for every $k \in \mathcal{T}[j] - \{j\}$.* ■

One possibility for implementing this test is to use the following result, which is obvious from an examination of Figure 17.

Theorem 5.4 *Let $\{c_1, c_2, \ldots, c_s\} = \Omega(\boldsymbol{A}^{\flat}_{i,*})$ with $c_0 < c_1 < c_2 < \cdots < c_s$ and $c_0 = 0$. Then vertex c_t is a leaf of $\mathcal{T}_r[i]$ if and only if $c_{t-1} < c_t - |\mathcal{T}[c_t]| + 1$.* ■

Figure 17. Illustration of Theorem 5.4.

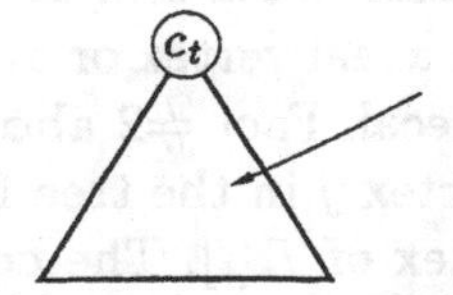

Vertices in the subtree are numbered $c_t - 1, c_t - 2, \ldots, c_t - |\mathcal{T}[c_t]| + 1$.

Theorem 5.4 has two requirements. The first is that it needs the numbers $|\mathcal{T}[i]|, 1 \leq i \leq n$. These are easy to obtain, as shown in Algorithm 6.

Algorithm 6 Algorithm to determine $|\mathcal{T}[j]|$.

$SIZE(1:n) \leftarrow 1$
for $vertex = 1, \ldots, n-1$ **do**
  $parent \leftarrow PARENT(vertex)$
  $SIZE(parent) = SIZE(parent) + SIZE(vertex)$
end for

Unfortunately, the second requirement of Theorem 5.4 is that the column subscripts $\Omega(A^{\llcorner}_{i,*})$ be sorted in increasing order. For efficiency, one would like to avoid doing this explicitly.

The sorting can be done *implicitly* using the following standard technique, which can be viewed as examining all the row subscript lists $\Omega(A^{\llcorner}_{i,*})$ "in parallel[2]." Examining $\Omega(A^{\urcorner}_{1,*})$ (in any order) reveals which rows have nonzeros in column 1; examining $\Omega(A^{\urcorner}_{2,*})$ next yields those rows having nonzeros in column 2, and so on. Thus, examining the row structures in order provides the nonzeros in each row in increasing order of column subscript. How this information is used, and whether some or all of it is retained as it emerges, will depend on the application.

With this technique in hand, the problem of discovering the leaf vertices of the row subtrees can now be solved. Theorem 5.4, together with the implicit sorting technique described above, allows one to find the leaf vertices as well as the internal vertices of each row subtree. This is presented as Algorithm 7, a "framework" which will be employed later.

6. Supernodes and why they are important

6.1. DEFINITION OF SUPERNODES AND FUNDAMENTAL SUPERNODES

A supernode is a consecutive set of columns having the "same" structure. More precisely, the *column subset* $S = \{s, s+1, \ldots, s+t-1\}$ is a supernode of the matrix L if and only if it is a maximal contiguous column subset satisfying $\Omega(L^{\llcorner}_{*k}) = \Omega(L_{*,k+1}), s \leq k \leq s+t-2$. Figure 18 depicts a supernode and the relevant parameters in the definition above.

The columns of a supernode S have a *dense* diagonal block and identical column structure below row $s+t-1$. Indeed, it is this property that makes finding supernodes important: data structures and computational

[2] It can also be viewed as a somewhat disguised "bin sort" of the column subscripts in each row[1, p. 274].

Algorithm 7 Algorithm to determine whether a vertex is a leaf.

```
PREVNONZ(1 : n) ← 0
for j = 1, ..., n do
    for {i | i > j and a_ij ≠ 0} do
        k ← PREVNONZ(i)
        if k < j − SIZE(j) + 1 then
            print "vertex j is a leaf of T_r[i] "
        else
            print "vertex j is an internal vertex of T_r[i] "
        end if
        PREVNONZ(i) ← j
    end for
end for
```

algorithms exploit this structure to gain high efficiency[4, 29, 30]. For example, only one copy of the row subscripts for the columns of a supernode need to be stored. Also, it should be clear that if it is known where each supernode begins, the symbolic factorization algorithm can be improved significantly. It is possible to design symbolic factorization algorithms that run in time proportional to $\sum_{k \in F} \eta(L_{*k})$, where F is the set of first columns of superodes[11, 28].

Figure 18. Illustration of a supernode.

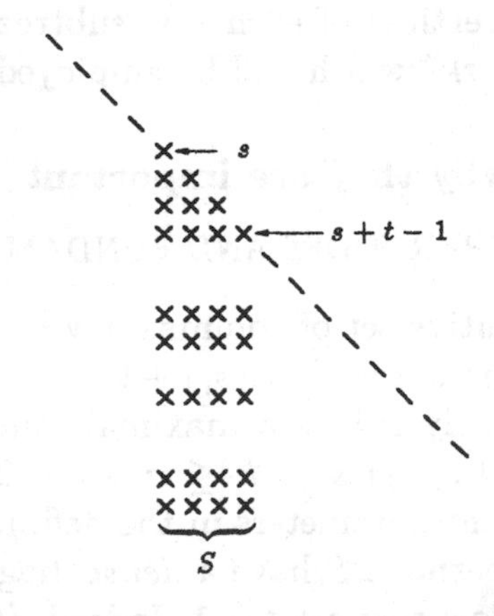

If the number of nonzeros in each column of L is known, the problem of finding supernodes is easy: if $\ell_{k+1,k} \neq 0$ and $\eta(L^{\llcorner}_{*,k}) = \eta(L_{*,k+1})$, then column k and column $k+1$ belong to the same supernode. This provides motivation for having a fast algorithm to find the row and/or column counts of L. This problem is considered later.

Note that $\ell_{k+1,k} \neq 0$ if and only if vertex k is a child of vertex $k+1$ in the elimination tree. Thus, the column counts of L plus the elimination tree allow one to identify the supernodes in L.

In practice, it is beneficial to adopt a slightly restricted definition of supernode. A *fundamental supernode* is a supernode as defined above with the additional restriction that vertex $s+i-1$ is the *only* child of $s+i$ in the elimination tree for $i = 1, \ldots, t-1$. Thus the vertices in S form a chain in $\mathcal{T}$. Note that this restriction is exactly the assumption contained in Theorem 2.6.

This restriction does not change the number of supernodes much, but it has some practical advantages. In particular, the supernodes characterized in this way do not depend on the postordering used. All postorderings will produce the same supernode partitioning.

6.2. FINDING SUPERNODES

Much of the preliminary development has already been done in the previous sections where a way to find leaf vertices of the row subtrees of $\mathcal{T}$ was developed. The importance of finding leaf vertices is apparent from the following result taken from Liu et. al.[26]. Note that it is essentially equivalent to Theorem 2.6.

Theorem 6.1 *Column j is the first vertex in a fundamental supernode if and only if vertex j has two or more children in the elimination tree, or j is a leaf vertex of some row subtree of the elimination tree.* ■

It is easy to find those vertices that have two or more children, as shown in Algorithm 8.

Algorithm 8 Algorithm to find the number of children of each vertex.

$CHILDREN(1:n) \leftarrow 0$
for $vertex = 1, \ldots, n-1$ **do**
  $parent \leftarrow PARENT(vertex)$
  $CHILDREN(parent) = CHILDREN(parent) + 1$
end for

Earlier an economical way to determine leaf vertices of row subtrees was developed. Thus, it is now possible to present an efficient algorithm for finding supernodes. Algorithm 7 is modified so that it determines whether vertex j is a leaf of *at least one* row subtree in the elimination tree. The result is Algorithm 9 below, where $XSUPER(i)$ points to the beginning of the i-th supernode. The algorithm assumes that the $CHILDREN$ vector has already been computed.

Algorithm 9 Algorithm to find supernodes.

```
nsuper ← 0; PREVNONZ(1 : n) ← 0
for j = 1, ... , n do
    ISLEAF ← false
    for {i | i > j and a_ij ≠ 0} do
        k ← PREVNONZ(i)
        if k < j − SIZE(j) + 1 then
            ISLEAF ← true
        end if
        PREVNONZ(i) ← j
    end for
    if ISLEAF or CHILDREN(j) > 1 then
        nsuper ← nsuper + 1; XSUPER(nsuper) ← j
    end if
end for
```

7. Finding row and column counts for L

The objective of this section is to present a fast algorithm for finding the number of nonzeros in each row and column of L. The algorithm is due to Gilbert, Ng and Peyton[16]. The basic idea is to (implicitly) traverse each row subtree of the elimination tree, counting the number of distinct edges within it. The novelty of the algorithm is that it deals with the potential redundancies and inefficiencies hinted at earlier in the "key facts" about row subtrees. The result is an $O(\eta(A), \alpha(\eta(A), n))$ algorithm for finding row and column counts for L.

7.1. FINDING ROW COUNTS FOR L

In order to focus on one thing at a time, this subsection deals with the problem of obtaining row counts only. Obtaining the column counts is considered in the next subsection. Some of the required algorithmic machinery is already available: an efficient method to detect leaf vertices of each row subtree means that only those paths in $\mathcal{T}_r[i]$ beginning at leaf vertices will be traced. However, as yet, there is no way to avoid the inefficiency caused by the fact that some of the paths partially overlap, which leads to parts of

$\mathcal{T}_r[i]$ being examined more than once. Gilbert, Ng and Peyton[16] provide two key techniques; they show:

1. how to decompose $\mathcal{T}_r[i]$ into *disjoint paths*, thus eliminating the problem of tracing overlapping paths.
2. how to avoid explicitly tracing the (disjoint) paths by *computing* their length.

The method for decomposing $\mathcal{T}_r[i]$ into disjoint paths depends on the following theorem, which is a slightly weakened (but adequate for the purpose of this paper) version of one in [16]. Here and elsewhere $lca(p, q)$ denotes the least common ancestor of vertices p and q in $\mathcal{T}$, and is defined to be the vertex furthest from the root of $\mathcal{T}$ which is an ancestor of both p and q. For example, in Figure 19, $lca(1,3) = 4$, $lca(4,10) = 12$, $lca(6,8) = 8$, and $lca(1,13) = 13$.

Theorem 7.1 *Let $p_1 < p_2 \cdots < p_{k-1} < p_k = i$ be the set of leaves together with the root of $\mathcal{T}_r[i]$ and let q_j be the least common ancestor of vertex p_j and $p_{j+1}, 1 \leq j < k$. Then each edge in $\mathcal{T}_r[i]$ is on the path from p_j to q_j for exactly one j.* ∎

An example of a row subtree and the path decomposition induced by its leaves and root according to Theorem 7.1 is given in Figure 19.

Figure 19. Decomposition of a row tree into disjoint paths using least common ancestors.

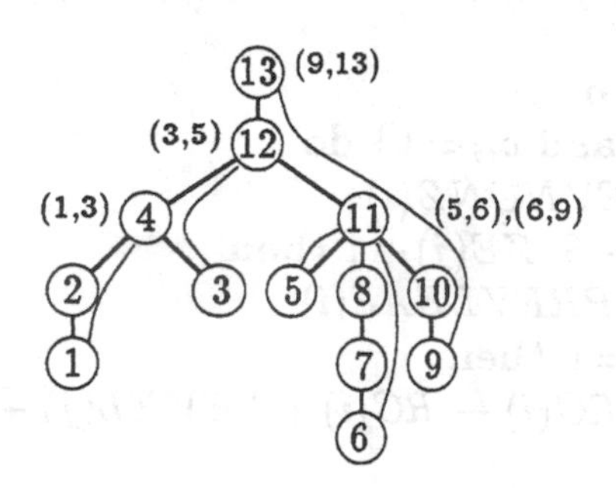

Suppose an algorithm is available to find the least common ancestor of any two vertices in $\mathcal{T}_r[i]$. The endpoints of a set of paths whose disjoint union is $\mathcal{T}_r[i]$ can then be obtained by computing $lca(p_1, p_2), lca(p_2, p_3), \cdots,$ $lca(p_{k-1}, i)$.

Rather than actually traversing the paths to count the number of edges in them, their lengths can be determined by computing the difference in the levels in the tree of the endpoints of the path. The *level* of vertex p in $\mathcal{T}$ is the distance from vertex p to the root (vertex n) of $\mathcal{T}$. That is, if vertex p is a descendent of vertex q in $\mathcal{T}$ then the length of the path joining them is $LEVEL(p) - LEVEL(q)$. The following simple algorithm uses the $PARENT$ vector to compute the level of each vertex in $\mathcal{T}$. Note that it is assumed that $PARENT(n) = 0$.

Algorithm 10 Algorithm to compute the level in $\mathcal{T}$ of each vertex.

```
LEVEL(0) ← 0
for vertex = n, n − 1, ..., 1 do
    LEVEL(vertex) = LEVEL(PARENT(vertex)) + 1
end for
```

It is beyond the scope of this paper to describe the details of an efficient algorithm for finding least common ancestors. It is sufficient to know that a very efficient algorithm due to Tarjan[35, 36, 37] is available to solve this problem.

The row count algorithm is presented as Algorithm 11 below. It assumes the existence of a function lca which returns the vertex number of the least common ancestor of its two arguments.

Algorithm 11 Algorithm to find $\eta(L_{i*}), i = 1, \ldots, n$.

```
PREVLEAF(1 : n) ← 0; PREVNONZ(1 : n) ← 0
RC(1 : n) ← 1
for j = 1, ..., n do
    for {i | i > j and a_ij ≠ 0} do
        k ← PREVNONZ(i)
        if k < j − SIZE(j) + 1 then
            p ← PREVLEAF(i)
            if p = 0 then
                RC(i) ← RC(i) + LEVEL(j) − LEVEL(i)
            else
                q = lca(p, j)
                RC(i) ← RC(i) + LEVEL(j) − LEVEL(q)
            end if
            PREVLEAF(i) ← j
        end if
        PREVNONZ(i) ← j
    end for
end for
```

7.2. FINDING COLUMN COUNTS FOR L

Let $c_j = \eta(L_{*j}), 1 \leq j \leq n$. The objective in this section is to develop a way of determining the c_j's using the framework developed for finding the row counts.

Suppose the *difference* Δ_j between c_j and the sum of the c's for the child vertices of vertex j is known. For definiteness, let the children of vertex j have labels s_1, s_2, and s_3. Then

$$\Delta_j = c_j - (c_{s_1} + c_{s_2} + c_{s_3}), \tag{1}$$

which means

$$c_j = c_{s_1} + c_{s_2} + c_{s_3} + \Delta_j. \tag{2}$$

Thus, Δ_j can be regarded as a "correction term" which needs to be added to $c_{s_1} + c_{s_2} + c_{s_3}$ in order to obtain c_j. When j has no children, $\Delta_j = c_j$. Given the Δ_j's, the c_j's can be computed using Algorithm 12.

Algorithm 12 Algorithm to compute c_j, given Δ_j.

$CC(1:n) \leftarrow DELTA(1:n)$
for $j = 1, \ldots, n$ **do**
  $CC(PARENT(j)) \leftarrow CC(PARENT(j)) + CC(j)$
end for

Recall Theorem 2.5, which with reference to (2) above says that the structure of column j is the structure of column j of A together with the union of the structures *below the diagonal* of columns s_1, s_2, and s_3. Thus Δ_j must account for two components: it must adjust for the fact that the column counts c_{s_1}, c_{s_2}, and c_{s_3} include the diagonal elements. In addition, Δ_j must account for the extent to which the structures of columns s_1, s_2, and s_3 overlap.

Let Δ_j be initially set to 1 for $1 \leq j \leq n$, and consider how the Δ_j's should be adjusted in the process of (implicitly) traversing the row trees using the scheme in the previous subsection.

Adjusting for the diagonal elements is straightforward: for all k satisfying $PARENT(k) = j$, Δ_j should be decremented by 1.

Dealing with the column overlap issue is somewhat more complicated; consider $\mathcal{T}_r[i]$ and any vertex j in $\mathcal{T}[i]$ having $d' \geq d$ child vertices. Here are the possibilities:

1. $j \notin \mathcal{T}_r[i]$
   Row i contributes nothing to c_j or to any of the children of vertex j. Thus, there is no contribution to Δ_j from row i.
2. j is a leaf of $\mathcal{T}_r[i]$ (hence, with no children in $\mathcal{T}_r[i]$)
   Row i contributes one to c_j and nothing to any child of vertex j. Thus, row i contributes one to Δ_j.
3. j and $d \leq d'$ of its children are in $\mathcal{T}_r[i]$
   Row i contributes one to c_j and d units to d of j's children. Thus, row i contributes $1 - d$ units to Δ_j. (See (1) above.)

The key point needed to complete the development is the observation that the Δ_j's change only for those vertices that are explicitly involved in the *lca* computations. In particular:

1. if $j \notin \mathcal{T}_r[i]$, Δ_j does not change.
2. if j is a leaf of $\mathcal{T}_r[i]$, then Δ_j increases by 1.
3. if j and $d \leq d'$ of its children are in $\mathcal{T}_r[i]$, then Δ_j increases by $(1 - d)$. There are two key points to be made about this circumstance:
   - if $d = 1$, Δ_j does not change. Thus, internal vertices on paths in $\mathcal{T}_r[i]$ do not change Δ_j.
   - if $d > 1$, then j will be the least common ancestor of exactly $d - 1$ pairs of leaves of $\mathcal{T}_r[i]$. Thus the adjustment to Δ_j can be made by subtracting one from it each time j serves as a least common ancestor.

With this analysis available, the complete algorithm for finding the row and column counts for L is as given as Algorithm 13.

8. Concluding remarks

This paper has studied a number of fundamental algorithms used in analyzing the structure of Cholesky factors. The key structure is the elimination tree and its row subtrees which, together, capture all essential information about the factor, as well as information concerning whether some elements of the original matrix can be discarded without changing the structure of the factor. A fast algorithm for finding elimination trees was described, along with an efficient technique to find row subtrees.

These techniques were then used to solve two important problems: detecting supernodes in the Cholesky factor, and finding the row and column counts for the Cholesky factor. Being able to solve these problems before explicitly carrying out the symbolic factorization leads to exceptionally efficient symbolic factorization schemes, and aids in software design.

The reader will perhaps note that solving the row/column count problem renders the algorithm for finding supernodes redundant, since the elimi-

Algorithm 13 Algorithm to find row and column counts of L.

```
PREVLEAF(1 : n) ← 0; PREVNONZ(1 : n) ← 0
DELTA(1 : n) ← 1; RC(1 : n) ← 1
for j = 1, ..., n do
    if j < n then
        DELTA(PARENT(j)) ← DELTA(PARENT(j)) − 1
    end if
    for {i | i > j and a_ij ≠ 0} do
        k ← PREVNONZ(i)
        if k < j − SIZE(j) + 1 then
            p ← PREVLEAF(i); DELTA(j) ← DELTA(j) + 1
            if p = 0 then
                RC(i) ← RC(i) + LEVEL(j) − LEVEL(i)
            else
                q = lca(p, j)
                RC(i) ← RC(i) + LEVEL(j) − LEVEL(q)
                DELTA(q) ← DELTA(q) − 1
            end if
            PREVLEAF(i) ← j
        end if
        PREVNONZ(i) ← j
    end for
end for
CC(1 : n) ← DELTA(1 : n)
for j = 1, ..., n do
    CC(PARENT(j)) ← CC(PARENT(j)) + CC(j)
end for
```

nation tree together with the column counts makes the problem of detecting supernodes trivial. Presenting the supernode finding algorithm first was for reasons of exposition; the row/column count algorithm and the supernode finding algorithm employ the same framework and similar ideas. Having the features of the latter algorithm in hand made the presentation of the row/column count algorithm much easier.

References

1. A. V. Aho, J. E. Hopcroft, and J. D. Ullman. *Data Structures and Algorithms*. Addison-Wesley, Reading, MA, 1983.
2. P.R. Amestoy and I.S. Duff. Vectorization of a multiprocessor multifrontal code.

Internat. J. Supercomp. Appl., 3:41–59, 1989.
3. C. Ashcraft. A vector implementation of the multifrontal method for large sparse, symmetric positive definite linear systems. Technical Report ETA-TR-51, Engineering Technology Applications Division, Boeing Computer Services, Seattle, Washington, 1987.
4. C. Ashcraft and R. Grimes. The influence of relaxed supernode partitions on the multifrontal method. *ACM Trans. Math. Software*, 15:291–309, 1989.
5. R.E. Benner, G.R. Montry, and G.G. Weigand. Concurrent multifrontal methods: shared memory, cache, and frontwidth issues. *Internat. J. Supercomp. Appl*, 1:26–44, 1987.
6. E.C.H. Chu and A. George. Sparse orthogonal decomposition on a hypercube multiprocessor. *SIAM J. Sci. Stat. Comput.*, 11:453–465, 1990.
7. I.S. Duff. Parallel implementation of multifrontal schemes. *Parallel Computing*, 3:193–204, 1986.
8. I.S. Duff and J.K. Reid. A comparison of sparsity orderings for obtaining a pivotal sequence in Gaussian elimination. *J. Inst. Maths. Appl.*, 14:281–291, 1974.
9. I.S. Duff and J.K. Reid. The multifrontal solution of indefinite sparse symmetric linear equations. *ACM Trans. Math. Software*, 9:302–325, 1983.
10. S.C. Eisenstat, M.H. Schultz, and A.H. Sherman. Software for sparse Gaussian elimination with limited core storage. In I. S. Duff and G. W. Stewart, editors, *Sparse Matrix Proceedings*, pages 135–153. SIAM Press, 1978.
11. A. George and J. W-H. Liu. An optimal algorithm for symbolic factorization of symmetric matrices. *SIAM J. Comput.*, 9:583–593, 1980.
12. A. George and J. W-H. Liu. *Computer Solution of Large Sparse Positive Definite Systems.* Prentice-Hall Inc., Englewood Cliffs, New Jersey, 1981.
13. A. George, J. W-H. Liu, and E. G-Y. Ng. Communication reduction in parallel sparse Cholesky factorization on a hypercube. In M. T. Heath, editor, *Hypercube Multiprocessors*, pages 576–586, Philadephia, PA., 1987. SIAM Publications.
14. A. George and E. G-Y. Ng. Symbolic factorization for sparse Gaussian elimination with partial pivoting. *SIAM J. Sci. Stat. Comput.*, 8:877–898, 1987.
15. J. R. Gilbert and E. G. Y. Ng. Predicting structure in nonsymmetric sparse matrix factorizations. In Alan George, John R. Gilbert, and Joseph W.H. Liu, editors, *Graph Theory and Sparse Matrix Computation*, volume IMA #56, pages 107–140. Springer-Verlag, 1993.
16. John R. Gilbert, Esmond G. Y. Ng, and Barry W. Peyton. An efficient algorithm to compute row and column counts for sparse Cholesky factorization. *SIAM J. Matrix Anal. Appl.*, 15(4):1075–1091, 1994.
17. J.R. Gilbert and H. Hafsteinsson. Parallel symbolic factorization of sparse linear systems. *Parallel Computing*, 14:151–162, 1990.
18. J. W-H. Liu. A compact row storage scheme for Cholesky factors using elimination trees. *ACM Trans. Math. Software*, 12:127–148, 1986.
19. J. W. H. Liu. On the storage requirement in the out-of-core multifrontal method for sparse factorization. *TOMS*, 12:249–264, 1986.
20. J. W. H. Liu. Equivalent sparse matrix reordering by elimination tree rotations. *SISSC*, 9:424–444, 1988.
21. J. W. H. Liu. A tree model for sparse symmetric indefinite matrix factorization. *SIMAX*, 1:26–39, 1988.
22. J. W-H. Liu. The multifrontal method and paging in sparse Cholesky factorization. *ACM Trans. Math. Software*, 15:310–325, 1989.
23. J. W. H. Liu. Reordering sparse matrices for parallel elimination. *Parallel Comput.*, 11:73–91, 1989.
24. J. W-H. Liu. The role of elimination trees in sparse factorization. *SIAM J. Matrix Anal. Appl.*, 11:134–172, 1990.
25. J. W-H. Liu. The multifrontal method for sparse matrix solution: theory and practice. *SIAM Review*, 34:92–109, 1992.

26. J.W.H. Liu, E. Ng, and B.W. Peyton. On finding supernodes for sparse matrix computations. *SIAM J. Matrix Anal. Appl.*, 14:242–252, 1993.
27. C.B. Moler. Matrix computations with Fortran and paging. *Comm. Assoc. Comput. Mach.*, 15:268–270, 1972.
28. E. Ng. Supernodal symbolic Cholesky factorization on a local-memory multiprocessor. *Parallel Computing*, 19:153–162, 1993.
29. E. Ng and B.W. Peyton. Block sparse Cholesky algorithms on advanced uniprocessor computers. *SIAM J. Sci. Comput.*, 14:1034–1056, 1993.
30. E. Ng and B.W. Peyton. A supernodal Cholesky factorization algorithm for shared-memory multiprocessors. *SIAM J. Sci. Comput.*, 14:761–769, 1993.
31. S.V. Parter. The use of linear graphs in Gaussian elimination. *SIAM Review*, 3:364–369, 1961.
32. F.J. Peters. *Sparse Matrices and Substructures.* Mathematisch Centrum, Amsterdam, The Netherlands, 1980. Mathematical Centre Tracts 119.
33. R. Schreiber. A new implementation of sparse Gaussian elimination. *ACM Trans. Math. Software*, 8:256–276, 1982.
34. A.H. Sherman. *On the efficient solution of sparse systems of linear and nonlinear equations.* PhD thesis, Yale University, 1975.
35. R.E. Tarjan. Efficiency of a good but not linear set union algorithm. *J. ACM*, 22:215–225, 1975.
36. R.E. Tarjan. Application of path compression on balanced trees. *J. ACM*, 26:690–715, 1979.
37. R.E. Tarjan. *Data Structures and Network Algorithms.* CBMS-NSF Regional Conference Series in Applied Math, SIAM Publications, 1983.

THE GO-AWAY ALGORITHM FOR BLOCK FACTORIZATION OF A SPARSE MATRIX

Pedro R. Almeida Benítez
University of Las Palmas de Gran Canaria

José R. Franco Brañas
University of La Laguna

Abstract

In this paper we consider the problem of colouring the vertices of a graph, without getting two adjacent vertices with the same colour and we study an algorithm to do so. Next, we present the Go-Away algorithm for block factorization of a sparse matrix and we obtain a substantial reduction of the factorization costs by using Cholesky's algorithm.

1 Notation and definitions

We will use the term *simple graph* to refer to the pair (X, E) where X is the set of the vertices or nodes and E is the set of edges, which do not have multiple edges nor loops. A *complete graph* is a simple graph in which every pair of vertices is connected by an edge. K_n will stand for the complete graph of n vertices. *Regular graph* refers to the graph whose vertices have all the same degree.

The *degree* of a vertex x is the number of edges that originate from the vertex.

A set of vertices T of a graph G is a *dominant set* if any vertex of G is either in T or adjacent to a vertex in T. Let us consider the graph of the following figure:

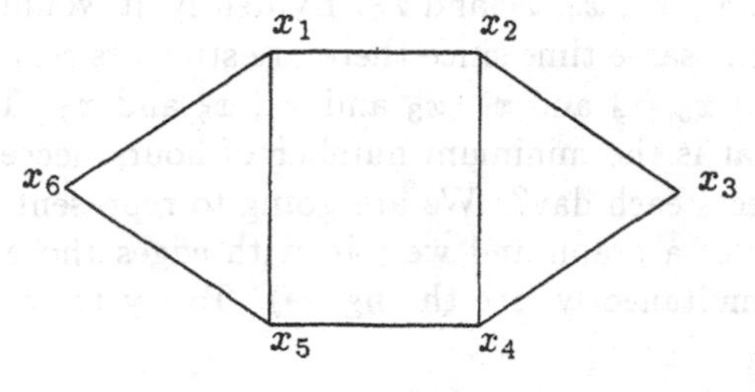

G. Winter Althaus and E. Spedicato (eds.), Algorithms for Large Scale Linear Algebraic Systems, 107–117.

For example, the set $\{x_6, x_3\}$ is dominant. However, the set $T = \{x_1, x_5, x_6\}$ is not, because x_3 is not in T nor is it adjacent to any vertex in T.

2 Sparse systems

Let A be a symmetric and positive defined nxn matrix. We want to solve the linear equations system

$$Ax = b$$

in which b is an nx1 vector of constants and x is a vector of nx1 variables. Since A is symmetric and positive defined, we can factorize it in the following way (known as *Cholesky's factorization*):

$$A = LL^t$$

where L is a subtriangular matrix. Finally, we will resolve the triangular systems $Ly = b$ and $L^t x = y$.

If A is dense, the number of multiplications/divisions required for the *Cholesky* factorization is of the order of $O(n^3/6)$ (see [1]). If A is sparse, it will be possible to save time and storage memory by manipulating the zeros. An important difficulty that we will face is that the factorization process is going to produce *fill-in*, that is, positions that in the matrix A were zero, they may not be so in L.

There are different methods which attempt to limit the *fill-in* effect. Some do by means of a local minimzation of the *fill-ins*. These are the strategies of *Markowitz*, *Minimum Degree*, etc. Other methods involve reducing the bandwidth: *Cuthill-McKee*, *King*, etc. Finally, others try to divide the graph of matrix A in blocks without *fill-ins* between nodes of differents blocks: *One-Way*, *Nested Dissection*, etc.

3 Colouring of graphs

Let's assume that we want to prepare a timetable for a set of six subjects that we will name x_1, x_2, x_3, x_4, x_5 and x_6. Evidently, it would not be right to schedule them all at the same time since there are students registered in x_1 and x_2, x_1 and x_4, x_2 and x_6, x_4 and x_5, x_3 and x_5, x_5 and x_6. The question we ask ourselves is: What is the minimum number of hours neccesary to be able to teach all the subjects each day?. We are going to represent the subjects by means of the vertices of a graph and we join with edges those subjects which can not be taught simultaneosly (see the figure). To try to solve the problem, we will assign a colour to each vertex so that there are not two adjacent vertices with the same colour. Each colour will represent an hour.

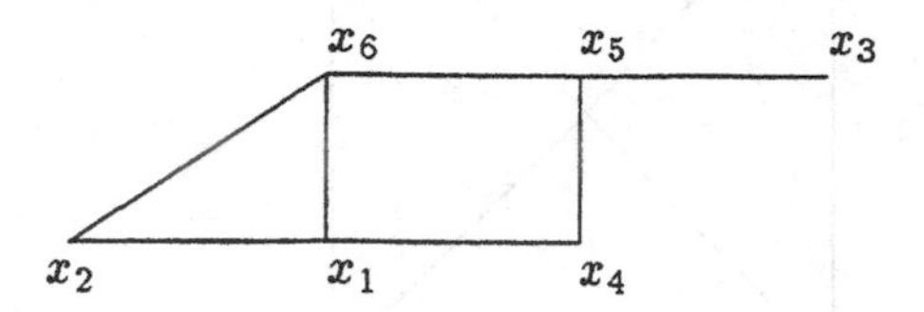

If we have colours 1,2,3,4,..., available, we can colour the 6 vertices in many ways. For example, one of them could be:

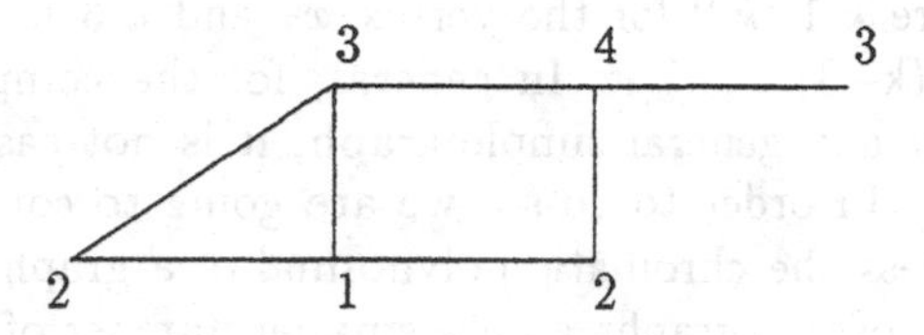

We have used four colours and we wonder if it would be possible to use less than four. Without difficulty we can find a solution using only three colours:

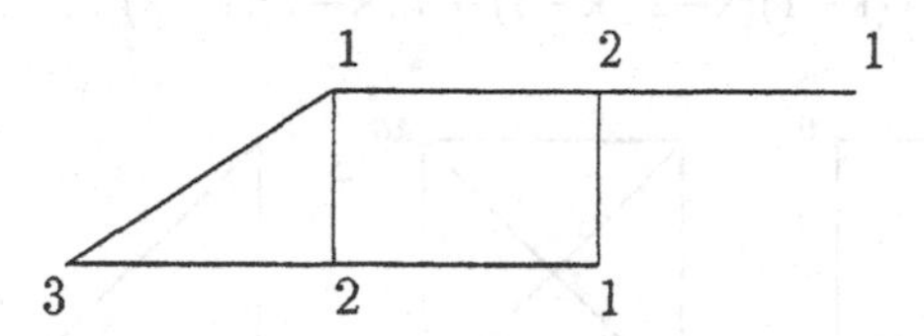

The lowest number of different colours that allows us to colour the vertices of a graph G, so that there are no two adjacent vertices with the same colour, is named the "chromatic number" of G. In the preceding example, the chromatic number is 3.

4 Number and chromatic polynomial

Let $G = (X, E)$ be a simple graph and $P(k)$ the number of possible ways of colouring the X vertices of the graph with k colours so that there are not two adjacent vertices with the same colour. We will call $P(k)$ the chromatic polynomial of G. For example, let's consider the complete graph K_4:

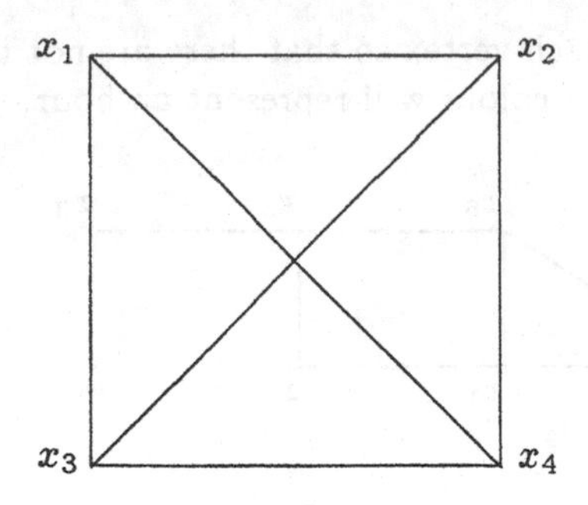

If k colours are available to us, for the vertex x_1 there are k possibilities,for the vertex x_2 there are k-1, k-2 for the vertex x_3 and k-3 for the x_4. In all, $P(k) = k(k-1)(k-2)(k-3) = V_{k,4}$. In general, for the complete graph K_n, $P(k) = V_{k,n}$. But, for any general simple graph, it is not easy to obtain the chromatic polynomial. In order to do so, we are going to consider a theorem that allows us to express the chromatic polynomial of a graph in terms of the chromatic polynomials of two graphs with a smaller number of vertices.

Theorem.- Let G be a simple graph, and let G_1 and G_2 be the graphs obtained from graph G eliminating and contracting an edge. It can be shown that $P_G(k) = P_{G_1}(k)+P_{G_2}(k)$. For example, for the graph of figure, the theorem tells us that: $P_G(k) = k(k-1)(k-2)(k-3) + k(k-1)(k-2)$.

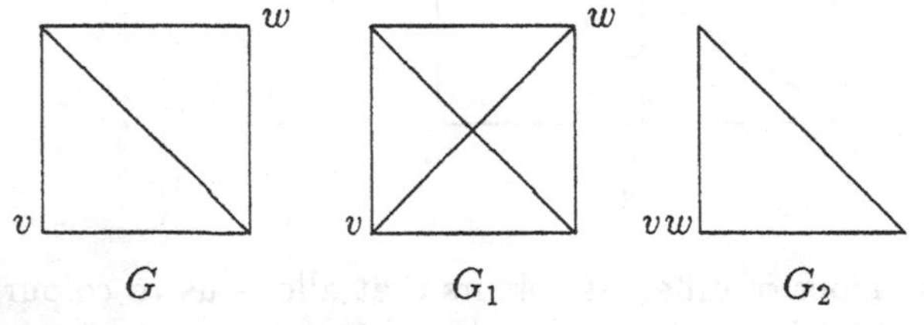

To see the proof, refer to [6].

If we known the chromatic number of a graph G, $P_G(k)$, we can calculate the chromatic number, since such a number will be the least natural number k for which $P_G(k) > 0$. On the other hand, the chromatic polynomial can be obtained applying the preceding theorem, eliminating and contracting edges, replacing the initial graph G with smaller graphs. However, for a slightly complex graph, the computation of the chromatic polynomial is a hard task, using the preceding theorem. We are going to study an algorithm that produces dominant sets with non-adjacent vertices.

5 Algorithm for colouring a graph

The purpose of this algorithm is to colour a graph using a small number of colours. The objective is to get a partition of the graph with the least possible number of classes in such a partition [6,8]:

The strategy we use is the following:

1. We choose a vertex of maximum degree x. Let A_1={x}.

2. Among the non-adjacent vertices to set A_1, we choose a vertex of maximum degree which we added to A_1.

3. In the same way, we continue adding vertices to A_1 (non-adjacent and of maximum degree) until there are no more vertices left.

4. With the remaining vertices, we repeat the process, forming a new set A_2 and so on.

5. We repeat the process until we finish all the vertices of the graph. Finally, to the nodes of each set A_i we assign the colour i.

What we have done has been to leave for the end the vertices of less degree. In this way, the vertices of the last A_i have few connections and this contributes to the disminishing number of A_i's.

We are going to study the graph of figure:

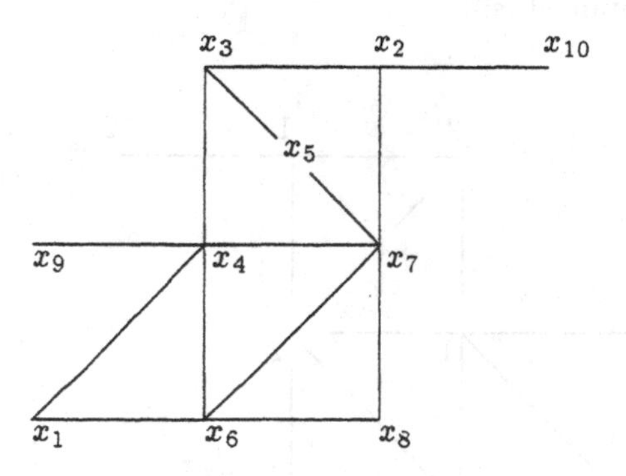

We will procede in the following manner:

vertex	*adj.vert.*	*degree*
x_1	x_4, x_6	2
x_2	x_3, x_7, x_{10}	3
x_3	x_2, x_4, x_5	3
x_4	x_1, x_3, x_6, x_7, x_9	5
x_5	x_3, x_7	2
x_6	x_1, x_4, x_7, x_8	4
x_7	x_2, x_4, x_5, x_6, x_8	5
x_8	x_6, x_7	2
x_9	x_4	1
x_{10}	x_2	1

We choose a vertex of maximum degree: x_4. Let $A_1 = \{x_4\}$.

Among the non-adjacent vertices to A_1 we choose a vertex of maximum degree: x_2. So, $A_1 = \{x_4, x_2\}$.

Among the remaining non-adjacent vertices, we select one of maximum degree: x_5. So, $A_1 = \{x_4, x_2, x_5\}$.

In the same way, we choose x_8 and we have: $A_1 = \{x_4, x_2, x_5, x_8\}$.

As there are no more vertices left, we begin by choosing a vertex of maximum degree among those remaining: x_7. We make: $A_2 = \{x_7\}$ and repeat the process, obtaining $A_2 = \{x_7, x_3, x_1, x_9, x_{10}\}$.

Finally: $A_3 = \{x_6\}$.

The colouring obtained is:

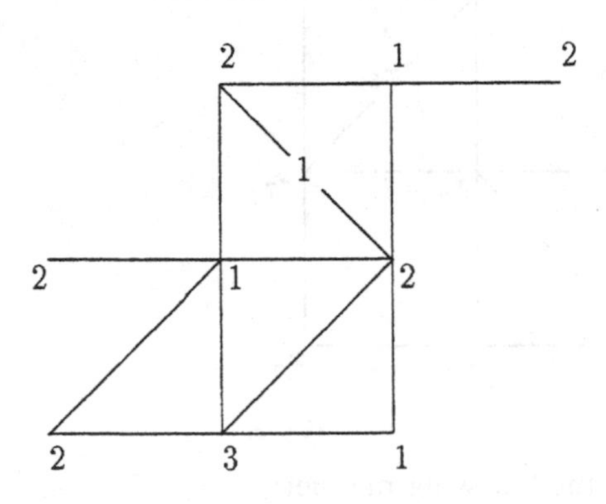

6 The Go-Away algorithm for sparse matrices

Now we will present an algorithm for block factorization of a sparse matrix and obtain an important reduction in the factorization costs. Let's consider the sparse matrix which corresponds to the graph of figure:

$$\begin{pmatrix}
x_1 & & & * & & * & & & & \\
& x_2 & * & & & & * & & & * \\
& * & x_3 & * & * & & & & & \\
* & & * & x_4 & & * & * & & * & \\
& & * & & x_5 & & * & & & \\
* & & & * & & x_6 & * & * & & \\
& * & & * & * & * & x_7 & * & & \\
& & & & & * & * & x_8 & & \\
& & & * & & & & & x_9 & \\
& * & & & & & & & & x_{10}
\end{pmatrix}$$

By applying the preceding algorithm we get a reordering of the nodes

$$A_1 = \{x_4, x_2, x_5, x_8\}, A_2 = \{x_7, x_3, x_1, x_9, x_{10}\}, A_3 = \{x_6\}$$

so that the preceding matrix turns out like this:

$$\begin{pmatrix}
x_4 & & & & * & * & * & * & & * \\
& x_2 & & & * & * & & & * & \\
& & x_5 & & * & * & & & & \\
& & & x_8 & * & & & & & * \\
* & * & * & * & x_7 & & & & & * \\
* & * & * & & & x_3 & & & & \\
* & & & & & & x_1 & & & * \\
* & & & & & & & x_9 & & \\
& * & & & & & & & x_{10} & \\
* & & & * & * & & * & & & x_6
\end{pmatrix}$$

We see that there are no connections within each diagonal block, since the nodes of A_1 are mutually independent, and so are the nodes of A_2.

If we have a sparse system $Ax = b$, the descomposition which is obtained with this algorithm can be useful in solving the system through a direct block method. On the other hand, the nodes of each block are numbered in descending order which could cause an increase of the *fill-in* in the factorization. To avoid this, we renumber within each block the nodes in asecnding order, that is, in an opposite way to what we have just done. This is the basic idea of *Go-Away* algorithm in block factorization a sparse matrix.

Let A be a symmetric and positive defined squared matrix of order n to which we have applied the preceding algorithm, which transforms into a matrix of rxr blocks which we will assume to be of equal size s:

$$\begin{pmatrix} D_{11} & A_{21}^t & A_{31}^t & \cdots & A_{r1}^t \\ A_{21} & D_{22} & A_{32}^t & \cdots & A_{r2}^t \\ A_{31} & A_{32} & D_{33} & \cdots & A_{r3}^t \\ \cdots & \cdots & \cdots & \cdots & \cdots \\ A_{r1} & A_{r2} & A_{r3} & \cdots & D_{rr} \end{pmatrix}$$

We are able to apply *Cholesky's factorization*, where L is a rxr blocks subtriangular matrix and D_{ii} are diagonal matrices. $L.L^t$:

$$\begin{pmatrix} L_{11} & & & & \\ L_{21} & L_{22} & & & \\ L_{31} & L_{32} & L_{33} & & \\ \cdots & \cdots & \cdots & & \\ L_{r1} & L_{r2} & L_{r3} & \cdots & L_{rr} \end{pmatrix} \begin{pmatrix} L_{11}^t & L_{21}^t & L_{31}^t & \cdots & L_{r1}^t \\ & L_{22}^t & L_{32}^t & \cdots & L_{r2}^t \\ & & L_{33}^t & \cdots & L_{r3}^t \\ & & & & L_{rr}^t \end{pmatrix}$$

where the matrices L_{ii} are subtriangular.

The computation of L_{11} is simple. Since the matrix D_{11} is diagonal, the elements of L_{11} are the square roots of the corresponding elements of D_{11}.

On the other hand,

$$L_{21}.L_{11}^t = A_{21} \rightarrow L_{21} = A_{21}.L_{11}^{-1}$$

But $L_{11}^t = L_{11}$. Besides, the inverse of a diagonal matrix is easy to calculate since its elements will be the inverse of the elements of such a matrix.

In the same way, we calculate the first column of blocks:

$$L_{i1}.L_{i1}^t = A_{i1} \rightarrow L_{i1} = A_{i1}.L_{11}^{-1}$$

Next, we calculate the subtriangular matrix L_{22}:

$$L_{21}.L_{21}^t + L_{22}.L_{22}^t = D_{22} \rightarrow L_{22}.L_{22}^t = D_{22} - L_{21}.L_{21}^t$$

we use *Cholesky's factorization* of $D_{22} - L_{21}.L_{21}^t$ and we obtain L_{22}.

In general, we can find the diagonal subtriangular matrices L_{ii} by *Cholesky* factorizing the difference (fixed i):

$$D_{ii} - \sum_{k=1}^{i-1} L_{ik}.L_{ik}^t$$

Next, we find L_{32} bearing in mind that: "L_{22}^t is a subtriangular matrix and its inverse is easy to calculate".

Continuing in this way, we can calculate the second column of blocks L_{i2}:

$$L_{i2} = [A_{i2} - L_{i1}.L_{21}^t].(L_{22}^t)^{-1}$$

In general, we can calculate L_{ij} by:

$$L_{ij} = [A_{ij} - \sum_{k=1}^{j-1} L_{ik}.L_{jk}^t].(L_{jj}^t)^{-1}$$

In this way we calculate the columns of subdiagonal blocks.

Finally, the system $Ax = b$ takes on the form $LL^t x = b$, which we can solve making:

$Ly = b$ (descending) and L^tx=y (ascending)

7 The Go-Away algorithm for regular grids

Let us consider a graph in grid form with mxn nodes, as the shown in the figure:

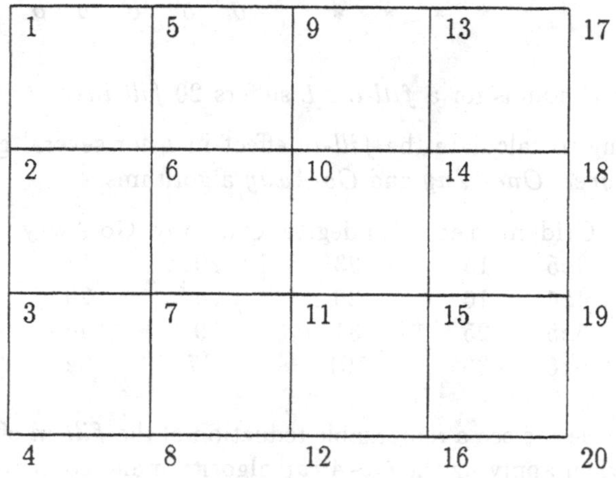

1 5 9 13 17
2 6 10 14 18
3 7 11 15 19
4 8 12 16 20

where the nodes are the corners of the squares of the network and corresponds to the matrix A of a system $Ax = b$ which may be factorized in four blocks using the *Go-Away* algorithm (the nodes can be coloured with 2 non-adjacent colours):

$$\begin{pmatrix} D_{11} & A_{21}^t \\ A_{21} & D_{22} \end{pmatrix}$$

observe that matrices D_{11} and D_{22} are diagonal. On the other hand, the *fill-in* effect becomes confined in the matrix D_{22}. For a 4x4 grid:

$$
\begin{pmatrix}
x_{16} & & & & & & & & & & * & * & & & & \\
& x_1 & & & & & & & & & & & * & * & & \\
& & x_{14} & & & & & & * & & * & & & & * & \\
& & & x_9 & & & & & * & & & & * & & * & \\
& & & & x_8 & & & & & * & & * & & & & * \\
& & & & & x_3 & & & & * & & & & * & & * \\
& & & & & & x_{11} & & & & * & * & & & * & * \\
& & & & & & & x_6 & & & & & * & * & * & * \\
& & * & * & & & & & x_{13} & & o & & o & & o & \\
& & & & * & * & & & & x_4 & & o & & o & & o \\
* & & * & & & & * & & o & & x_{15} & o & o & & o & o \\
* & & & & * & & * & & & o & o & x_{12} & o & o & o & o \\
& * & & * & & & & * & o & & o & o & x_5 & o & o & o \\
& * & & & & * & & * & & o & & o & o & x_2 & o & o \\
& & * & * & & & * & * & o & & o & o & o & o & x_{10} & o \\
& & & & * & * & * & * & & o & o & o & o & o & o & x_7
\end{pmatrix}
$$

where the "o" stands for a *fill-in*. L suffers 20 *fill-ins*.

We are going to calculate the *fill-in* effect in L for several grids using the *Minimum Degree*, *One-Way* and *Go-Away* algorithms:

Grid	num.ec.	Min.degree	One-Way	Go-Away
$3x5$	15	23	20	17
$4x4$	16	26	24	20
$5x5$	25	81	59	46
$6x6$	36	191	117	94

In the table we can see a remarkable reduction of the *fill-in* effect for several regular grids when applying the *Go-Away* algorithm and comparing it with the *Minimum Degree* and *One-Way* algorithms. On the other hand, the implementation of the algorithm is staightforward as we saw above. The decrease of the *fill-in* effect implies an important reduction of the computational effort and a shorter roundoff propagation.

References

[1] **R.L.BURDEN et J.L.FAIRES**.-*Análisis Numérico.* E. Iberoamérica. México. 1985.

[2] **J.J.DONGARRA ET AL.**.-*Solving Linear Systems on Vector and Shared Memory Computers.* SIAM. Philadelphia. 1991.

[3] **K.A.GALLIVAN ET AL.**.-*Parallel Algorithms for Matrix Computations.* SIAM. Philadelphia. 1991.

[4] **A.JENNINGS AND J.J.MCKEOWN**.-*Matrix Computation.* John Wiley and Sons. Chichester. 1992.

[5] **D.J.A.WELSH ET M.B.POWELL**.-*An upper bound on the chromatic number of a graph and its applications to timetabling problems.* The Computer. 1967.

[6] **R.J.WILSON**.-*Introducción a la teoría de grafos.* Alianza Editorial. Madrid. 1983.

[7] **R.J.WILSON ET J.J.WATKINS**.-Graphs-An Introductory Appoach. John Wiley and Sons. New York. 1990.

[8] **D.C.WOOD**.-A technique for colouring a graph applicable to large scale timetabling problems. The Computer. 1969.

RENUMBERING SPARSE MATRICES BY SIMULATED ANNEALING.

G. Winter*, M. Galán*, I. Sánchez*

Abstract

This paper deals with Simulated Annealing algorithm and their application to reduce simultaneously the band of a sparse matrix and their total number of null terms between the first non-null term in a row of the lower triangle and the diagonal term. The t arget of this application is to increase in practice, performance of ILU factorization as preconditioner. This problem is a multi-objective optimization problem with two terms in the utility function cost.

1 Simulated annealing as renumbering procedure.

The problem of finding strong preconditioners have increased its importance in later times due to the high order of the systems that appear in industrial applications and for several other considerations. In the set preconditioners, we can consider two kinds:

- Local preconditioners, which address the problem of local ill-condition of the matrix.
- Global preconditioners, which improve the global conditioning of the matrix.

Among global ones, it seems very efficient the preconditioners SSOR and ILU. These two try to approximate the LU factorization of the matrix. Obviously, these two preconditioners will be as good as the approximation of A^{-1} performed by them.

If we restrict ourselves to the ILU preconditioner, it is clear that it will approximate the full LU factorization better when the "fill in" effect is minimized. Nevertheless, we must not forget that we are doing is an incomplete LU factorization; meaning that, to a certain extent, the objective is not exactly the same in the incomplete LU case than in the " full" LU decomposition.

Borrowing from the direct solvers we can consider two possibilities:

* *Universidad de Las Palmas de Gran Canaria*

G. Winter Althaus and E. Spedicato (eds.), Algorithms for Large Scale Linear Algebraic Systems, 119–129.

Once that the "stiffness" matrix is assembled, we can consider the possibility of performing some kind of pivoting to eliminate the deleterious effect of nearly null pivots.

Other possibility is that, previously to the assembly stage, we perform renumeration algorithms. These techniques were devised originally to minimize the storage of the matrix, however it seems natural that they will increase the strength of ILU as preconditioner, as it will approximate better the LU factorization.

There are well known renumeration algorithms, namely Cunthill-McKee[?] Reverse CM, Gibbs[?], Minimum Degree, Minimum Neighbouring , Minimum fill-in, Snake, etc...

There are several studies about the performance of these algorithms when combined the ILU factorization(i.e. see L.C. Dutto [?]). It seems that the effect on the performance of ILU factorization as preconditioner is increased drastically by the use of "almost" any of these reordering algorithms.

This approach seems to be more suitable treated by stochastic "improvers".

The method of simulated annealing (i.e. Press *et al* [?]) is a suitable technique for optimization of large scale problems, especially ones when a global extreme is hidden among many local extremes. The method has been used successfully for designing complex integrated circuits (i.e. see M.P. Vecchi & Kirpatrick [?] and R.H. Otten & van Ginneken [?]) and to solve several other combinatorial problems.

We have an objective function to be minimized but, the space over which that function is defined is not simply the N-dimensional of N continuously variable parameters. The number of elements in the configuration space is factorially large, so that they cannot be explored exhaustively, , stochastic approach tries random steps in a process that is described mathematically as *directed multidimensional Markov process.*

Simulated annealing a family of heuristic optimization techniques was derived by a natural analogy with the statistical mechanics when a liquid freezes and crystallizes or metal cools and anneals. At high temperatures, the molecules of a liquid move freely with respect to one another. If the liquid is cooled slowly, thermal mobility is lost. A number of atoms of order 10^{23} per cubic centimeter are able to line themselves up and form a pure crystal that is completely ordered over a distance up to billions of times the size of an individual atom in all directions. The crystal is the state of minimum energy. The fact is that for slowly cooled system, nature is able to find this minimum energy state through intermediate random states. Rapid cooling causes defects inside the material and for this reason a slow and controlled cooling ensures highly ordered crystaline state.

The essence of the process is *slow* cooling, allowing ample time for redistribution of the atoms as they lose mobility. This is the technical definition of *annealing*, and it is essential for ensuring that a low energy will be achieved.

The Boltzmann probability expresses the idea that a system in thermal equilibrium at temperature T has its energy probabilistically distributed among all different energy states. Even at low temperture ther e is a chance of a system to get out of a local energy minimum in favour of finding a better, more global, one:

$$\mathtt{Prob}(E) = exp\left(\frac{-E}{kT}\right)$$

E is the energy, which depending on atomic positions of the system, that is to say E is the energy of the configuration and the quantity k is a constant that relates temperature to energy (Boltzmann's constant). The simulated annealing approach makes the followings identifications between the parameter of the real problem (i.e. optimization problem) and a fluid:

Fluid	*Solving a real optimization problem*
internal energy	objective function
energy state	configuration
atomic locations	variables
cooling slowly towards crystaline state	find a near optimal with an annealing schedule
temperature	a control parameter
Boltzmann's constant	a control parameter of behaviour of the convergence (scaling factor)

These kinds of principles were first incorporated by Metropolis and coworkers [39] in 1953 into numerical calculations. Generally if $\Delta E \leq 0$, $\Delta E = (x_{i+1}) - f(x_i)$, then it is accepted the new configuration and $x_{i+1} = x$ and otherwise it is considered the new configuration with a probab ility of

$$P(\Delta E) = P(\Delta f) = e^{-\Delta f/kT}$$

To make use of the Metropolis algorithm for other than thermodynamic systems, one must provide the following elements:

1. A description of possible system configuration.
2. A generator of random changes in the configuration; these changes are "options" presented to the system.
3. An objective function E (analog energy) whose minimization is the goal of the procedure.
4. A control parameter T (analog temperature) and an *annealing schedule* which tells how it is lowered from high to low values. The meaning of "high" and "low" in this context and the assignment of a schedule,

may require physical insight and /or trial-and-error experiments. In applications to solve optimization problems is necessary to start with an "*adequate high* " effective temperature T_0.

It is important to know that the quality of the final solution is not affected by the initial candidates (initial solutions). We have a sparse matrix and we try to reduce, in a first approach, the number of "intermediate" null entries in the matrix; this means that the null entries that exist between the first non-null element in a row of the lower triangle of the matrix and the diagonal.

This problem belongs to the class known as *NP-complete* problems, whose computation time for an *exact* solution increases with N *as* $\exp(const.N)$, becoming rapidly prohibitive in cost as N increases. This problem also belongs to a class of minimization problems for which the objective function E has many local minima and many global minima. In practice, one desires to find a good solution, not the global minima, but being acceptable as a solution which is significantly better than many others to be found.

To test the flexibility of the algorithm we add another criterion to the "cost function": as an instance we have taken the variance of the length of the semi-rows of the matrix, that is, the number of entries in each row of the lower triangle of the matrix between the first non-null entry and the diagonal.

Our simulated annealing approach has been handled as follows:

We build the connection graph of the matrix which describes its non-null entries:

1. Configuration: The nodes are numbered i=1,2,...N, a configuration is a permutation of the number 1, 2, ..., N, interpreted as the order in which the nodes of the graph of the matrix are arranged.
2. Rearrangement: An efficient set of moves has been suggested by Lin[?]. These moves consist of two types:

   (a) A section of the permutation vector is removed and then is replaced with the same section in the opposite order, for example,

1	2	3	4	5	6	7	8	9	10	11
2	1	3	8	**6**	**4**	**5**	**7**	11	9	10

1	2	3	4	5	6	7	8	9	10	11
2	1	3	8	**7**	**5**	**4**	**6**	11	9	10

   (b) A section of the permutation vector is removed and then is replaced inbetween two contiguous positions on another randomly chosen part of the vector.

For example,

1	2	3	4	5	6	7	8	9	10	11
2	1	3	8	**7**	**5**	**4**	6	11	9	10

1	2	3	4	5	6	7	8	9	10	11
2	1	3	8	6	11	**7**	**5**	**4**	9	10

3. Objective function: E is proposed as the cost function:

$$E = \alpha.NC + \beta.\sigma^2$$

Being NC the total number of intermediate entries and σ the standard deviation of the length.

4. Annealing schedule: we first generate some random rearrangements and use them to determine the range of values of ΔE that will be encountered from move to move.Choosing a starting value for the parameter T which is considerably larger than the largest ΔE encountered, we proceed downward in multiplicative steps each amounting to a 10 percent decrease in T. We hold each new value of T constant for κN reconfigurations or for L successful reconfigurations, whichever comes first. When efforts to reduce E become sufficiently discouraging, we stop.

   The algorithm of simulated annealing for minimizing non null entries and band in a sparse matrix is:

```
begin
   t←0
   Initialize temperature T by generating some random
   rearrangements and the evaluation of ΔE that is produced
   Select a new rearrangement X_c
   Evaluate E for X_c
   repeat
      repeat
         Generate a new rearrangement X_n
            Move type (a) or Move type (b)
         Evaluate E for X_n
      until (κN evaluations or L successes)
      decrement T by a selected percent
   until (T is a very small or we get no gain in the inner loop)
```

2 Numerical experiments

We consider a test for a regular mesh in a rectangular domain. The first numeration (figure1) gives the best distribution for a banded matrix storage scheme. This distribution has 36 "intermediate" null entries. If we apply our criterion with *k=50* and *L=10* then the renumeration in figure 2 is obtained, which has 31 "intermediate" null entries. If we take *k=20* and keep *L=10* then we obtain the mesh in figure 3 which also has 32 intermediate nulls. Figures 4,5, and 6 correspond to the matricial representation of the renumeration cases respectively in accordance with figures 1, 2, 3. Now we present several results about renumeration of systems with the following orders: 318, 742, 950 and 1023.

3 Concluding remarks

- Simulated annealing procedure for searching a permutation of the matrix is proposed.

- We propose the possible application of stochastic rnumbering procedure for te improvement of preconditioner effectiveness for incomplete ILU factorization associated to GMRES method or system resolution.

- Numerical experiments show that in the first iteration the simulated annealing procedure, a drastic reduction in the cost function is achivied. This fact could indicate the possibility of performing a "*quick and dirty* " improvement doing only a few iterations in the procedure.

- It could ve interesting o achieve a much better optimization as the time consumed in such optimization would be shared among the many linear resolutions; leading, this way, to an "*economic* " and robust non linear solver to solve a non-linear evolutionary intefral operator which involves the resolution of "*many* " linear systems, all of them sharing the same matrix.

- This approach we believe that will produce a scheme of resolution with the possibility of using less memory allocation due to the dimension of Krylov subspace reduction.

- Another possible alternative to the presented simulated annealing method could be the implementation by Genetic Algorithm associated to deterministic reordering methods; by which could choose among "*pseoudoperipherical* " nodes by the stochast ic procedure and with flexible criteria on the fitness function.

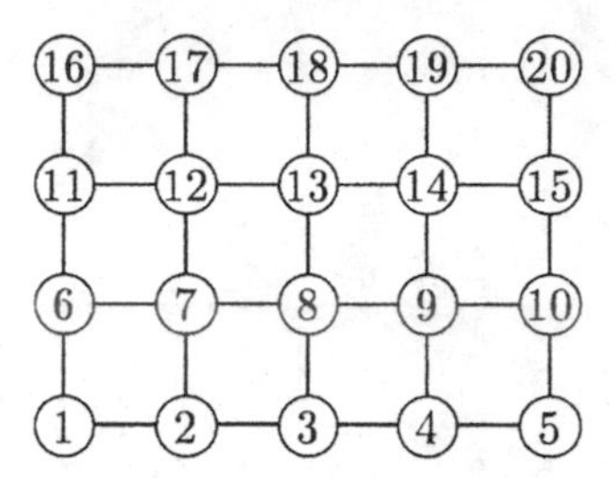

Figure 1: Initial Ordering

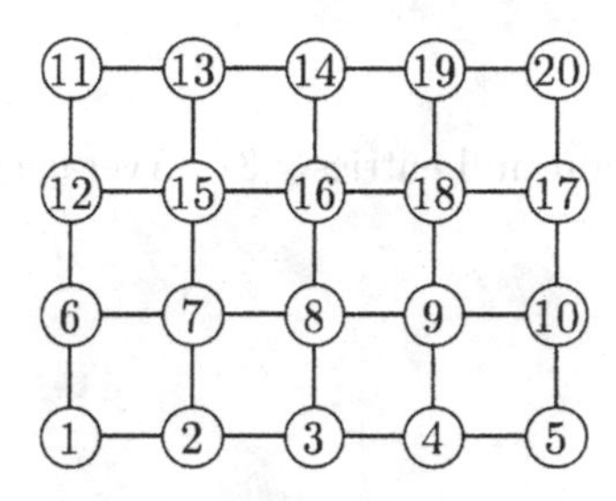

Figure 2: Renumerated (1)

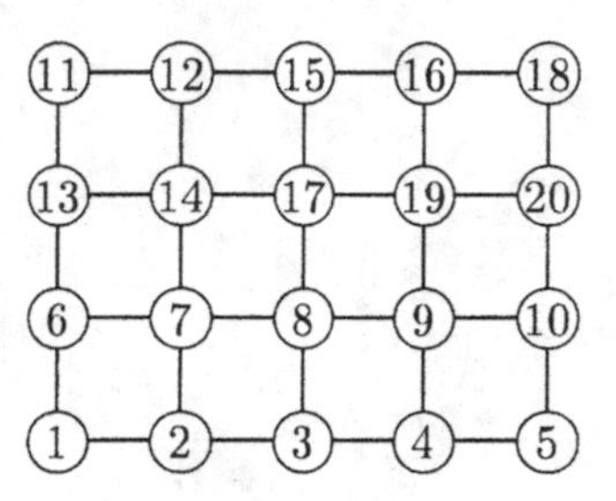

Figure 3: Renumerated (2)

$$
\begin{pmatrix}
x & x & & & & x & x & & & & & & & & & & & & & \\
x & x & x & & & x & x & x & & & & & & & & & & & & \\
 & x & x & x & & & x & x & x & & & & & & & & & & & \\
 & & x & x & x & & & x & x & x & & & & & & & & & & \\
 & & & x & x & & & & x & x & & & & & & & & & & \\
x & x & & & & x & x & & & & x & x & & & & & & & & \\
x & x & x & & & x & x & x & & & x & x & x & & & & & & & \\
 & x & x & x & & & x & x & x & & & x & x & x & & & & & & \\
 & & x & x & x & & & x & x & x & & & x & x & x & & & & & \\
 & & & x & x & & & & x & x & & & & x & x & & & & & \\
 & & & & & x & x & & & & x & x & & & & x & x & & & \\
 & & & & & x & x & x & & & x & x & x & & & x & x & x & & \\
 & & & & & & x & x & x & & & x & x & x & & & x & x & x & \\
 & & & & & & & x & x & x & & & x & x & x & & & x & x & x \\
 & & & & & & & & x & x & & & & x & x & & & & x & x \\
 & & & & & & & & & & x & x & & & & x & x & & & \\
 & & & & & & & & & & x & x & x & & & x & x & x & & \\
 & & & & & & & & & & & x & x & x & & & x & x & x & \\
 & & & & & & & & & & & & x & x & x & & & x & x & x \\
 & & & & & & & & & & & & & x & x & & & & x & x
\end{pmatrix}
$$

Figure 4: N^o non null entries =36. Average width=4.55

$$\begin{pmatrix}
x&x&&& & x&x&&& & &&&& & &&&& \\
x&x&x&& & x&x&x&& & &&&& & &&&& \\
&x&x&x& & &x&x&x& & &&&& & &&&& \\
&&x&x&x & &&x&x&x & &&&& & &&&& \\
&&&x&x & &&&x&x & &&&& & &&&& \\
x&x&&& & x&x&&& & &x&&&x & &&&& \\
x&x&x&& & x&x&x&& & &x&&&x & x&&&& \\
&x&x&x& & &x&x&x& & &&&&x & x&&x&& \\
&&x&x&x & &&x&x&x & &&&& & x&x&x&& \\
&&&x&x & &&&x&x & &&&& & &x&x&& \\
&&&& & &&&& & x&x&x&&x & &&&& \\
&&&& & x&x&&& & x&x&x&&x & &&&& \\
&&&& & &&&& & x&x&x&x&x & x&&&& \\
&&&& & &&&& & &&x&x&x & x&&x&x& \\
&&&& & x&x&x&& & x&x&x&x&x & x&&&& \\
&&&& & &x&x&x& & &&x&x&x & x&&x&x& \\
&&&& & &&&x&x & &&&& & &x&x&x&x \\
&&&& & &&x&x&x & &&&x& & x&x&x&x&x \\
&&&& & &&&& & &&&x& & x&x&x&x&x \\
&&&& & &&&& & &&&x& & x&x&x&x&x
\end{pmatrix}$$

Figure 5: N^o non null entries =31. Average width =4.3

$$\begin{pmatrix}
x&x&&& & x&x&&& & &&&& & &&&& \\
x&x&x&& & x&x&x&& & &&&& & &&&& \\
&x&x&x& & &x&x&x& & &&&& & &&&& \\
&&x&x&x & &&x&x&x & &&&& & &&&& \\
&&&x&x & &&&x&x & &&&& & &&&& \\
x&x&&& & x&x&&& & &&x&x& & &&&& \\
x&x&x&& & x&x&x&& & &&x&x& & &x&&& \\
&x&x&x& & &x&x&x& & &&&x& & &x&&x& \\
&&x&x&x & &&x&x&x & &&&& & &x&&x&x \\
&&&x&x & &&&x&x & &&&& & &&&x&x \\
&&&& & &&&& & x&x&x&x& & &&&& \\
&&&& & &&&& & x&x&x&x&x & &x&&& \\
&&&& & x&x&&& & x&x&x&x& & &&&& \\
&&&& & x&x&x&& & x&x&x&x&x & &x&&& \\
&&&& & &&&& & &x&&x&x & x&x&&x& \\
&&&& & &&&& & &&&&x & x&x&x&x&x \\
&&&& & &x&x&x& & &x&&x&x & x&x&&& \\
&&&& & &&&& & &&&& & x&&x&x&x \\
&&&& & &&x&x&x & &&&&x & x&x&x&x&x \\
&&&& & &&&x&x & &&&& & x&&x&x&x
\end{pmatrix}$$

Figure 6: N^o non null entries =32. Average width = 4.35

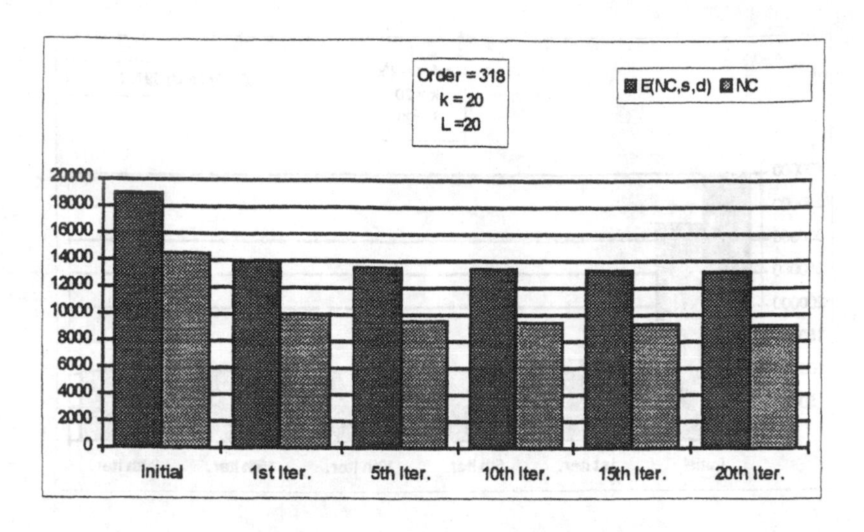
Order = 318
k = 20
L =20
E(NC,s,d)
NC
20000
18000
16000
14000
12000
10000
8000
6000
4000
2000
0
Initial
1st Iter.
5th Iter.
10th Iter.
15th Iter.
20th Iter.

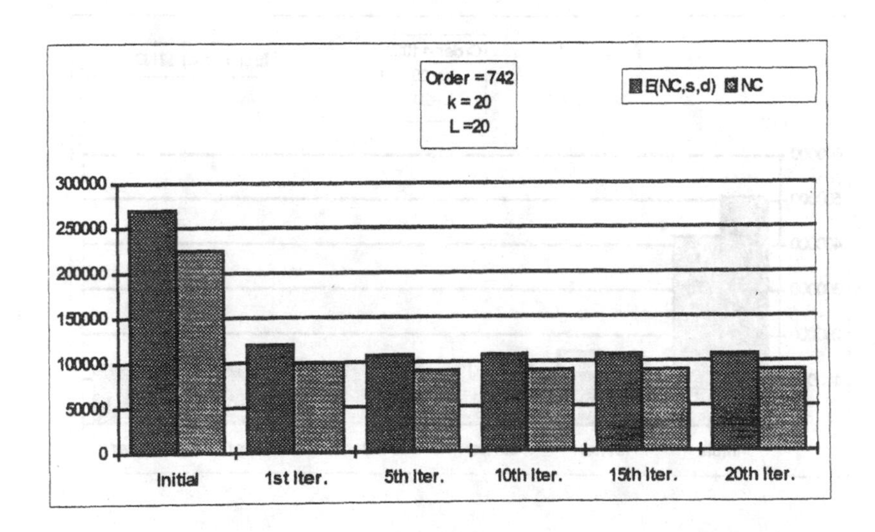
Order = 742
k = 20
L =20
E(NC,s,d)
NC
300000
250000
200000
150000
100000
50000
0
Initial
1st Iter.
5th Iter.
10th Iter.
15th Iter.
20th Iter.

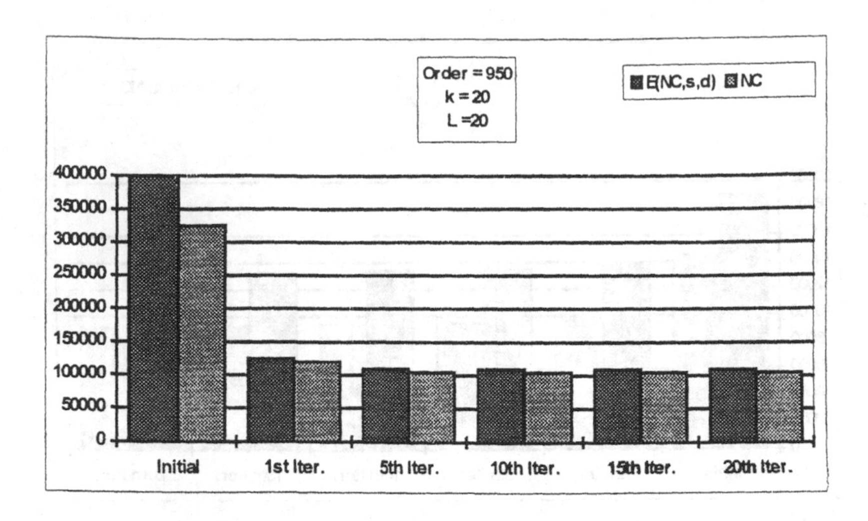
Order = 950
k = 20
L =20
E(NC,s,d)
NC
400000
350000
300000
250000
200000
150000
100000
50000
0
Initial
1st Iter.
5th Iter.
10th Iter.
15th Iter.
20th Iter.

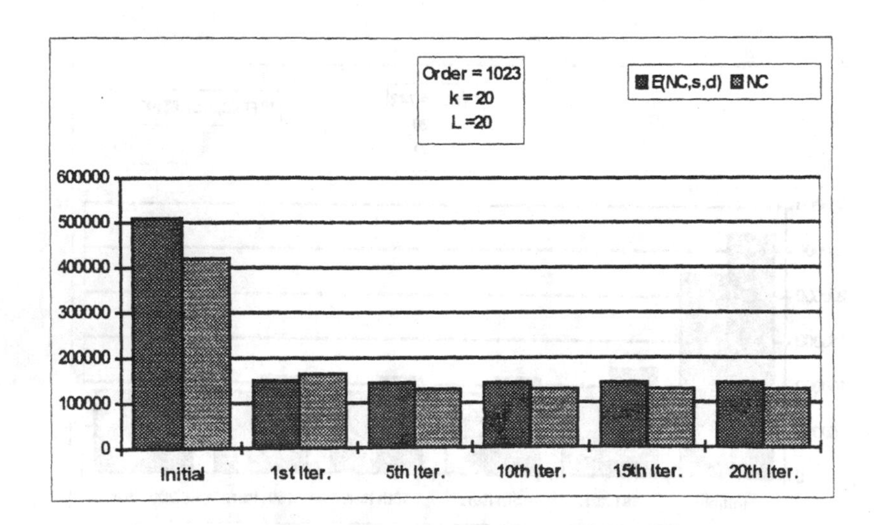
Order = 1023
k = 20
L =20
E(NC,s,d)
NC
600000
500000
400000
300000
200000
100000
0
Initial
1st Iter.
5th Iter.
10th Iter.
15th Iter.
20th Iter.

References

[1] Cunthill, E.H. and J.M.McKee (1969), "*Reducing the band width of sparse symmetric matrices*", Proccedings 24th National conference or the Association for Computing Machenery, 157-172.

[2] Dutto, L.C. (1993) "*The effect of ordering on preconditioned GMRES algorithm for solving the comprensible Navier-Stokes equations*". International Journal for Methods in Engineering (36) (457-497).

[3] Gibbs, N.E., Poole, W.G. Jr and Stockmeyer (1976), "*An Algorithm for reducing the band width and profile of a sparse matrix*", SIAM J. Numer. Anal., (13) (236-250). Rawlins (Ed.).

[4] Kahaner, D., Moler C., Nash S. (1989) "*Numerical Methods and Software*" Prentice Hall.

[5] Kirpatric, S., Stoll, E. (1981) "*A Very fast Shift-Register Sequence Random Number Generator*", Journal of computational Physics, V.40.

[6] Kirpatrick, S. (1984) Journal of Statistical Physics, (34), (975-986).

[7] Kirpatrick, S., Gelatt, C.D., Vecchi, M.P. (1983) Science, (220), (671-680).

[8] Knuth, D.E. (1981) "*Seminumerical Algorithms: The art of Computer Programming*" Addison-Wesley.

[9] Lin, S. (1965) Bell system Technical Journal, (44), (2245-2269).

[10] Metropolis, N., Rosenbluth, A. Rosenbluth, M., Teller, A. and Teller, E. (1953) Journal of Chemical Physics (21) (1087-1092).

[11] Otten, R.H. and van Ginneken, L.P. (1989) "*The Annealing Algorithm*", Boston: Kluwer.

[12] Press , W., Teukolsky S., Vetterling, W. and Frannery, B. (1992) "*Numerical Recipes in C: The Art of Scientific Computing*". Cambridge Univ. Press.

[13] Schrage, L. (1979) "*ACM Transactions on Mathematical Software*", (5), (132-138).

[14] Vecchi, M.P. and Kirkpatrick, S. (1983) IEEE Transactions on Computer Aided Design, (CAD-2), (215-222).

PRECONDITIONED KRYLOV SUBSPACE METHODS

Y. SAAD
University of Minnesota, Department of Computer Science
200 Union Street S.E., Minneapolis, MN 55455

Abstract. Preconditioned Krylov subspace techniques have had a substantial impact on the numerical solution of engineering and scientific problems. These methods can be mandatory for large problems arising from 3-dimensional models. However, although iterative methods are rapidly gaining ground, there are still a number of open questions and unresolved issues related to their use in representative applications. In this paper we give an overview of the most commonly used techniques, and discuss open questions and current trends.

Key Words: Sparse Linear Systems; Krylov Subspace Methods; Iterative techniques; Preconditioning.

1. Introduction

Iterative methods for solving linear systems in scientific and engineering applications have had a varying level of success over the past few decades and across different disciplines. New algorithms are sometimes discovered and then abandoned because of a particular weakness, only to be rehabilitated years later. One difficulty stems from the changing nature of the problems that are tackled. Until relatively recently most simulations were two-dimensional and direct solvers were generally preferred. Direct methods have also been preferred for some types of problems which are known to yield extremely ill-conditioned matrices as for example in structural engineering applications. Prior to the middle-50's most methods used for solving linear systems were either direct methods – or relaxation type techniques. The work of Lanczos [35, 36] and Hestenes and Stiefel [26], began the era of conjugate gradient-type techniques, which matured into sophisticated and efficient methods only in the early to the middle seventies. Krylov subspace methods for both linear systems and eigenvalue problems, were promoted by the studies of Paige [39], Reid [40], Godunov and Prokopov [22], and

G. Winter Althaus and E. Spedicato (eds.), Algorithms for Large Scale Linear Algebraic Systems, 131–149.

others at roughly the same time, who established that loss of orthogonality is not a serious flaw of these methods, in that it does not prevent the algorithm to converge toward the desired solution. Then, preconditioning techniques started to play a major role in the 1960s and 1970s. Preconditioning ideas were present in many of the early papers on Krylov subspace techniques, see e.g., the survey [23]. The paper by Meijerink and van der Vorst [38] marked a turning point. The authors defined a general purpose technique for constructing a preconditioner based on the original coefficient matrix.

While iterative techniques for matrix problems have made substantial inroads into industrial applications, there are still a few serious stumbling blocks. Ironically, part of the problem is a result of the success of iterative techniques and more generally numerical solution techniques. Indeed, as these techniques are becoming capable of solving larger problems, fluid dynamicists are attempting harder simulations which may involve heat transfer and chemical reactions. These models lead to much harder linear systems characterized by very poor conditioning and highly indefinite coefficient matrices.

Iterative techniques based on Krylov subspace projection coupled with suitable preconditioners offer a good compromise between efficiency and robustness. There are two ingredients in the use of a preconditioned Krylov subspace approach. First, the user must select a preconditioning matrix M for the problem. The matrix M must be an approximation to A, and yet it must be not too expensive to compute $M^{-1}v$ for an arbitrary v. The original linear system

$$Ax = b \tag{1}$$

is then preconditioned by, for example, transforming it into the "left-preconditioned" equivalent system

$$M^{-1}Ax = M^{-1}b.$$

One can also solve the right-preconditioned system,

$$AM^{-1}(Mx) = b. \tag{2}$$

Here, the system $AM^{-1}z = b$ is solved for the unknown $z \equiv Mx$, and the final x is obtained through the post-transformation $x = M^{-1}z$. The main difference between the left and right preconditioning approaches is that the actual residual norm is available at each step of the iterative process in the right-preconditioned case whereas only the preconditioned residual is available in the left preconditioning case.

The second ingredient in the method is the "accelerator", typically a conjugate gradient-like method, based on projections on Krylov subspaces. Next, is a survey of these techniques.

2. KRYLOV SUBSPACE METHODS

Let x_0 be an initial guess to the linear system (1), where A is a large sparse real matrix, and let r_0 be the associated initial residual vector $r_0 = b - Ax_0$. A Krylov subspace method finds an approximate solution $\tilde{x}$ from the affine subspace $x_0 + K_m(A, r_0)$ where $K_m(A, r_0)$ is the Krylov subspace

$$K_m(A, r_0) = \text{ span } \{r_0, Ar_0, \ldots, A^{m-1}r_0\}$$

by imposing a number of conditions, such as optimality, of the approximate solution. If there is no ambiguity we will denote by K_m the subspace $K_m(A, r_0)$. The residual vector $r_m = b - A\tilde{x}$ of the approximate solution is of the form $r_m = \phi_m(A)r_0$, where ϕ_m is a certain polynomial of degree $\leq m$. The polynomial ϕ_m is referred to as the *residual polynomial* and methods which lead to residuals which have this form are referred to as *polynomial acceleration methods.*

The canonical way of extracting an approximation from K_m, is to impose a Petrov-Galerkin condition of the form:

$$b - A\tilde{x} \perp L_m \tag{3}$$

where L_m is another subspace of dimension m.

The different versions of Krylov subspace methods arise from different choices of the subspaces K_m and L_m and from the ways in which the system is preconditioned.

2.1. ORTHOGONAL BASES METHODS

This class of techniques are based on constructing orthogonal bases of the Krylov subspace and extracting an approximate solution by exploiting this bases. Note that the orthogonality of the basis may be relative to a non-Euclidean inner product.

A number of algorithms in this class can be derived from the Arnoldi procedure. Given a vector v_1 this procedure constructs an orthogonal basis of the Krylov subspace $K_m(A, v_1)$.

ALGORITHM 2.1 *Arnoldi*

1. *Choose a vector* v_1 *of norm 1.*
2. *For* $j = 1, 2, \ldots, m$ *do:*
3. *Compute* $w_j := Av_j$
4. *For* $i = 1, \ldots, j$ *do*
5. $h_{ij} = (w_j, v_i)$
6. $w_j := w_j - h_{ij}v_i$
7. *EndDo*
8. $h_{j+1,j} = \|w_j\|_2$. *If* $h_{j+1,j} = 0$ *Stop.*
9. $v_{j+1} = w_j / h_{j+1,j}$.

This algorithm is essentially a modified Gram-Schmidt procedure in which at each step the new vector to be orthogonalized is obtained by multiplying the most recent vector in the sequence by A. We use the following notation. The $n \times m$ matrix whose column vectors are the vectors $v_1, \ldots, v_m$ is denoted by V_m. The $(m+1) \times m$ upper Hessenberg (quasi-upper triangular) matrix whose coefficients h_{ij} are defined by the algorithm is denoted by $\bar{H}_m$. We also denote by H_m the $m \times m$ matrix obtained from $\bar{H}_m$ by deleting its last row. We have the important relations:

$$AV_m = V_{m+1}\bar{H}_m, \quad (4)$$

$$V_m^T A V_m = H_m \, . \quad (5)$$

FOM. Suppose that we take $v_1 = r_0/\|r_0\|_2$ in Arnoldi's method, and set $\beta = \|r_0\|_2$. An approximate solution from K_m can be written as

$$\tilde{x} = x_0 + V_m y \, . \quad (6)$$

If we impose the orthogonality condition $r_m \perp K_m$ we immediately get $y = H_m^{-1}(\beta e_1)$ and the approximate solution,

$$x_m = x_0 + V_m H_m^{-1}(\beta e_1) \, .$$

The algorithm related to this approximate solution is referred to as the Full Orthogonalization Method [41]. Mathematically equivalent methods have been proposed by Axelsson [4] and Young and Jea [29]. Note that computing the approximate solution $\tilde{x}$ requires storing the m Arnoldi vectors $v_1, \ldots, v_m$ and this may become a big burden. One solution is to restart the algorithm, i.e., if x_m is not accurate enough we restart FOM with x_0 set to the most recent approximate solution obtained.

GMRES. We can also exploit the Arnoldi basis to obtain an optimal solution from the Krylov subspace. Specifically, we can seek an approximation of the form (6) whose residual norm is minimal. If we take again $v_1 = r_0/\|r_0\|$, we observe that we have,

$$b - A\tilde{x} = V_{m+1}[\beta e_1 - \bar{H}_m y] \, . \quad (7)$$

Using the orthonormality of the column-vectors of V_{m+1} we immediately obtain that y must minimize

$$\|\beta e_1 - \bar{H}_m y\|_2 \quad (8)$$

over all y's in $\mathbf{R}^m$. The algorithm leading to a solution obtained in this manner is termed the Generalized Minimum Residual algorithm [48]. It can be sketched as follows.

ALGORITHM **2.2** *GMRES*

1. $r_0 := b - Ax_0$, $v_1 := r_0/(\beta := \|r_0\|_2)$
2. *Compute* V_m, $\bar{H}_m$ *by Arnoldi's method.*
3. *Compute* y_m *which minimizes (8)*
4. $x_m = x_0 + V_m y_m$.

In practical situations, one must restart the Algorithm, just as for FOM.

GCR and ORTHOMIN. An alternative implementation of the full GMRES algorithm is to exploit a basis of the Krylov subspace different from the Arnoldi basis. Specifically, if $p_0, p_1, \ldots, p_{m-1}$ is a basis of the Krylov subspace $K_m(A, r_0)$ which is $A^T A$-orthogonal, i.e., such that

$$(Ap_i, Ap_j) = 0 \quad \text{for } i \neq j,$$

then it is known that the approximate solution x_m which has the smallest residual norm in the affine space $x_0 + K_m$ can be 'updated' from x_{m-1} by,

$$x_m = x_{m-1} + \frac{(r_{m-1}, Ap_{m-1})}{(Ap_{m-1}, Ap_{m-1})} p_{m-1}\,, \tag{9}$$

in which $r_i = b - Ax_i$. The idea then is to progressively compute a basis of K_m which is $A^T A$ - orthonormal and to update the iterate by the above formula each time a new p vector is computer. One such procedure is the Generalized Conjugate Residual (GCR) algorithm [17].

ALGORITHM **2.3** *GCR*

1. *Compute* $r_0 = b - Ax_0$. *Set* $p_0 = r_0$.
2. *For* $j = 0, 1, \ldots,$ *until convergence do:*
3. $\quad \alpha_j = \frac{(r_j, Ap_j)}{(Ap_j, Ap_j)}$
4. $\quad x_{j+1} = x_j + \alpha_j p_j$
5. $\quad r_{j+1} = r_j - \alpha_j Ap_j$
6. $\quad$ *For* $i = 0, \ldots, j$ *compute:*
7. $\quad\quad \beta_{ij} = -\frac{(Ar_{j+1}, Ap_i)}{(Ap_i, Ap_i)}$
8. $\quad p_{j+1} = r_{j+1} + \sum_{i=0}^{j} \beta_{ij} p_i$
9. *EndDo*

A variant of this algorithm, known as ORTHOMIN(k) and suggested by Vinsome [54], consists of truncating the recurrence for computing the p-vectors, replacing it with one involving only the k most recent p_i's.

A few other algorithms that are mathematically equivalent to the above schemes, or similar in spirit, have been developed in recent years [29], [17], [4], [3], [18], [49].

2.2. BI-ORTHOGONAL BASES METHODS

Recent research on accelerators has been focused mainly on methods derived from the non-Hermitian Lanczos algorithm. There are essentially two sub-classes of methods in this category: variants of the bi-conjugate gradient algorithm and transpose-free alternatives.

Lanczos, BiCG, and QMR Given two vectors v_1 and w_1, the non-symmetric Lanczos algorithm builds simultaneously a basis $\{v_i\}$ of $K_m(A, v_1)$ and a basis $\{w_j\}$ of $K_m(A^T, w_1)$ which have the property of being bi-orthogonal to each other, i.e., such that

$$(v_i, w_j) = \delta_{ij}$$

where δ_{ij} is the Kronecker symbol.

ALGORITHM **2.4** *Lanczos*

Select v_1, w_1 *with* $(v_1, w_1) = 1$.
Set $\beta_1 = \delta_1 \equiv 0$, $w_0 = v_0 \equiv 0$.
For $j = 1, 2, \ldots, m$ *Do:*
$\alpha_j = (Av_j, w_j)$
$\hat{v}_{j+1} = Av_j - \alpha_j v_j - \beta_j v_{j-1}$
$\hat{w}_{j+1} = A^T w_j - \alpha_j w_j - \delta_j w_{j-1}$
$\delta_{j+1} = |(\hat{v}_{j+1}, \hat{w}_{j+1})|^{1/2}$
$\beta_{j+1} = (\hat{v}_{j+1}, \hat{w}_{j+1})/\delta_{j+1}$
$w_{j+1} = \hat{w}_{j+1}/\beta_{j+1}$
$v_{j+1} = \hat{v}_{j+1}/\delta_{j+1}$
EndDo

A few techniques exploit this algorithm for obtaining an approximate solution based on a projection process onto $K_m(A, r_0)$ and orthogonally to a subspace of the form $K_m(A^T, w_1)$, where w_1 is an arbitrary vector. For consistency with the Arnoldi process, we denote by $\bar{H}_m$ the $(m+1) \times m$ tridiagonal matrix with diagonal entries α_j, $j = 1, \ldots, m$, superdiagonal entries β_j, $j = 1, \ldots, m$, and subdiagonal entries δ_j, $j = 1, \ldots, m+1$. Then it is clear that the relation (4) is still satisfied, while (5) is replaced by

$$W_m^T A V_m = H_m \,.$$

If we take $v_1 = w_1 = r_0/\|r_0\|_2$ and seek an approximate solution of the form (6) whose residual is orthogonal to the subspace $K_m(A^T, r_0)$, we would find the same expression as in FOM for the approximate solution x_m except that H_m is now a tridiagonal matrix. A procedure based on this approach is not appealing because it requires that we store all the vectors v_i in order to compute the solution. However, it is possible to obtain the approximate solution in a progressive manner, i.e., it is possible to obtain x_m from x_{m-1} by using auxiliary vectors which satisfy a short term recurrence. This is achieved by the Bi-Conjugate gradient algorithm proposed by Fletcher [19].

ALGORITHM 2.5 *Bi-CG*

Compute $r_0 := b - Ax_0$.
Choose r_0^* *such that* $(r_0, r_0^*) \neq 0$.
Set, $p_0 := r_0$, $p_0^* := r_0^*$.
For $j = 0, 1, \ldots,$ *until convergence Do:*

$$\alpha_j := (r_j, r_j^*)/(Ap_j, p_j^*)$$
$$x_{j+1} := x_j + \alpha_j p_j$$
$$r_{j+1} := r_j - \alpha_j A p_j$$
$$r_{j+1}^* := r_j^* - \alpha_j A^T p_j^*$$
$$\beta_j := (r_{j+1}, r_{j+1}^*)/(r_j, r_j^*)$$
$$p_{j+1} := r_{j+1} + \beta_j p_j$$
$$p_{j+1}^* := r_{j+1}^* + \beta_j p_j^*$$

EndDo

The bi-CG algorithm does not obey any simple optimality properties in general and the behavior of the residual norm that it produces can be quite erratic. To remedy this, Freund and Nachtigal [21] proposed a middle ground solution based on the Lanczos algorithm. Indeed, the vectors produced by the Lanczos algorithm 2.4 satisfy a relation which is identical with (4) except that the matrix $\bar{H}_m$ is $(m+1) \times m$ tridiagonal instead of Hessenberg. Therefore, the residual vector r_m associated with an approximation $\tilde{x} = x_0 + V_m y$ also satisfies the relation (7). We could attempt to obtain the vector y which minimizes the residual norm $b - A\tilde{x}$ as was done for GMRES. This is rendered difficult by the fact that the column vectors of V_{m+1} are no longer orthonormal. As an alternative, the Quasi Minimal Residual algorithm [21] ignores the orthonormality of these vectors and produces the approximate solution (6) in which y minimizes the norm

$$\|\beta e_1 - \bar{H}_m y\|_2$$

just as in GMRES. Thus, the formal description of the QMR algorithm differs from that of algorithm 2.2 only in step 2 which computes the Arnoldi basis. In general, the resulting algorithm does tend to produce a convergence behavior which is much smoother than that of BiCG.

CGS, BiCGSTAB, and TFQMR. A significant development in Lanczos-based algorithms was the realization that it is possible to avoid the use of A^T in the BiCG algorithm and gain a faster convergence for the same number of matrix-vector products. Indeed, what Sonneveld [51] discovered in the early 80's (although the final article appeared much later) is that the operations with A^T in BiCG can in essence be replaced by operations with matrix-vector products with A to produce a new sequence of iterates whose residual polynomial is the square of that realized by the BiCG algorithm. A whole class of methods based on these techniques have recently been discovered. Among them are the BiCGSTAB algorithm [52] and the Transpose-Free QMR algorithm [20]. These are discussed in full detail in another contribution of these proceedings.

Normal equations and related techniques. Solving the normal equations

$$A^T A x = A^T b \,, \tag{10}$$

is often advocated as an alternative to using a Krylov method on the original system (1). This is a symmetric positive definite linear system whose solution is the same as that of the system (1). Therefore, a preconditioned conjugate gradient algorithm can used. There are two difficulties with this approach. First, the condition number of the resulting system can be very large, being the square of that of the original system, and this can render the CG iteration ineffective. A second, somewhat related difficulty is that it is not easy to precondition the coefficient matrix. We will return to this issue in the next section.

There are cases where some form of the normal equations occurs naturally. For example, when solving least-squares problems

$$\min \|b - Ax\|_2$$

A is $n \times m$ with $m < n$, then no matter which method is used it will compute a solution to the normal equations (10). Therefore, methods for solving the normal equations are important by themselves.

3. PRECONDITIONING TECHNIQUES

A critical component, possibly the most important, in the success of iterative methods is the preconditioner. The idea of preconditioning is to transform the original system, into one which will be easier to solve by a Krylov subspace method. For example, when the preconditioner M is applied to the right, we will be solving instead of (1), the preconditioned linear system (2). In some applications it can be important to allow the preconditioner to vary from step to step in the inner Krylov subspace iteration. We start by discussing a class of accelerators which accommodate these variations.

3.1. FLEXIBLE ACCELERATORS

We consider here a situation which is gaining importance because of the impact of parallel processing in particular. Sometimes a preconditioner can be itself defined from an iterative process, such as a relaxation scheme. In this situation, M is not a constant matrix and can instead be viewed as an operator which changes at each step. When right preconditioning is used, it is possible to obtain a solution which is optimal in a certain sense

Step 4 of the GMRES algorithm 2.2 applied to the system (2) will compute the approximation via the modified formula,

$$x_m = x_0 + M^{-1} V_m y_m \ .$$

where M is the preconditioner. This forms a linear combination of the preconditioned vectors $z_i = M^{-1} v_i, i = 1, \ldots, m$, where v_i are the Arnoldi vectors. Because these vectors are all obtained by applying the same preconditioning matrix M^{-1} to the v's, we need not save them. We only need to apply M^{-1} to the linear combination of the $v's$, i.e., to $V_m y$. If we allowed the preconditioner to vary at every step, i.e., if z_j is now defined by

$$z_j = M_j^{-1} v_j$$

we may save these vectors and use them when computing x_m. This observation yields the following 'flexible' preconditioning GMRES algorithm [43].

ALGORITHM 3.1 *FGMRES*

Until convergence do :

$r_0 = b - Ax_0;\ \beta := \|r_0\|_2;\ v_1 = r_0/\beta.$

For $j = 1, \ldots, m$ *do:*

$z_j := M_j^{-1} v_j$

$w := Az_j$

Do $i = 1, \ldots, j,\ \left\{ \begin{array}{l} h_{i,j} := (w, v_i) \\ w := w - h_{i,j} v_i \end{array} \right.$

$v_{j+1} = w/(h_{j+1,j} := \|w\|_2)$

Define $Z_m := [z_1, \ldots, z_m]$.

Compute $x_m = x_0 + Z_m y_m$ where

$y_m = \text{argmin}_y \|\beta e_1 - \bar{H}_m y\|_2.$

EndDo

Set $x_0 \leftarrow x_m$.

EndDo

The solution obtained by the algorithm minimizes the residual norm over the affine space $x_0 + \text{span}\{Z_m\}$. The algorithm may be useful if we wish to mix the effects of short range coupling and long range coupling as is done with multi-level preconditioners. In this case one can alternate between two types of preconditioners. For additional details, see reference [43]. We note that there has been at least two other methods developed recently which allow variable preconditioners. Axelsson and Vassilevski [5] developed a version from the block Generalized Conjugate Gradient viewpoint while Van der Vorst and Vuik [53] developed a rank-1 update-type algorithm, similar to the Broyden approach.

3.2. STANDARD PRECONDITIONERS

The simplest way to define a preconditioning is to take the matrix M to be the diagonal or block diagonal of A. This is referred to as Jacobi (or diagonal), or block Jacobi (block diagonal) preconditioning and is effective only in special cases, e.g., for transient solutions. However, there is currently renewed interest in Block Jacobi preconditioners in certain applications mainly because of the impact of parallel processing, see for example [16].

SOR and SSOR Preconditioners. We can define a rather important class of preconditioners based on point or block relaxation techniques. The best known of these is the SSOR preconditioning matrix defined by

$$M = (D - \omega E)D^{-1}(D - \omega F) \tag{11}$$

in which $-E$ is the strict lower part of A, $-F$ its strict upper part, D its diagonal, and ω is the relaxation parameter.

In the context of preconditioning, it is common to simply take $\omega = 1$ as the gains from selecting an optimal ω are typically small. Note also that an optimal ω for the SOR iteration is unlikely to be the optimal ω for the combination SOR/GMRES. However, it is clear that one can use different values of ω at each step of FGMRES and this can open up the possibility of using heuristics to determine the best ω dynamically, by simply monitoring convergence.

Consider now any general splitting of the form

$$A = M - N ,$$

with which is associated the iteration,

$$u_{k+1} = M^{-1}(Nu_k + b) . \tag{12}$$

We can define a preconditioning matrix associated with applying $k+1$ steps of this iterative process by noticing that

$$u_{k+1} = u_0 + \left[\sum_{j=0}^{k}(M^{-1}N)^j\right] M^{-1}[b - Au_0]$$

which means that $k+1$ steps of (12) amount to approximating the exact solution $x = u_0 + A^{-1}[b - Au_0]$ by u_{k+1} defined above. Hence, the preconditioning matrix associated with this iteration is

$$M_k = \left[\sum_{j=0}^{k}(M^{-1}N)^j\right] M^{-1} . \tag{13}$$

This preconditioning matrix is referred to as the k-step preconditioner associated with the splitting $A = M - N$. Adams [1, 2] studied such preconditioners in the context of the conjugate gradient method and showed in particular that for the SSOR splitting, the preconditioning matrix is positive definite under certain conditions. Clearly, the same techniques can be applied to nonsymmetric matrices as well, although the theory is not as well understood. An interesting observation here is that in the non-Hermitian case, there is little attraction in using SSOR as opposed to the simpler SOR. In fact we found that for our model problems SSOR(k) was rarely superior to SOR(k).

Currently there is renewed interest in multi-step relaxation-type preconditioners, see for example [44, 46, 13, 56, 50]. There are several reasons for this.

- These preconditioners require little or no extra storage.
- Their application is relatively easy to parallelize. Multicoloring schemes can be used to extract maximum parallelism.
- Often, the underlying codes have used relaxation schemes and an 'expertise' may have been acquired on the use of these codes over the years.

On the first point, we note that the scalar versions require no extra storage while the block versions require some additional storage to save the factors of the blocks. Storage is often the most important obstacle in realistic applications so this is a very important advantage of relaxation-based preconditioners.

In the late 60's and early 70's the idea came about to use an M which has the same form as that of the SSOR preconditioner (11) with $\omega = 1$ but with a D that is defined recursively to ensure that the diagonal elements of M and A are the same. This leads to the ILU(0) preconditioner in the special case of 5-point matrices.

ILU(0). If we denote by $NZ(A)$ the nonzero structure of A, i.e., the set of all pairs (i, j) such that $a_{ij} \neq 0$, then ILU(0) can be described as follows.

Algorithm 3.2 *ILU(0)*

For $i = 1, \ldots, N$ *Do:*
    For $k = 1, \ldots, i-1$
        and if $(i,k) \in NZ(A)$ *Do:*
        Compute $a_{ik} := a_{ik}/a_{kj}$
        For $j = k+1, \ldots$
            and if $(i,j) \in NZ(A)$, *Do:*
            compute $a_{ij} := a_{ij} - a_{ik}a_{k,j}$.
        EndDo
    EndDo
EndDo

This algorithm is nothing but an (i,k,j) version of Gaussian elimination [24] which is essentially restricted to the $NZ(A)$ part of the matrix. It can be generalized to any preset nonzero pattern. In particular, one can classify the fill-ins introduced in a Gaussian Elimination procedure by assigning them a level defined recursively from the parents from which the fill-ins are generated [28]. For diagonally dominant matrices, the higher the level-of-fill the smaller the element. Once the level of fill of each element is defined we can execute an algorithm similar to the one above, in which $NZ(A)$ is replaced by $NZ_p(A)$ which is the set of all elements whose level-of-fill does not exceed p. This defines the ILU(p) factorization. The ILU(p) factorization of A requires more storage but is more accurate than ILU(0), so the number of steps of the resulting preconditioned Krylov subspace iteration may be reduced dramatically.

3.3. ILUT AND ILUTP

The elements that are dropped in the ILU(p) factorizations depend only on the pattern of A and not on the values. These level-based ILU factorizations work well for M-Matrices or diagonally dominant matrices but they may fail otherwise. An alternative strategy is to use the same general structure of the ILU factorization, namely the i,k,j variant of Gaussian Elimination, but to drop elements according to their magnitude. One such strategy defined in [45], is referred to as ILU with threshold (ILUT). This incomplete LU factorization uses two parameters, an integer parameter p which sets a limit on the number of fill-ins allowed in L and U, and a tolerance parameter ϵ which serves for filtering out small elements.

Algorithm 3.3 *ILUT(p,ϵ)*

For $i = 1, \ldots, N$ *Do:*
    Compute $\epsilon_i := \epsilon \|a_{i,:}\|_2$
    For $k = 1, \ldots, i-1$ *and if* $a_{i,k} \neq 0$ *Do:*

Compute $a_{ik} := a_{ik}/a_{kj}$
If $|a_{ik}| \geq \epsilon_i$ *Then*
For $j = k+1, \ldots$ *Do:*
compute $a_{ij} := a_{ij} - a_{ik}a_{k,j}$.
If $|a_{ij}| \leq \epsilon_i$ *then* $a_{ij} := 0$
EndDo
EndIf
Keep p *largest elements in L-part of* $a_{i,:}$ *and and* p *largest elements in the U-part of* $a_{i,:}$;
EndDo
EndDo

An advantage of ILUT is that it allows to control the amount of fill-in. When $\epsilon = 0$, then the higher the parameter p, the more accurate the factorization. Reordering for reducing fill-in can help improve the quality of the factorization, and we refer to [14] and [15] for similar experiments performed with the ILU(0) factorization.

The ILUT factorization can be used to solve indefinite problems, and does work for a much broader set of matrices than ILU(0). However, there may be problems computing the factorization itself, because a zero pivot can be encountered. An obvious solution for cases where ILUT fails to yield a good incomplete factorization is to perform some form of pivoting. We can perform a form of pivoting and obtain an algorithm similar, in simplicity and cost, to ILUT. Because of the structure of ILUT, it is not very practical to perform row pivoting. However, a column pivoting variant is easy to develop. The algorithm uses a permutation array *perm* as well as its reverse permutation array to hold the new orderings of the variables. The corresponding algorithm is termed ILUTP (ILUT with Pivoting). The complexity of the ILUTP procedure is virtually identical with that of ILUT.

3.4. PRECONDITIONING THE NORMAL EQUATIONS

Since the normal equations are typically poorly conditioned it is even more important to precondition them before using a Krylov subspace method. If a good preconditioner can be found for A itself, then one can solve the normal equations associated with the preconditioned system. However, in this situation using a Krylov subspace technique on the direct equations is generally preferable. Preconditioning the normal equations can be attractive when the coefficient matrix A is so indefinite that ILU – type preconditioners on the original matrix A fail.

Incomplete Cholesky. One of the first ideas that was suggested for handling symmetric positive definite sparse linear systems $Ax = b$, that are not diagonally dominant, was to use an incomplete Cholesky factorization on the 'shifted' matrix $A + \alpha I$, see, e.g., [31, 37]. The shift is necessary since the IC(0) factorization does not necessarily exist for positive definite matrices. The Preconditioned Conjugate Gradient (PCG) can then be used to solve the preconditioned system. This idea can be applied to the normal equations. Thus, we can use the special forms of PCG to solve the system $A^T Ax = A^T b$ (CGNR variant), preconditioned with a matrix $M = LL^T$ which is the IC(0) factorization of the shifted matrix $A^T A + \alpha I$, see e.g., [47].

An issue which is not satisfactorily resolved is to find good values for the shift α. There is no easy and well-founded solution to this problem for irregularly structured symmetric sparse matrices. One idea is to select the smallest possible α which makes the shifted matrix diagonally dominant. However, this shift tends to be too large in general since, as was observed in [37], IC(0) may exist for much smaller values of α. We can also try to determine the smallest α for which the IC(0) factorization exists. This is unfortunately not the best strategy. Typically, the number of steps required for convergence starts decreasing as α increases, and then it increases again. It is our experience that preconditioning the normal equations with a row projection type method, e.g., using a Kaczmarz technique [30, 8], is more reliable and often performs comparatively with the best incomplete Cholesky approach. We recall that the Kaczmarz approach is equivalent to an SSOR iteration on the normal equations [7, 30, 8].

ILQ and CIMGS A different class of techniques for preconditioning the normal equations is based on the intimate relation between the LQ factorization of A and the Cholesky factorization of AA^T. Consider a general sparse matrix A and let us denote by $a_1, a_2, ..., a_n$ its rows. The (complete) LQ factorization of A is defined by

$$A = LQ$$

where L is a lower triangular matrix and Q is unitary, i.e., $Q^T Q = I$. As can be easily seen, the L factor in the above factorization is identical with the Cholesky factor of the matrix $B = AA^T$. This relationship can be exploited to obtain preconditioners for the normal equations.

The Incomplete LQ (ILQ) approach proposed in [42] consists of using the Gram-Schmidt process, combined with a drop strategy. This idea may seem undesirable at first because of potential numerical difficulties when orthogonalizing a large number of vectors. However, because the rows remain very sparse in the incomplete LQ factorization, any given row of A

will be typically orthogonal to most of the previous rows of Q, and, as a result the Gram-Schmidt process is much less prone to numerical difficulties. Recently, this idea was extended to the modified Gram-Shmidt process [55]. It turns out the Modified Gram Schmidt process can be implemented without the columns of Q. An important by-product of this observation is that it is possible to define a nonzero pattern for AA^T in advance, such that *the incomplete Cholesky based on this pattern is guaranteed to exist.*

3.5. APPROXIMATE INVERSES

There is currently a growing interest in techniques which seek to precondition a linear system by exploiting a sparse approximation to the inverse of A [6, 25, 12, 27, 34, 33, 11, 9] motivated in part by parallel processing and in part by the failure of the ILU-type preconditioners for indefinite problems.

One approach consists of finding a matrix M such that AM is close to the identity matrix. Then the preconditioned system is of the form

$$AMy = b, \quad x = My \ .$$

Each column m_j of M can be obtained by approximately solving the linear system $Am_j = e_j$. In [11] we exploited the observation that the iterates produced by algorithms such as GMRES or MR stay sparse if the initial guess is sparse. Initially, the approximate inverse M is taken to be the identity matrix. Rough approximations can be obtained inexpensively by performing a very small number of steps starting with the columns of the current M. Once a first approximation of M is obtained we can improve it by incorporating it in an outer loop which takes as initial guess the columns of the most current M and improves it with a few steps of sparse-sparse mode iterative technique. A potential improvement can be obtained by using the most recent approximate inverse to precondition the system solved when approximating a column. We refer to this as *self-preconditioning.* The idea of sparse-sparse mode computating was also exploited in a different context to compute incomplete factorizations for sparse skyline matrices [10].

However, the most promising use of approximate inverse techniques is in combining them with other factorizations, such as block factorizations. A number of techniques based on this viewpoint were developed in [32] and, more recently, in [9]. The main idea in [9] is to exploit blockings of the original linear system in the block form,

$$\begin{pmatrix} B & F \\ E & C \end{pmatrix} \begin{pmatrix} x \\ y \end{pmatrix} = \begin{pmatrix} f \\ g \end{pmatrix}. \tag{14}$$

This blocking may originate from a domain decomposition partitioning for example or may be the original structure of A, as in the Navier-Stokes equations.

The above blocking can be exploited in several different ways to define preconditioners for A. One of the simplest and most effective techniques is to approximate the block LU factorization

$$M = LU$$

in which

$$L = \begin{pmatrix} B & 0 \\ E & S \end{pmatrix} \quad \text{and} \quad U = \begin{pmatrix} I & B^{-1}F \\ 0 & I \end{pmatrix}$$

where S is the Schur complement

$$S = C - EB^{-1}F.$$

Since F is a sparse block, the general approach of approximate inverse techniques can be effectively used to construct a sparse approximation Y to $B^{-1}F$. Once Y is obtained, then S is replaced by $M_S = C - EY$ in the L factor and $B^{-1}F$ is replaced by Y in the U factor. This yields an approximate block-factorization. Interestingly enough, this technique is remarkably robust, possibly due to the inherent decoupling of the block structure.

CONCLUDING REMARKS

There are currently a number of practical difficulties that emerge when incorporating preconditioned Krylov subspace methods in typical industrial applications. A major stumbling block is that we do not know too well how to precondition linear systems that are strongly indefinite. The normal equation approach may work well sometimes. However, in most cases the condition number is typically so large that these techniques, or those related to preconditioned CGNE, are bound to fail. It is often possible to solve these hard linear systems by using more memory, in the form of a very accurate ILU factorization. This middle-ground solution between direct and iterative solvers is not always appealing to many practitioners because of the demand on memory.

Preconditioning large indefinite systems is likely to remain a challenging problem for some time to come. Iterative methods still suffer from their origins, since they have been mostly developed with simple elliptic PDEs in mind. These are hardly representative models of the applications world.

References

1. L. M. Adams. *Iterative algorithms for large sparse linear systems on parallel computers.* PhD thesis, Applied Mathematics, University of Virginia, Charlottsville, VA, 1982. Also NASA Contractor Report 166027.

2. L. M. Adams. M-step preconditioned conjugate gradient methods. *SIAM J. Sci. Statist. Comp*, 6:452–463, 1985.
3. O. Axelsson. Conjugate gradient type-methods for unsymmetric and inconsistent systems of linear equations. *Linear Algebra and its Applications*, 29:1–16, 1980.
4. O. Axelsson. A generalized conjugate gradient, least squares method. *Numerische Mathematik*, 51:209–227, 1987.
5. O. Axelsson and P. S. Vassilevski. A block generalized conjugate gradient solver with inner iterations and variable step preconditioning. *SIAM Journal on Matrix Analysis and Applications*, 12, 1991.
6. M. W. Benson and P. O. Frederickson. Iterative solution of large sparse linear systems arising in certain multidimensional approximation problems. *Utilitas Math.*, 22:127–140, 1982.
7. A. Björck and T. Elfving. Accelerated projection methods for computing pseudo-inverse solutions of systems of linear equations. *BIT*, 19:145–163, 1979.
8. R. Bramley and A. Sameh. Row projection methods for large nonsymmetric linear systems. *SIAM Journal on Scientific and Statistical Computing*, 13:168–193, 1992.
9. E. Chow and Y. Saad. Approximate inverse techniques for block-partitioned matrices. *SIAM Journal on Scientific Computing*, 1997. To appear.
10. E. Chow and Y. Saad. ILUS: an incomplete LU factorization for matrices in sparse skyline format. Technical Report UMSI-95-78, University of Minnesota, Supercomputer Institute, Minneapolis, MN 55415, April 1995.
11. E. Chow and Y. Saad. Approximate inverse preconditioners for general sparse matrices. Technical Report UMSI 94-101, University of Minnesota Supercomputer Institute, Minneapolis, MN, May 1994.
12. J. D. F. Cosgrove, J. C. Díaz, and A. Griewank. Approximate inverse preconditioning for sparse linear systems. *Intl. J. Comp. Math.*, 44:91–110, 1992.
13. M. A. DeLong and J. M. Ortega. SOR as a preconditioner. Technical Report CS-94-43, Department of Computer Science, University of Virginia, Charlottesville, VA, 1994.
14. I. S. Duff and G. A. Meurant. The effect of reordering on preconditioned conjugate gradients. *BIT*, 29:635–657, 1989.
15. L. C. Dutto. The effect of reordering on the preconditioned GMRES algorithm for solving the compressible Navier-Stokes equations. *International Journal for Numerical Methods in Engineering*, 36:457–497, 1993.
16. L. C. Dutto, W. G. Habashi, M. P. Robichaud, and M. Fortin. A method for finite element parallel viscous compressible flow calculations. *International Journal for Numerical Methods in Fluids*, 19:275–294, 1994.
17. S. C. Eisenstat, H. C. Elman, and M. H. Schultz. Variational iterative methods for nonsymmetric systems of linear equations. *SIAM Journal on Numerical Analysis*, 20:345–357, 1983.
18. H. C. Elman. *Iterative Methods for Large Sparse Nonsymmetric Systems of Linear Equations*. PhD thesis, Yale University, Computer Science Dept., New Haven, CT., 1982.
19. R. Fletcher. Conjugate gradient methods for indefinite systems. In G. A. Watson, editor, *Proceedings of the Dundee Biennal Conference on Numerical Analysis 1974*, pages 73–89, New York, 1975. Springer Verlag.
20. R. W. Freund. A Transpose-Free Quasi-Minimal Residual algorithm for non-Hermitian linear systems. *SIAM Journal on Scientific Computing*, 14(2):470–482, 1993.
21. R. W. Freund and N. M. Nachtigal. QMR: a quasi-minimal residual method for non-Hermitian linear systems. *Numerische Mathematik*, 60:315–339, 1991.
22. S. K. Godunov and G. P. Propkopov. A method of minimal iteration for evaluating the eigenvalues of an elliptic operator. *Zh. Vichsl. Mat. Mat. Fiz.*, 10:1180–1190, 1970.
23. G. H. Golub and D. P. O'Leary. Some history of the conjugate gradient and Lanczos

algorithms: 1948-1976. *SIAM review*, 31:50–102, 1989.

24. G. H. Golub and C. Van Loan. *Matrix Computations*. The John Hopkins University Press, Baltimore, 1989.
25. M. Grote and H. D. Simon. Parallel preconditioning and approximate inverses on the connection machine. In R. F. Sincovec, D. E. Keyes, L. R. Petzold, and D. A. Reed, editors, *Parallel Processing for Scientific Computing – vol. 2*, pages 519–523. SIAM, 1992.
26. M. R. Hestenes and E. L. Stiefel. Methods of conjugate gradients for solving linear systems. *Journal of Research of the National Bureau of Standards*, Section B, 49:409–436, 1952.
27. T. Huckle and M. Grote. A new approach to parallel preconditioning with sparse approximate inverses. Technical Report SCCM-94-03, Stanford University, Scientific Computing and Computational Mathematics Program, Stanford, California, 1994.
28. J. W. Watts III. A conjugate gradient truncated direct method for the iterative solution of the reservoir simulation pressure equation. *Society of Petroleum Engineers Journal*, 21:345–353, 1981.
29. K. C. Jea and D. M. Young. Generalized conjugate gradient acceleration of nonsymmetrizable iterative methods. *Linear Algebra and its Applications*, 34:159–194, 1980.
30. C. Kamath and A. Sameh. A projection method for solving nonsymmetric linear systems on multiprocessors. *Parallel Computing*, 9:291–312, 1988.
31. D. S. Kershaw. The incomplete Choleski conjugate gradient method for the iterative solution of systems of linear equations. *Journal of Computational Physics*, 26:43–65, 1978.
32. L. Yu Kolotilina and A. Yu Yeremin. On a family of two-level preconditionings of the incomplete block factorization type. *Soviet Journal of Numerical Analysis and Mathematical Modeling*, 1:293–320, 1986.
33. L. Yu. Kolotilina and A. Yu. Yeremin. Factorized sparse approximate inverse preconditionings II. solution of 3-D FE systems on massively parallel computers. Technical Report EM-RR 3/92, Elegant Mathematics, Inc., Bothell, Washington, 1992.
34. L. Yu. Kolotilina and A. Yu. Yeremin. Factorized sparse approximate inverse preconditionings I. Theory. *SIAM Journal on Matrix Analysis and Applications*, 14:45–58, 1993.
35. C. Lanczos. An iteration method for the solution of the eigenvalue problem of linear differential and integral operators. *Journal of Research of the National Bureau of Standards*, 45:255–282, 1950.
36. C. Lanczos. Solution of systems of linear equations by minimized iterations. *Journal of Research of the National Bureau of Standards*, 49:33–53, 1952.
37. T. A. Manteuffel. An incomplete factorization technique for positive definite linear systems. *Mathematics of Computations*, 34:473–497, 1980.
38. J. A. Meijerink and H. A. van der Vorst. An iterative solution method for linear systems of which the coefficient matrix is a symmetric M-matrix. *Mathematics of Computations*, 31(137):148–162, 1977.
39. C. C. Paige. *The computation of eigenvalues and eigenvectors of very large sparse matrices*. PhD thesis, London University, Institute of Computer Science, London, England, 1971.
40. J. K. Reid. On the method of conjugate gradients for the solution of large sparse systems of linear equations. In J. K. Reid, editor, *Large Sparse Sets of Linear Equations*, pages 231–254. Academic Press, 1971.
41. Y. Saad. Krylov subspace methods for solving large unsymmetric linear systems. *Mathematics of Computation*, 37:105–126, 1981.
42. Y. Saad. Preconditioning techniques for indefinite and nonsymmetric linear systems. *Journal of Computational and Applied Mathematics*, 24:89–105, 1988.
43. Y. Saad. A flexible inner-outer preconditioned GMRES algorithm. *SIAM Journal on Scientific and Statistical Computing*, 14:461–469, 1993.

44. Y. Saad. Highly parallel preconditioners for general sparse matrices. In G. Golub, M. Luskin, and A. Greenbaum, editors, *Recent Advances in Iterative Methods, IMA Volumes in Mathematics and Its Applications*, volume 60, pages 165–199, New York, 1994. Springer Verlag.
45. Y. Saad. ILUT: a dual threshold incomplete ILU factorization. *Numerical Linear Algebra with Applications*, 1:387–402, 1994.
46. Y. Saad. ILUM: a parallel multi-elimination ILU preconditioner for general sparse matrices. *SIAM Journal on Scientific Computing*, 17(4):830–847, 1996.
47. Y. Saad. *Iterative Methods for Sparse Linear Systems*. PWS publishing, New York, 1996.
48. Y. Saad and M. H. Schultz. GMRES: a generalized minimal residual algorithm for solving nonsymmetric linear systems. *SIAM Journal on Scientific and Statistical Computing*, 7:856–869, 1986.
49. Y. Saad and K. Wu. DQGMRES: a direct quasi-minimal residual algorithm based on incomplete orthogonalization. Technical Report UMSI-93/131, Minnesota Supercomputer Institute, University of Minnesota, Minneapolis, MN, 1993.
50. J. Shadid and R. Tuminaro. A comparison of preconditioned Krylov methods on large scale MIMD machines. *SIAM Journal on Scientific and Statistical Computing*, 15:440–459, 1994.
51. P. Sonneveld. CGS, a fast Lanczos-type solver for nonsymmetric linear systems. *SIAM Journal on Scientific and Statistical Computing*, 10(1):36–52, 1989.
52. H. A. van der Vorst. Bi-CGSTAB: A fast and smoothly converging variant of Bi-CG for the solution of non-symmetric linear systems. *SIAM Journal on Scientific and Statistical Computing*, 12:631–644, 1992.
53. H. A. van der Vorst and C. Vuik. GMRESR: a family of nested GMRES methods. *Numerical Linear Algebra with Applications*, 1:369–386, 1994.
54. P. K. W. Vinsome. Orthomin, an iterative method for solving sparse sets of simultaneous linear equations. In *Proceedings of the Fourth Symposium on Resevoir Simulation*, pages 149–159. Society of Petroleum Engineers of AIME, 1976.
55. X. Wang, K. Gallivan, and R. Bramley. CIMGS: An incomplete orthogonal factorization preconditioner. Technical Report 394, Indiana University at Bloomington, Bloomington, IN, 1993.
56. C. H. Wu. A multicolor SOR method for the finite-element method. *Journal of Computational and Applied Mathematics*, 30:283–294, 1990.

PRECONDITIONING KRYLOV METHODS

A. SUAREZ, D. GARCIA, E. FLOREZ AND G. MONTERO
Department of Mathematics, Las Palmas de Gran Canaria University.
Edificio de Informatica y Matematicas, Campus Univ, de Tafira,
Las Palmas de Gran Canaria 35017, Spain.

1. Introduction

The application of any discretization technique to differential equations problems leads to an algebraic system of equations, generally large and sparse. Recently, several methods based on the Krylov subspaces have been developed and the most popular ones seem to be BICGSTAB [36], QMRCGSTAB [3] and VGMRES [8], [9], [26] and [27].

The equivalence between the different preconditioning patterns and an adequate choice of the started vector for methods like CGS [32] and BICGSTAB is well known. Thus, the field of possibilities is opened far from the left, right and both side preconditioning. The first goal of this paper is to study the convergence behavior of these algorithms when several alternatives are used.

On the other hand, an efficient resolution of sparse linear systems of equations does not only depend on the iteration type, but also on the preconditioning technique. Here, results of convergence for BICGSTAB, QMRCGSTAB and VGMRES methods are presented using different preconditioners. These are Diagonal, SSOR(w), ILU(0) and several steps of one iterative method. Also, the effect of ordering techniques, as Reverse Cuthill-Mckee and Minimum Neighboring algorithms, on the preconditioners is studied to improve the convergence.

2. Preconditioning

The convergence of Krylov subspace methods is improved using preconditioning techniques. These generally consist in changing the original system $\mathbf{Ax} = \mathbf{b}$ by another one $\bar{\mathbf{A}}\bar{\mathbf{x}} = \bar{\mathbf{b}}$ in such a way that $\bar{\mathbf{A}}$ has a lower condition number or a better eigenvalues distribution than $\mathbf{A}$. To perform the preconditioning, a matrix

G. Winter Althaus and E. Spedicato (eds.), Algorithms for Large Scale Linear Algebraic Systems, 151–174.

M may be introduced as follows:

$$\begin{array}{ll} \mathbf{M}^{-1}\mathbf{A}\mathbf{x} = \mathbf{M}^{-1}\mathbf{b} & \text{(Left preconditioning)} \\ \mathbf{A}\mathbf{M}^{-1}\mathbf{M}\mathbf{x} = \mathbf{b} & \text{(Right preconditioning)} \\ \mathbf{M}_1^{-1}\mathbf{A}\mathbf{M}_2^{-1}\mathbf{M}_2\mathbf{x} = \mathbf{M}^{-1}\mathbf{b} & \text{(Both sides preconditioning)} \end{array} \quad (1)$$

if **M** can be splitted as $\mathbf{M} = \mathbf{M}_1\mathbf{M}_2$.

A good preconditioner should be a Matrix **M**, approximation of **A**, which allows to compute the $\mathbf{M}^{-1}$ by vector product (often involving some backward substitutions) at a reasonable cost.

The massive parallel machines provide new opportunities in this topic. Grote and Simon [12] propose to obtain $\mathbf{M}^{-1}$as an approximation of $\mathbf{A}^{-1}$ minimizing the Frobenius Norm

$$\left\|\mathbf{A}\mathbf{M}^{-1} - \mathbf{I}\right\|_F \quad (2)$$

that means to solve n small least squared problems which are independent each other (n is the matrix order). Also in [28] and [29] **M** is built as a polynomial $P(\mathbf{A})$ and in [30] as least-squares polynomials.

Element-By-Element (EBE) preconditioners have been geared toward finite element problems and allow to avoid assembling finite element matrices. The global stiffness matrix **A** is the sum of matrices $\mathbf{A}^e$ associated with each element e,

$$\mathbf{A} = \sum_{e=1}^{Nel} \mathbf{A}^e \quad (3)$$

Each finite element matrix is transformed (see i.e. Montero et al.[17]) using the diagonal of **A**, regularized (i.e. using Wingler regularization) and factorized (i.e. using LU factorization). Thus, the preconditioner is defined as

$$\mathbf{M} = \prod_{e=1}^{Nel} \mathbf{L}^e \times \prod_{e=Nel}^{1} \mathbf{U}^e \quad (4)$$

where $\mathbf{L}^e$ and $\mathbf{U}^e$ are the LU factorization of transformed and regularized $\bar{\mathbf{A}}^e$.

Many other preconditioners have been developed, but the most exploited are the well-known Diagonal, SSOR and ILU. In this paper, we will consider these last ones.

DIAGONAL PRECONDITIONER

It arises from comparing the Richardson iteration applied to the preconditioned system with Jacobi method applied to the non-preconditioned system. This preconditioning matrix is diagonal and its entries are the ones of the diagonal of **A**.

SSOR PRECONDITIONER

Now, if we compare with SSOR method, another preconditioning matrix $\mathbf{M}$ is obtained. Let consider a splitting of $\mathbf{A}$ as $\mathbf{A} = \mathbf{D} - \mathbf{E} - \mathbf{F}$ ($\mathbf{D}$ is the diagonal of $\mathbf{A}$ and $\mathbf{E}$, $\mathbf{F}$ lower and upper triangular matrices, respectively), then $\mathbf{M}$ stands as:

$$\mathbf{M} = \frac{1}{w(2-w)}(\mathbf{D} - w\mathbf{E})\mathbf{D}^{-1}(\mathbf{D} - w\mathbf{F}) \tag{5}$$

where w is the relaxation parameter.

For symmetric systems it is splitted as

$$\mathbf{M} = \left[\frac{(\mathbf{D} - w\mathbf{E})\mathbf{D}^{-\frac{1}{2}}}{\sqrt{w(2-w)}}\right]\left[\frac{(\mathbf{D} - w\mathbf{E})\mathbf{D}^{-\frac{1}{2}}}{\sqrt{w(2-w)}}\right]^T \tag{6}$$

and for nonsymmetric systems:

$$\mathbf{M} = (\mathbf{I} - w\mathbf{E}\mathbf{D}^{-1})\left(\frac{\mathbf{D} - w\mathbf{F}}{w(2-w)}\right) \tag{7}$$

ILU(0) PRECONDITIONER

It arises from the approximation of $\mathbf{A}$ by an Incomplete LU factorization, keeping the same null entries in the triangular matrices,

$$\mathbf{A} = \mathbf{LU} \approx \mathbf{ILU}(0) = \mathbf{M} \tag{8}$$

where m_{ij} are the entries of $\mathbf{M}$ such that,

$$m_{ij} = 0 \quad \text{if} \quad a_{ij} = 0 \tag{9}$$

$$\{\mathbf{A} - \mathbf{LU}\}_{ij} = 0 \quad \text{if} \quad a_{ij} \neq 0 \tag{10}$$

Diagonal preconditioner requires less computational effort than the others. It is directly built from matrix $\mathbf{A}$, and the inverse matrix by vector products are carried out straightforwardly. Nevertheless, there are many ill-conditioned problems where this preconditioner does not get convergence in a competitive way. SSOR preconditioner is also built directly from $\mathbf{A}$, but the product of inverse matrix by vector means two backward substitutions. ILU(0) preconditioner has an added cost due to the incomplete factorization, with respect to SSOR.

3. Storage scheme

The matrices of the linear systems that we are solving are sparse and the number of non-null entries is much lower than the total number of entries in the matrix. The storage scheme of these matrices should be taken into account in order to reduce the computational cost and the storage requirements.

Nowadays, there are several types of storage schemes that improve those like *skyline* or *band* schemes, which involve null entries to be stored. On the other hand, concerning with the Element-By-Element preconditioning, the finite element matrices can be stored, performing the matrix operations without assembling them into the global matrix **A**. This storage is less competitive when a parallel machine is not available.

The storage scheme used here for the sparse matrix **A** of dimension n is a version of the Ellpack-Itpack format [31]. Two rectangular arrays of dimension $n \times N_d$ are required, one real and another integer, where N_d is the maximum number of non-null entries ($N_d << n$). The first column of the real array contains the diagonal terms of the matrix followed by the others entries of each row. In the first column of the integer array, the number of non-null entries per row is stored, followed by the column positions of the respective entries. In both arrays the rows are completed by zeros as necessary. This storage scheme has interesting advantages for the matrix-by-vector product in vector machines and allows to use the same integer array of **A** for ILU(0) and SSOR preconditioners.

4. Ordering

ILU(0) preconditioner is an approximation of matrix **A** by a product of two triangular matrices. As this is an incomplete factorization, an error is involved if we consider that this product of matrices is approximation of LU factorization of **A**.

If an ordering technique is used, we obtain a new matrix whose band width or profile is smaller. Also, the fill-in effect given in the LU factorization may be diminished. This is a simple way to understand why ILU(0) preconditioner is more effective after reordering the matrix of the system. Similar effect occurs with SSOR, since it is another approximation of a factorization of **A**.

Two ordering techniques have been considered in the numerical experiments: the Reverse Cuthill-Mckee (RCM) and the Minimum Neighboring (MN) algorithms. These algorithms are given below.

RCM Algorithm
Build the graph associated to matrix **A**, $g(x) = \langle V, E \rangle$, where V is the set of nodes and $E = \{\{a, b\} : a \neq b \ / \ a, b \in V\}$.

Find an initial (pseudo-peripheral) node and reorder it as x_1.
Reorder the nodes connected to x_i from the nodes of lowest degree to those of highest degree.
Carry out the reverse ordering.

MN Algorithm
Build the graph associated to matrix $\mathbf{A}$, $g(x) = \langle V, E \rangle$, where V is the set of nodes and $E = \{\{a,b\} : a \neq b \ / \ a,b \in V\}$.
Do while $V \neq \emptyset$:

Choose a node v of minimum degree in $g(x) = \langle V, E \rangle$ and reorder it as the followed node.

Define:

$$V_v = V - \{v\}, \quad E_v = \{\{a,b\} \in E \ / \ a,b \in V_v\}.$$

Make

$$V = V_v, \quad E = E_v \text{ and } g(x) = \langle V, E \rangle.$$

EndDo

The choice of the initial node in the step 2 of the above algorithms is carried out using the algorithm of George [11], which allows to begin from a pseudo-peripheral node.

5. Preconditioned Krylov Algorithms

Krylov subspace methods are considered the most important iterative tools for solving large and sparse linear systems of equations ($\mathbf{Ax} = \mathbf{b}$), such as those which arise from finite element discretization. In this work, we have considered the classical Conjugate Gradient method (CG), used for symmetric problems, and more recent others like BICGSTAB, QMRCGSTAB and VGMRES for the non-symmetric case.

Once the original system is transformed in the preconditioned one, this is going to be solved using Krylov methods. Some changes should be introduced in the algorithms taking into account the general characteristics that the preconditioning matrix must hold, in order to assure that they are applicable in practice:

- Evidently the preconditioning matrix must be never multiplied by the system matrix $\mathbf{A}$.

- Though we are solving the preconditioned system $\bar{\mathbf{A}}\bar{\mathbf{x}} = \bar{\mathbf{b}}$, the algorithm must provide the solution of the original system $\mathbf{x}$.

- The stop criterion must be defined from the residual vector of the unpreconditioned system ($\mathbf{r} = \mathbf{b} - \mathbf{Ax}$).

In principle, there should be three preconditioned algorithms corresponding to the preconditioning patterns (left, right and both sides). To obtain them, we will establish the relations between the matrices, the residual vectors and the solution before and after preconditioning for each case. So, for example, the both sides preconditioned system is:

$$\bar{\mathbf{A}}\bar{\mathbf{x}} = \bar{\mathbf{b}} \qquad \begin{cases} \bar{\mathbf{A}} = \mathbf{L}^{-1}\mathbf{A}\mathbf{U}^{-1} \\ \bar{\mathbf{x}} = \mathbf{U}\mathbf{x} \\ \bar{\mathbf{b}} = \mathbf{L}^{-1}\mathbf{b} \end{cases} \tag{11}$$

and the relations between the residual vectors are:

$$\begin{array}{ll} \bar{\mathbf{r}}_i = \mathbf{L}^{-1}\mathbf{r}_i & \text{corresponding to original system } (\mathbf{r}_i = \mathbf{b} - \mathbf{A}\mathbf{x}_i) \\ \bar{\mathbf{r}}_i^* = \mathbf{U}^{-T}\mathbf{r}_i^* & \text{corresponding to auxiliary system } (\mathbf{r}_i^* = \mathbf{b} - \mathbf{A}^T\mathbf{x}_i) \end{array} \tag{12}$$

5.1. CG ALGORITHM

Nowadays, it is well known that Standard Conjugate Gradient method is the fastest strategy for solving symmetric linear systems of equations. Here we present the preconditioned version which will be used in the numerical experiment.

Preconditioned CG algorithm
Initial guess $\mathbf{x}_0$. $\mathbf{r}_0 = \mathbf{b} - \mathbf{A}\mathbf{x}_0$;
$\tilde{\rho}_0 = 1$; $\mathbf{p}_0 = \mathbf{0}$;
Do while $\| \mathbf{r}_{i-1} \| / \| \mathbf{r}_0 \| \geq \varepsilon \quad (i = 1, 2, 3, ...)$,
    solve $\mathbf{M}\mathbf{z} = \mathbf{r}_{i-1}$;
    $\tilde{\rho}_i = \mathbf{r}_{i-1}^t \mathbf{z}$;
    $\tilde{\beta}_i = \tilde{\rho}_i / \tilde{\rho}_{i-1}$;
    $\mathbf{p}_i = \mathbf{z} + \tilde{\beta}_i \mathbf{p}_{i-1}$;
    $\mathbf{v}_i = \mathbf{A}\mathbf{p}_i$;
    $\tilde{\alpha}_i = \frac{\tilde{\rho}_i}{\mathbf{p}_i^t \mathbf{v}_i}$;
    $\mathbf{x}_i = \mathbf{x}_{i-1} + \tilde{\alpha}_i \mathbf{p}_i$;
    $\mathbf{r}_i = \mathbf{r}_{i-1} - \tilde{\alpha}_i \mathbf{v}_i$;
EndDo

5.2. CGS ALGORITHM

Now, if we apply CGS algorithm to the system given in (11) and make some suitable transformations, then we will obtain the corresponding both sides preconditioned CGS algorithm. If the system is preconditioned by the left or by the right, and some similar transformations are made, both left and right preconditioned algorithms are obtained.

However, the only different thing between these three algorithms is the initial vector $\hat{\mathbf{r}}_0$. This vector influences to each iteration (through $\hat{\rho}_i$ and $\tilde{\alpha}_i$ parameters).

The relation of $\hat{\mathbf{r}}_0$ with the initial residual vector of the preconditioned auxiliary system is:

$$\begin{array}{ll} \hat{\mathbf{r}}_0 = \bar{\mathbf{r}}_0^* & \text{for right preconditioning} \\ \hat{\mathbf{r}}_0 = \mathbf{M}^{-t}\bar{\mathbf{r}}_0^* & \text{for left preconditioning} \\ \hat{\mathbf{r}}_0 = \mathbf{L}^{-t}\bar{\mathbf{r}}_0^* & \text{for both sides preconditioning} \end{array} \tag{13}$$

As $\bar{\mathbf{r}}_0^*$ is arbitrary, it can be chosen in a different way such that one algorithm may be transformed in another. We can conclude that the different patterns of preconditioning may be considered equivalent to the different selections of the initial vector $\hat{\mathbf{r}}_0$ in a single algorithm. This idea opens a wider field of possibilities for preconditioning. For example, we can choose:
- $\hat{\mathbf{r}}_0$ equal to the product of certain matrix by $\mathbf{r}_0$. This matrix may be the preconditioning one, or even any of its factors if it is splitted.
- $\hat{\mathbf{r}}_0$ equal to the residual vector computed after a small number of steps of an iterative method applied to $\bar{\mathbf{A}}\bar{\mathbf{x}} = \bar{\mathbf{b}}$.

If we used the relations between the residual vectors corresponding to the non-preconditioned and preconditioned systems:

$$\begin{array}{ll} \bar{\mathbf{r}}_0^* = \mathbf{M}^{-t}\mathbf{r}_0^* & \text{right preconditioning} \\ \bar{\mathbf{r}}_0^* = \mathbf{r}_0^* & \text{left preconditioning} \\ \bar{\mathbf{r}}_0^* = \mathbf{U}^{-t}\mathbf{r}_0^* & \text{both sides preconditioning} \end{array} \tag{14}$$

we would obtain three new algorithms again, which only differs in the initial vector. So the single algorithm may be written as follows:

Preconditioned CGS algorithm
Initial guess $\mathbf{x}_0$. $\mathbf{r}_0 = \mathbf{b} - \mathbf{A}\mathbf{x}_0$;
$\hat{\mathbf{r}}_0$ is arbitrary, such that $\hat{\mathbf{r}}_0^t\mathbf{r}_0 \neq 0$, e.g., $\hat{\mathbf{r}}_0 = \mathbf{r}_0$;
$\hat{\rho}_0 = 1$; $\mathbf{p}_0 = \mathbf{q}_0 = \mathbf{0}$;
Do while $\| \mathbf{r}_{i-1} \| / \| \mathbf{r}_0 \| \geq \varepsilon \quad (i = 1, 2, 3, ...)$,
   solve $\mathbf{M}\mathbf{z} = \mathbf{r}_{i-1}$;
   $\hat{\rho}_i = \hat{\mathbf{r}}_0^t\mathbf{z}$;
   $\tilde{\beta}_i = \hat{\rho}_i / \hat{\rho}_{i-1}$;
   $\mathbf{u} = \mathbf{z} + \tilde{\beta}_i\mathbf{q}_{i-1}$;
   $\mathbf{p}_i = \mathbf{u} + \tilde{\beta}_i(\mathbf{q}_{i-1} + \tilde{\beta}_i\mathbf{p}_{i-1})$;
   $\mathbf{y} = \mathbf{A}\mathbf{p}_i$;
   solve $\mathbf{M}\mathbf{v}_i = \mathbf{y}$;
   $\tilde{\alpha}_i = \frac{\hat{\rho}_i}{\hat{\mathbf{r}}_0^t\mathbf{v}_i}$;
   $\mathbf{q}_i = \mathbf{u} - \tilde{\alpha}_i\mathbf{v}_i$;
   $\mathbf{w} = \mathbf{u} + \mathbf{q}_i$;
   $\mathbf{x}_i = \mathbf{x}_{i-1} + \tilde{\alpha}_i\mathbf{w}$;
   $\mathbf{r}_i = \mathbf{r}_{i-1} - \tilde{\alpha}_i\mathbf{A}\mathbf{w}$;
EndDo

Though this algorithm is equivalent to the preconditioned CGS proposed by Sonneveld [32] for $\hat{\mathbf{r}}_0 = \mathbf{M}^{-t}\hat{\mathbf{r}}_{0_{Sonneveld}}$ in exact arithmetic, they have a different convergence behavior for the same selection of this vector in both algorithms, and in fact the one proposed here often leads to smoother convergence properties.

5.3. BICGSTAB ALGORITHM

BICGSTAB [36] method is a smoothing variant of BCG. It introduces a parameter $\tilde{\omega}$ which minimizes the residual. For each preconditioning pattern, let assume:

$$\begin{array}{ll} \tilde{\omega} = \frac{\mathbf{t}^t\mathbf{s}}{\mathbf{t}^t\mathbf{t}} & \text{right preconditioning} \\ \tilde{\omega} = \frac{(\mathbf{M}^{-1}\mathbf{t})^t(\mathbf{M}^{-1}\mathbf{s})}{(\mathbf{M}^{-1}\mathbf{t})^t(\mathbf{M}^{-1}\mathbf{t})} & \text{left preconditioning} \\ \tilde{\omega} = \frac{(\mathbf{L}^{-1}\mathbf{t})^t(\mathbf{L}^{-1}\mathbf{s})}{(\mathbf{L}^{-1}\mathbf{t})^t(\mathbf{L}^{-1}\mathbf{t})} & \text{both sides preconditioning} \end{array} \tag{15}$$

Thus, to compute $\tilde{\omega}$ involves one backward substitution per iteration for left preconditioning and two substitutions when both side preconditioning is used. However, all of these algorithms are equivalent to the right preconditioning one if $\tilde{\omega}$ is obtained from the minimization of the residual vector of the unpreconditioned system and the $\hat{\mathbf{r}}_0$ vector is chosen in a suitable way for each algorithm again. So, a single BICGSTAB algorithm is considered, and the preconditioning pattern depends on the selection of $\hat{\mathbf{r}}_0$ vector. Moreover, we may obtain another version of the BICGSTAB algorithm if we use the relations (14).

Preconditioned BICGSTAB algorithm

Initial guess $\mathbf{x}_0$. $\mathbf{r}_0 = \mathbf{b} - \mathbf{A}\mathbf{x}_0$;
$\hat{\mathbf{r}}_0$ is arbitrary, such that $\hat{\mathbf{r}}_0^t\mathbf{r}_0 \neq 0$, e.g., $\hat{\mathbf{r}}_0 = \mathbf{r}_0$;
$\hat{\rho}_0 = \tilde{\alpha}_0 = \tilde{\omega}_0 = 1$; $\mathbf{p}_0 = \mathbf{v}_0 = \mathbf{0}$;
Do while $\| \mathbf{r}_{i-1} \| / \| \mathbf{r}_0 \| \geq \varepsilon \quad (i = 1, 2, 3, ...)$,
    solve $\mathbf{M}\mathbf{z} = \mathbf{r}_{i-1}$;
    $\hat{\rho}_i = \hat{\mathbf{r}}_0^t\mathbf{z}$;
    $\tilde{\beta}_i = (\hat{\rho}_i/\hat{\rho}_{i-1})(\tilde{\alpha}_{i-1}/\tilde{\omega}_{i-1})$;
    $\mathbf{p}_i = \mathbf{z} + \tilde{\beta}_i(\mathbf{p}_{i-1} - \tilde{\omega}_{i-1}\mathbf{v}_{i-1})$;
    $\mathbf{y} = \mathbf{A}\mathbf{p}_i$;
    solve $\mathbf{M}\mathbf{v}_i = \mathbf{y}$;
    $\tilde{\alpha}_i = \frac{\hat{\rho}_i}{\hat{\mathbf{r}}_0^t\mathbf{v}_i}$;
    $\mathbf{s} = \mathbf{r}_{i-1} - \tilde{\alpha}_i\mathbf{y}$;
    $\mathbf{u} = \mathbf{z} - \tilde{\alpha}_i\mathbf{v}_i$;
    $\mathbf{t} = \mathbf{A}\mathbf{u}$;
    $\tilde{\omega}_i = \frac{\mathbf{t}^t\mathbf{s}}{\mathbf{t}^t\mathbf{t}}$;
    $\mathbf{x}_i = \mathbf{x}_{i-1} + \tilde{\alpha}_i\mathbf{p}_i + \tilde{\omega}_i\mathbf{u}$;
    $\mathbf{r}_i = \mathbf{s} - \tilde{\omega}_i\mathbf{t}$;
EndDo

5.4. QMRCGSTAB ALGORITHM

This method applies the quasi-minimization principle to the BICGSTAG method. The least squares minimization involved in the process is solved using QR decomposition of the Hessemberg matrix in an incremental way with the Givens rotations. In the following QMRCGSTAB algorithm, the Givens rotations are writing out explicitly as it was originally proposed by Chan et al.[3].

Preconditioned QMRCGSTAB algorithm
Initial guess $\mathbf{x}_0$. $\mathbf{r}_0 = \mathbf{b} - \mathbf{A}\mathbf{x}_0$;
$\hat{\mathbf{r}}_0$ is arbitrary, such that $\hat{\mathbf{r}}_0^t \mathbf{r}_0 \neq 0$, e.g., $\hat{\mathbf{r}}_0 = \mathbf{r}_0$;
$\hat{\rho}_0 = \tilde{\alpha}_0 = \tilde{\omega}_0 = 1$; $\theta_0 = \eta_0 = 0$; $\gamma = \| \mathbf{r}_0 \|$; $\mathbf{p}_0 = \mathbf{v}_0 = \mathbf{d}_0 = \mathbf{0}$;
Do while $\| \mathbf{r}_{i-1} \| / \| \mathbf{r}_0 \| \geq \varepsilon \quad (i = 1, 2, 3, ...)$,
    solve $\mathbf{M}\mathbf{z} = \mathbf{r}_{i-1}$;
    $\hat{\rho}_i = \hat{\mathbf{r}}_0^t \mathbf{z}$;
    $\tilde{\beta}_i = (\hat{\rho}_i / \hat{\rho}_{i-1})(\tilde{\alpha}_{i-1} / \tilde{\omega}_{i-1})$;
    $\mathbf{p}_i = \mathbf{z} + \tilde{\beta}_i(\mathbf{p}_{i-1} - \tilde{\omega}_{i-1}\mathbf{v}_{i-1})$;
    $\mathbf{y} = \mathbf{A}\mathbf{p}_i$;
    solve $\mathbf{M}\mathbf{v}_i = \mathbf{y}$;
    $\tilde{\alpha}_i = \frac{\hat{\rho}_i}{\hat{\mathbf{r}}_0^t \mathbf{v}_i}$;
    $\mathbf{s} = \mathbf{r}_{i-1} - \tilde{\alpha}_i \mathbf{y}$;
    $\bar{\theta}_i = \| \mathbf{s} \| / \gamma$;
    $c = 1 / \sqrt{1 + \bar{\theta}_i^2}$;
    $\bar{\gamma} = \gamma \bar{\theta}_i c$;
    $\bar{\eta}_i = c^2 \tilde{\alpha}_i$;
    $\bar{\mathbf{d}}_i = \mathbf{p}_i + \frac{\theta_{i-1}^2 \eta_{i-1}}{\tilde{\alpha}_i} \mathbf{d}_{i-1}$;
    $\bar{\mathbf{x}}_i = \mathbf{x}_{i-1} + \bar{\eta}_i \bar{\mathbf{d}}_i$;
    $\mathbf{u} = \mathbf{z} - \tilde{\alpha}_i \mathbf{v}_i$;
    $\mathbf{t} = \mathbf{A}\mathbf{u}$;
    $\tilde{\omega}_i = \frac{\mathbf{t}^t \mathbf{s}}{\mathbf{t}^t \mathbf{t}}$;
    $\mathbf{r}_i = \mathbf{s} - \tilde{\omega}_i \mathbf{t}$;
    $\theta_i = \| \mathbf{r}_i \| / \bar{\gamma}$;
    $c = 1 / \sqrt{1 + \theta_i^2}$;
    $\gamma = \bar{\gamma} \theta_i c$;
    $\eta_i = c^2 \tilde{\omega}_i$;
    $\mathbf{d}_i = \mathbf{s}_i + \frac{\bar{\theta}_i^2 \bar{\eta}_i}{\tilde{\omega}_i} \bar{\mathbf{d}}_i$;
    $\mathbf{x}_i = \bar{\mathbf{x}}_i + \eta_i \mathbf{d}_i$;
EndDo

5.5. VGMRES ALGORITHM

The variable GMRES algorithm is a variation of the original GMRES [27] or, if it is preferred, FGMRES [26]. The first change introduced is the direct solver for the least square problem [8] which appears in each iteration. On the other hand, a sub-tolerance value δ is fixed in order to use GMRES with an increasing Krylov subspace dimension k each step until the residual norm is higher or equal to δ and k is lower than the maximum Krylov subspace dimension k_{top}. From this point, the value of k is constant along the iterations (see Galán et al [9]).

Preconditioned VGMRES algorithm

Initial guess $\mathbf{x}_0$. $\mathbf{r}_0 = \mathbf{b} - \mathbf{A}\mathbf{x}_0$;
Choose k_{init}, k_{top}, $\delta \in [0,1]$, $k = k_{init}$
Do while $\| \mathbf{r}_{i-1} \| / \| \mathbf{r}_0 \| \geq \varepsilon \quad (i = 1, 2, 3, ...)$,
  $\beta_{i-1} = \| \mathbf{r}_{i-1} \|$;
  $\mathbf{v}_1 = \mathbf{r}_{i-1}/\beta_{i-1}$;
  If $\| \mathbf{r}_{i-1} \| / \| \mathbf{r}_0 \| \geq \delta$ and $k < k_{top}$ Do $k = k + 1$;
  For $j = 1, ..., k$ Do
    solve $\mathbf{M}\mathbf{z}_j = \mathbf{v}_j$;
    $\mathbf{w} = \mathbf{A}\mathbf{z}_j$;
    For $n = 1, ..., j$ Do
      $\{\mathbf{H}\}_{nj} = \mathbf{w}^t \mathbf{v}_n$;
      $\mathbf{w} = \mathbf{w} - \{\mathbf{H}\}_{nj} \mathbf{v}_n$;
    EndDo
    $\{\mathbf{H}\}_{j+1j} = \| \mathbf{w} \|$;
    $\mathbf{v}_{j+1} = \mathbf{w} / \{\mathbf{H}\}_{j+1j}$;
  EndDo
  solve $\mathbf{U}_k^t \bar{\mathbf{p}} = \mathbf{d}_k$ and $\mathbf{U}_k \mathbf{p} = \bar{\mathbf{p}}$; where $\begin{array}{l} \{\mathbf{d}_k\}_m = \{\mathbf{H}\}_{1m} \\ \{\mathbf{U}_k\}_{lm} = \{\mathbf{H}\}_{l+1m} \end{array} \quad l, m = 1, ..., k$;
  $\lambda_i = \frac{\beta_{i-1}}{1 + \mathbf{d}_k^t \mathbf{p}}$; $\mathbf{u}_k = \lambda_i \mathbf{p}$; $\mathbf{x}_i = \mathbf{x}_{i-1} + \mathbf{Z}_k \mathbf{u}_k$; being $\mathbf{Z}_k = [\mathbf{z}_1, \mathbf{z}_2, ..., \mathbf{z}_k]$;
  $\mathbf{r}_i = \mathbf{Z}_{k+1} \hat{\mathbf{r}}_i$; where $\begin{array}{l} \{\hat{\mathbf{r}}_i\}_1 = \lambda_i \\ \{\hat{\mathbf{r}}_i\}_{l+1} = -\lambda_i \{\bar{\mathbf{p}}\}_l \end{array} \quad l = 1, ..., k$;
EndDo

6. Numerical experiments

Four physical problems are considered in order to study the behavior of the algorithms in different cases. They were previously analyzed in Montero et al. [17].

All the examples are referred to a square of unit side $\Omega = \Omega_1 \cup \Omega_2 \cup \Omega_3$, of boundary $\Gamma = \Gamma_1 \cup \Gamma_2 \cup \Gamma_3 \cup \Gamma_4$.

EXAMPLE A

In our first experiment, we apply FEM to a problem of heat transfer in Ω, of equation $-K(u_{xx} + u_{yy}) = F$ with Dirichlet boundary conditions $u = 0$ on Γ_1 and Γ_3, and $u = 10^2$ on Γ_4, with a flow on Γ_2, $(-Ku_x = H(u - u_\infty))$. The values $K = 10^{-5}$, $F = 1$, $H = 20$ and $u_\infty = 10^3$ have been used. Our triangular mesh has produced a symmetric system of 16412 equations.

Figures 1-3 show the convergence curves of CG, BICGSTAB and QMR-CGSTAB respectively. Each of them have been implemented with different preconditioning strategies: non-preconditioning, Diagonal, ILU(0), SSOR, 5 steps of BICGSTAB with ILU(0) preconditioning and 5 steps of QMRCGSTAB with ILU(0) preconditioning too. The erratic behavior of unpreconditioned CG is evident. Also in this case, a faster convergence was found for SSOR and Diagonal Preconditioning in terms of CPU time. The cost of CG (which may be used here due to the symmetry and positivity) with respect to BICGSTAB or QMRCGSTAB is obviously lower, about a half. On the other hand, QMRCGSTAB iteration as preconditioner has not leaded to a competitive result in any case.

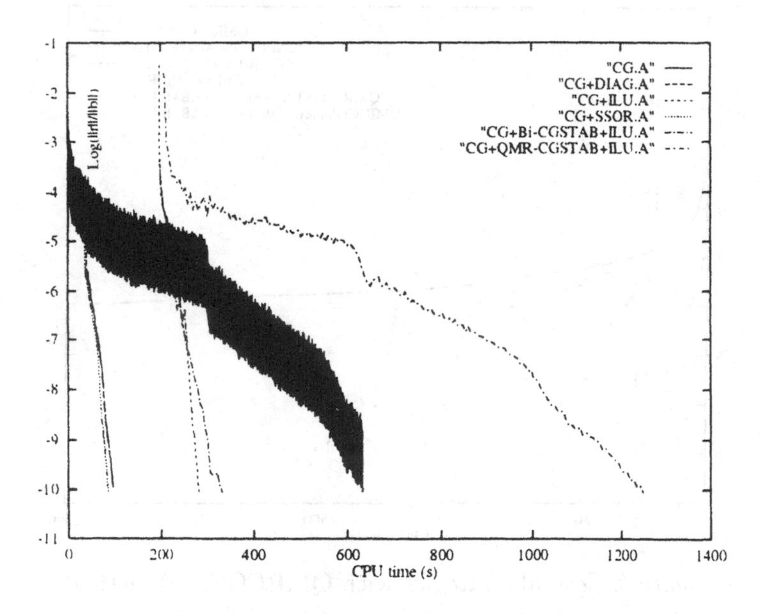

Figure 1. Several strategies with Conjugate Gradient method.

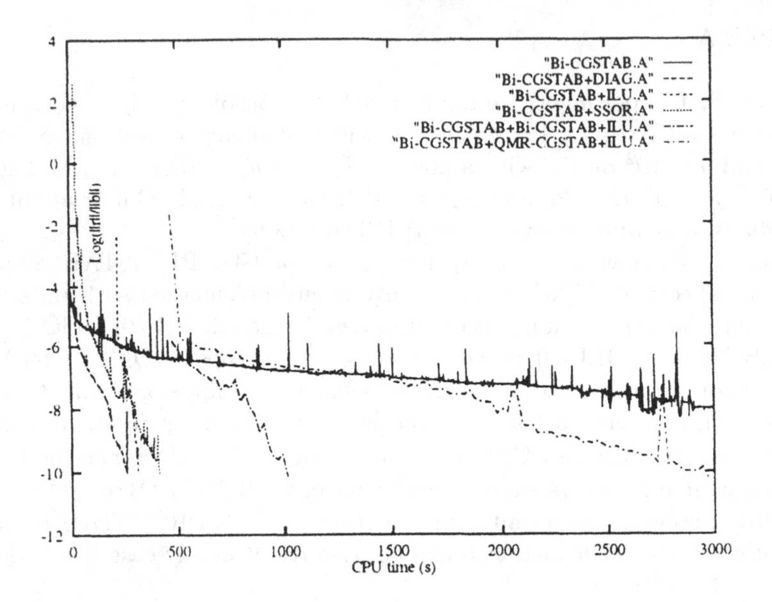

Figure 2. Several strategies with BICGSTAB method.

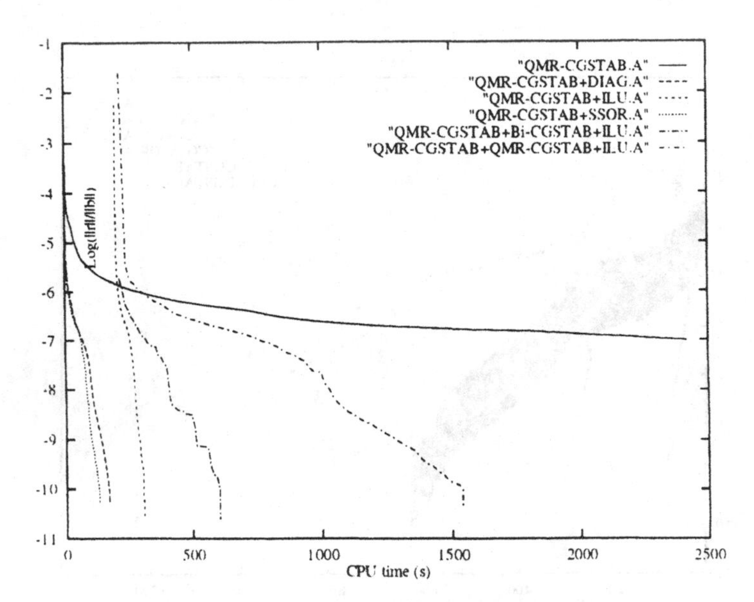

Figure 3. Several strategies with QMRCGSTAB method.

EXAMPLE B

The following problem corresponds to a steady convection-diffusion equation $v_1 u_x - K(u_{xx} + u_{yy}) = F$ in Ω, with horizontal velocities field defined by $v_1 = C\,(y - 1/2)\,(x - x^2)\,(1/2 - x)$. In this case, we have taken the value 10^{-5} for K in Ω_2 and 10^2 in the rest of the domain, and for F the value 10^3 in Ω_3 and 1 in the rest. The value of constant C is 10^4. We consider Dirichlet conditions on Γ_2 and $\Gamma_4 (u = 0$ and $u = 1)$, and Neuman conditions $\frac{\partial u}{\partial \mathbf{n}} = 0$ on the rest of the boundary. We have used two meshes which produces two nonsymmetric systems of 6585 and 13190 equations, respectively.

Figure 4 shows the smoother behavior of QMRCGSTAB towards BICGSTAB. In this nonsymmetric problem, we can see that a robust preconditioner is often required for convergence (for instance, the convergence is not obtained with Diagonal preconditioner).

In figure 5, how the number of BICGSTAB iterations (plus ILU(0) preconditioner) is obviously decreasing with the number of steps (also BICGSTAB plus ILU(0)) used to find $\hat{\mathbf{r}}_0$ is showed. Though this curve is monotonic, however the similar one in terms of CPU time presents a minimum for a small number of inner iterations (see figure 6). This effect has been contrasted in several other numerical experiments. The exact optimum is not trivial a priori, but it may be said that a good strategy is to use a few inner steps before fixing $\hat{\mathbf{r}}_0$.

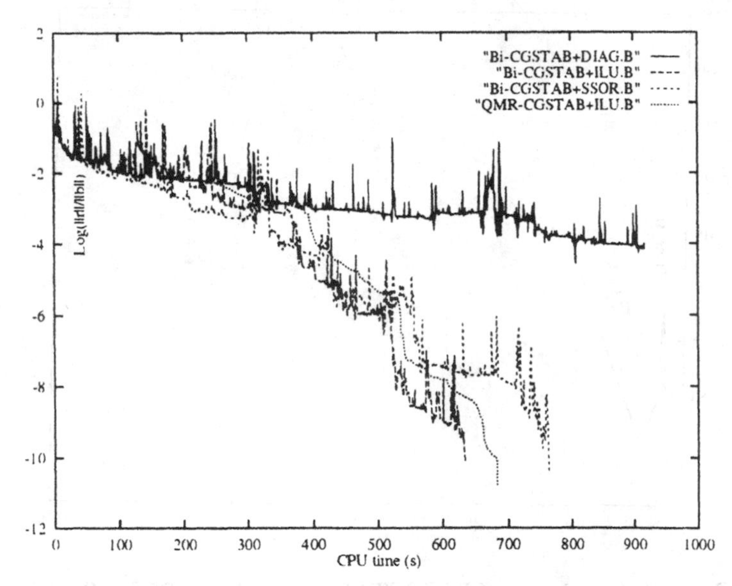

Figure 4. Effect of preconditioning in BICGSTAB and QMRCGSTAB.

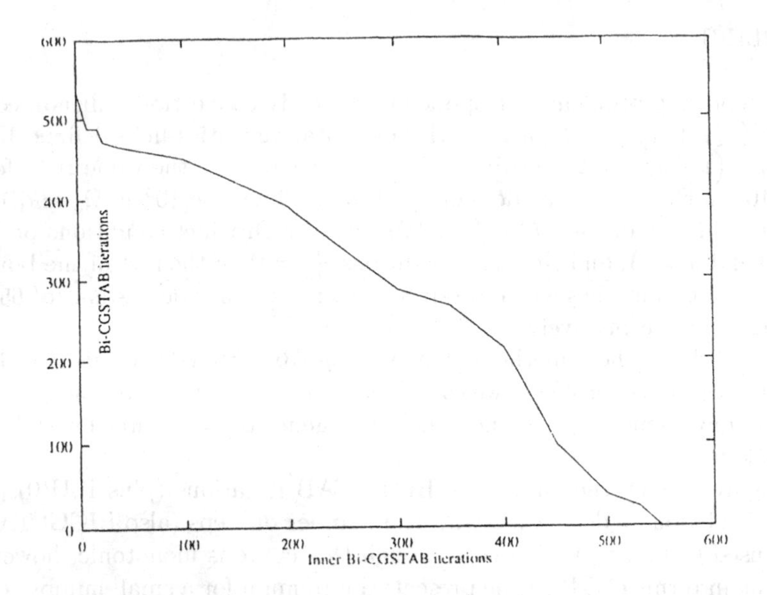

Figure 5. BICGSTAB iterations versus Inner BICGSTAB iterations.

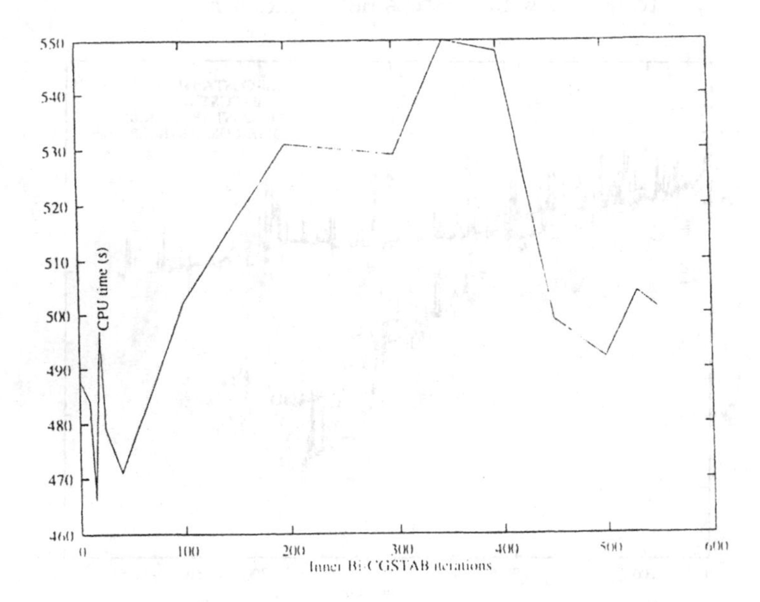

Figure 6. Computational cost of BICGSTAB versus Inner BICGSTAB iterations.

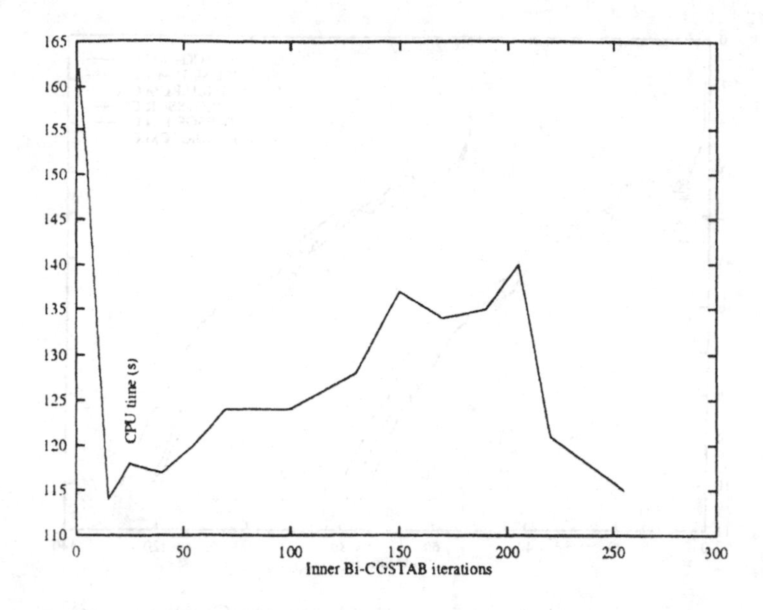

Figure 7. Computational cost of CGS versus Inner BICGSTAB iterations.

EXAMPLE C

Here, the deformation of a thin elastic slab submitted to a constant load in $\Gamma_2(L_2 = -1)$, and to another load which varies linearly in $\Gamma_3(L_3 = -\frac{x}{25})$, is studied. A condition of immobility is imposed in Γ_1 and in point $(1,1)$ [22]. Applying FEM with adaptive mesh refinement we obtain a symmetric system of 8434 equations.

In figures 8-11, similar results to problem A are presented, but also including the convergence histories of VGMRES. We can see that the reordering improves the convergence properties for ILU(0) and even sometimes for SSOR. In this case, VGMRES is not competitive with the other methods and the CG is evidently the fastest due to the symmetry of the problem.

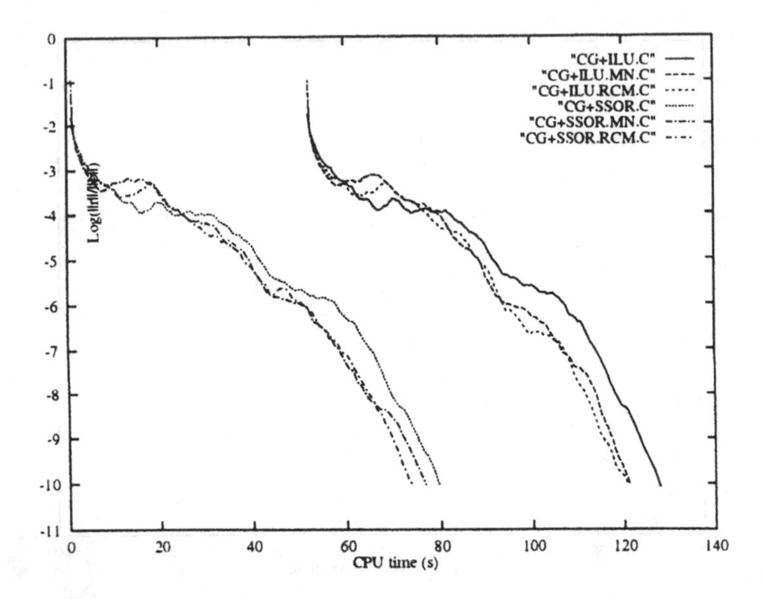

Figure 8. Several strategies with Conjugate Gradient method.

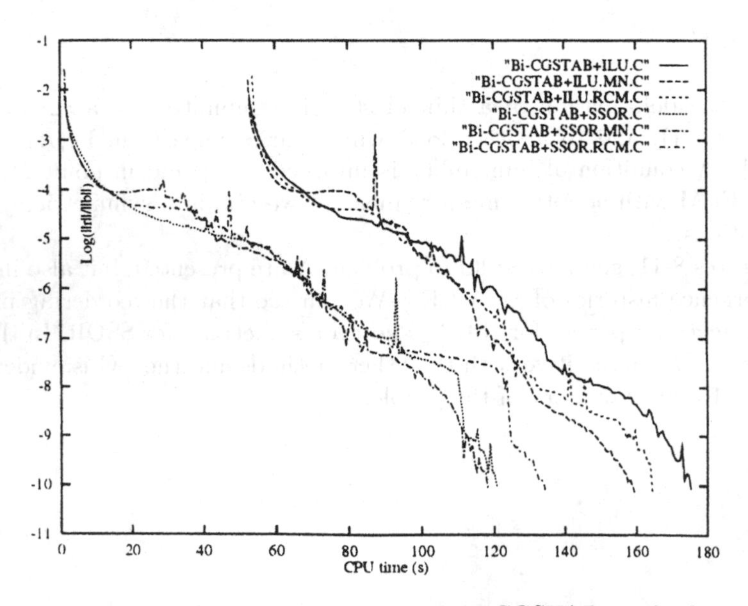

Figure 9. Several strategies with BICGSTAB method.

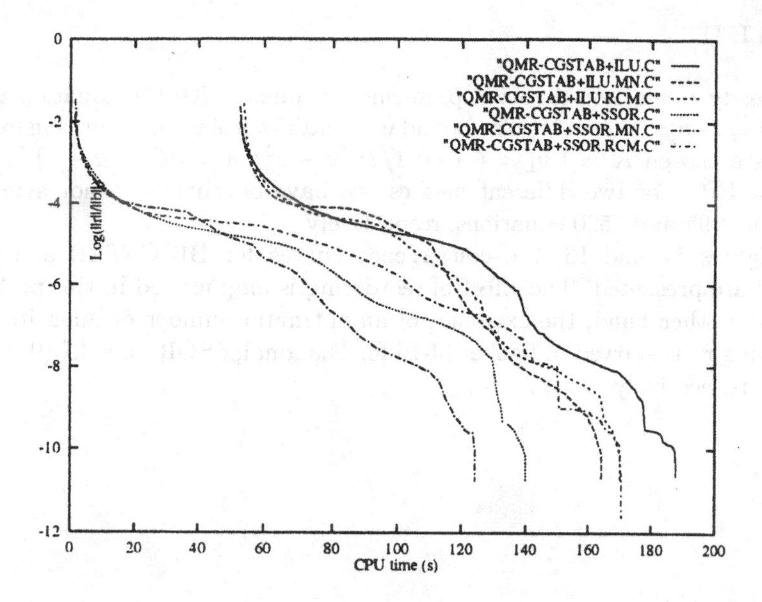

Figure 10. Several strategies with QMRCGSTAB method.

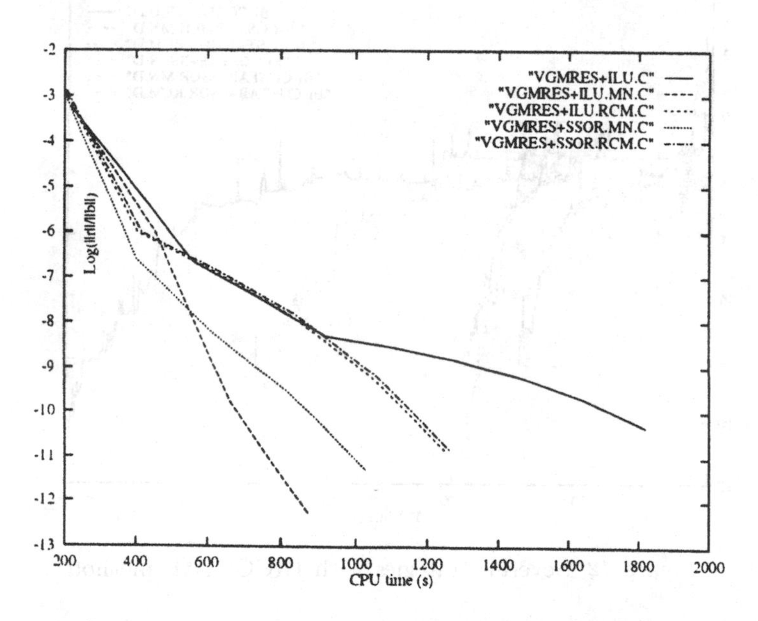

Figure 11. Several strategies with VGMRES method.

EXAMPLE D

A steady convection-diffusion problem is studied, with the equation $v_1 u_x + v_2 u_y - K(u_{xx} + u_{yy}) = 0$ in Ω. The boundary conditions are the same as in example B. We have chosen $K = 1$, $v_1 = C\,(y - 1/2)\,(x - x^2)$, $v_2 = C\,(1/2 - x)\,(y - y^2)$, with $C = 10^5$. For two different meshes, we have obtained two non-symmetric systems of 4095 and 7520 equations, respectively.

In figures 12 and 13, the convergence curves for BICGSTAB and QMR-CGSTAB are presented. The effect of reordering is emphasized in this problem.

On the other hand, the existence of an optimum number of inner iterations before fixing $\hat{\mathbf{r}}_0$ is showed in figures 14-16 for Diagonal, SSOR and ILU(0) preconditioners, respectively.

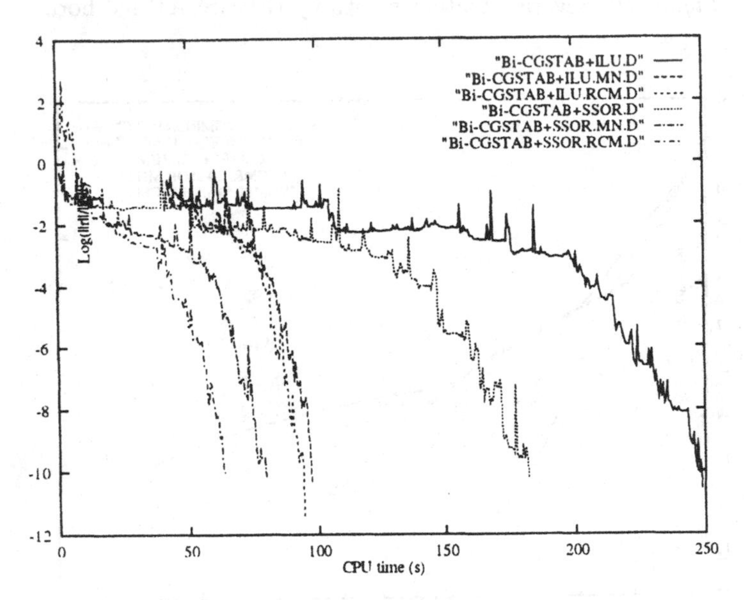

Figure 12. Several strategies with BICGSTAB method.

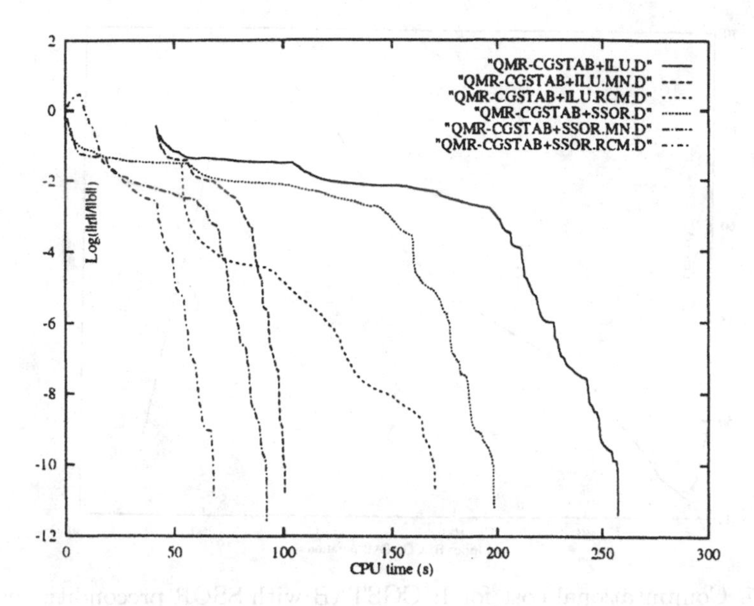

Figure 13. Several strategies with QMRCGSTAB method.

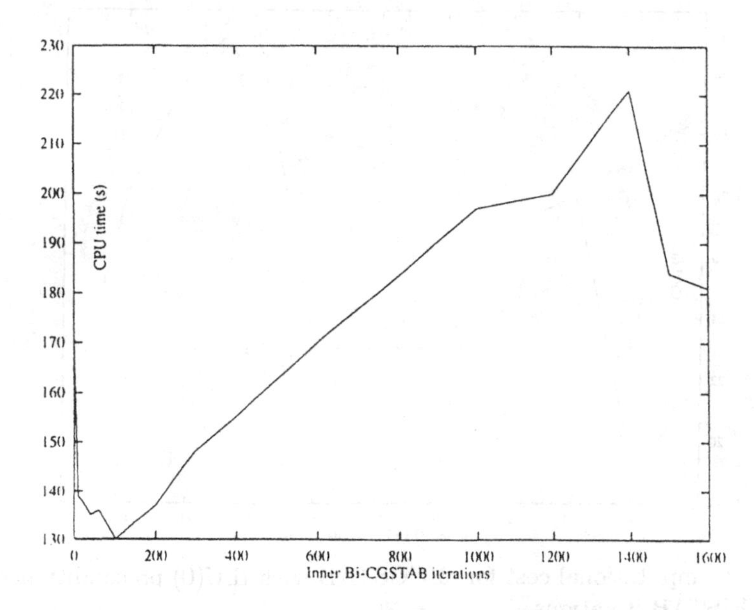

Figure 14. Computational cost for BICGSTAB with Diagonal preconditioner versus Inner BICGSTAB iterations.

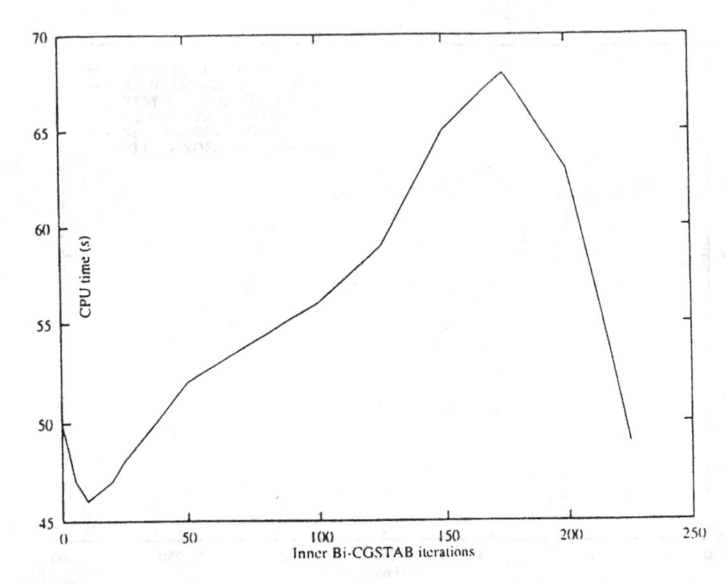

Figure 15. Computational cost for BICGSTAB with SSOR preconditioner versus Inner BICGSTAB iterations.

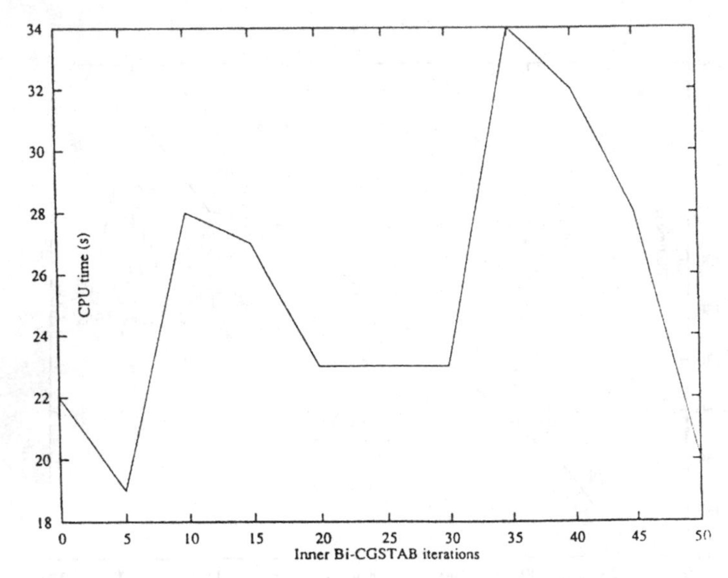

Figure 16. Computational cost for BICGSTAB with ILU(0) preconditioner versus Inner BICGSTAB iterations.

7. Conclusions

This work is a study of Krylov methods which analyses the effect of preconditioning and ordering in the convergence rate. SSOR preconditioner improve the convergence with a low cost. ILU(0) is a more robust preconditioner with a lower number of iterations but at higher computational cost.

The reordering generally improves the behavior of the preconditioners. This is more appreciable for nonsymmetric systems than symmetric ones. However, it must be study in depth also involving the relation with the condition number. In our numerical experiments, RCM algorithm often wins MN algorithm, but there is not much difference between them.

The results obtained for BICGSTAB and QMRCGSTAB algorithms are very near, nevertheless the last has a smoother curve of convergence. VGMRES generally involved more computational cost.

On the other hand, the preconditioning pattern for methods like CGS and BICGSTAB, or even TFQMR and QMRCGSTAB, is equivalent to the selection of the initial vector. However, the algorithms have only identical behavior in exact arithmetic, since they have different properties in finite arithmetic.

Finally, if we fix that initial vector after iterating the algorithm itself, one may wonder if there is a number of iteration which minimizes the total computational cost. Results show that this minimum often appears for a few steps.

8. References

[1] Ashby, S.F., Manteuffel, T.A. and Saylor, P.E. (1990) A taxonomy for conjugate gradient methods, *SIAM J. Numer. Anal.* **27**, 1542-1568.

[2] Barret, R., Berry, M., Chan, T.F., Demmel, J., Donato, J., Dongarra, J., Eijkhout, V., Pozo, R. Romine, C. and van der Vorst, H.A. (1994) *Templates for the solution of linear systems*, SIAM, Philadelphia.

[3] Chan, T.F., Gallopoulos, E., Simonsini, V., Szeto, T. and Tong, C.H. (1994) A Quasi-Minimal residual variant of the BI-CGSTAB algorithm for nonsymmetric systems, *SIAM J.Sci. Comput.* **15**, 2, 338-347.

[4] Dutto, L.C. (1993) The effect of ordering on preconditioned GMRES algorithm, *Int. Jour. Num, Meth. Eng.* **36**, 457-497.

[5] Fletcher, R. (1976) Conjugate gradient methods for indefinite systems, *Lectures Notes in Math.* **506**, 73-89.

[6] Freund, R.W. (1993) A transpose-free quasi-minimal residual algorithm for non-Hermitian linear systems, *SIAM J. Sci. Comput.* **14**, 470-482.

[7] Freund, R.W. and Nachtigal, N.M. (1991) QMR: a quasi-minimal residual method for non-Hermitian linear systems, *Numerische Math.* **60**, 315-339.

[8] Galán, M. Montero, G. and Winter, G. (1994) A direc solver for the least square problem arising from GMRES(k), *Com. Num. Meth. Eng.* **10**, 743-749.

[9] Galán, M., Montero, G. and Winter, G. (1994) Variable GMRES: an optimizing self-configuring implementation of GMRES(k) with dynamic memory allocation, *Technical Report of CEANI.*

[10] George, A. (1971) Computer implementation of the finite element method, *Report stan CS-71-208.*

[11] George, A. and Liu, J.W. (1989) The evolution of the minimum degree ordering algorithms, *SIAM Rev.* **31**, 1-19.

[12] Grote, M. and Simon, H.D. (1992) Parallel Preconditioning and Approximate Inverses on the Connection Machine, *Proceedings of the Sixth SIAM Conference on Parallel Processing for Scientific Computing*, **Vol. II**, 519-523, Philadelphia.

[13] Hestenes, M.R. and Stiefel, E. (1952) Methods of conjugate gradients for solving linear systems, *Jour. Res. Nat. Bur. Standards* **49**, 409-436.

[14] Jennings, A. and Malik, G.M. (1978) The solution of saparse linear equations by conjugate gradient method, *Int. Jour. Num. Meth. Eng.* **12**, 141-158.

[15] Martin, G. (1991) Methodes de preconditionnemen par factorisation incomplete, *Memoire de Maitrise*, Universite Laval, Quebec, Canada.

[16] Meijerink, J.A. and van der Vorst, H.A. (1977) An iterative solution method for linear systems of which the coefffecient matrix is a symmetric M-matrix, *Math. Comp.* **31**, 148-162.

[17] Montero, G. Montenegro, R. Winter, G. amd Ferragut, L. (1990) Aplicacion de esquemas EBE en procesos adaptativos, *Rev. Int. Met. Num. Cal. Dis. Ing.* **6**, 311-332.

[18] Montero, G. and Suárez, A. (1996) Efecto del precondicionamiento y reordenacion en los metodos CGS y BI-CGSTAB, *Proceeding of III Congreso de Metodos Numericos en Ingenieria*, Zaragoza, 1306-1315.

[19] Montero, G. and Suárez, A. (1995) Left-Right preconditioning versions of BCG-like methods, Neural, *Parallel & Scientific Computations* **3**, 4, 487-501.

[20] Nachttigal, N.M., Reddy, S.C. and Trefethen (1992) How fast are nonsymmetric matrix iterations?, *SIAM J. Matr. Anal. Appl.* **13**, 3, 778-795.

[21] Peters, A. (1993) Non-symmetric CG-like schemes and the finite solution of the advection-dispersion equation, *Int. Jour. Num. Meth. Fluids* **9**, 67-69.

[22] Prakhya. K.V.G. (1988) Some Conjugate Gradient methods for symmetric and nonsymmetric systems on a vector multiprocessor, *Com. Appl. Num. Meth.* **4**, 531-539.

[23] Radicati, G. and Robert, Y. (1986) Parallel conjugaate gradient-like algorithms for solving sparse non-symmetric systems on a vector multiprocessor, *Parallel Comput.* **11**, 223-239.

[24] Radicati, G. and Vitaletti, M. (1986) Sparse matrix-vector product and storage representations on the IBM 3090 with Vector Facility, *Report G513-4098, IBM-ECSEC*, Rome.

[25] Saad, Y. (1992) Highly parallel preconditioners for general sparse matrices, *Tech. Rep. 92-087, Army High Performance Computing Research Center*, Minneapolis, MN.

[26] Saad, Y. (1993) A flexible innner-outer preconditioned GMRES algorithm, *SIAM J.Sci. Statist. Comput.* **14**, 2, 461-469.

[27] Saad, Y. and Schultz, M.H. (1986) GMRES: A generalized minimal residual algorithm for solving nonsymmetric linear systems, *SIAM J.Sci. Statist. Comput.* **7**, 856-869.

[28] Saad, Y. (1983) Iterative solution of indefinite symmetric systems by methods using orthogonal polynomials over two disjoint intervals, *SIAM J. Num. Anal.* **20**, 784-811.

[29] Saad, Y. (1985) Practical use of polynomial preconditionings for the conjugate gradient method, *SIAM J. Sci. Statist. Comput.* **6**, 865-881.

[30] Saad, Y. (1987) Least squares polynomials in the complex plane and their use for solving sparse nonsymmetric linear systems, *SIAM J. Num. Anal.* **24**, 155-169.

[31] Saad, Y. (1996) *Iterative Methods for Sparse Linear Systems*, PWS PUBLISHING COMPANY, Boston.

[32] Sonneveld, P. (1989) CGS: a fast Lanczos-type solver for nonsymmetric linear systems, *J.Sci. Statist. Comput.* **10**, 36-52.

[33] Suárez, A. (1995) Contribucion a algoritmos de biortogonalizacion para la resolucion de sistemas de ecuaciones lineales, *Doct. Thesis*, Las Palmas de G.C. University.

[34] Mai, Tsun-zee (1993) Modified Lannczosmethod for solving large sparse linear systems, *Com. Num. Meth. Eng.* **9**, 67.79.

[35] van der Vorst, H.A. (1990) The convergence behaviour of preconditioned **CG** and **CG-S** in the presence of rounding errors, *Lectures Notes in Math.* **1457**, 126-136.

[36] van der Vorst, H.A. (1992) BI-CGSTAB: a fast and smoothly converging variant of BI-CG for the solution of nonnsymmetric linear systems, *SIAM J.Sci. Statist. Comput.* **13**, 631-644.

[37] van der Vorst, H.A. and Vuik, C. (1991) GMRESR: A family of nested GMRES methods, *Tech, Rep. 91-80, Delft University of Technology, Mathematics and Informatics*, Delft, The Netherlands.

CONVERGENCE AND NUMERICAL BEHAVIOUR OF THE KRYLOV SPACE METHODS

ZDENĚK STRAKOŠ
Institute of Computer Science, Academy of Sciences
Pod vod. věží 2, 182 07 Praha 8, Czech Republic
Email: na.strakos@na-net.ornl.gov

Abstract. Our contribution is devoted to two questions: How to describe the rate of convergence of the Krylov space methods and how is their behaviour affected by the effects of rounding errors. We review some classical results and emphasize connections with the related topics such as orthogonal polynomials and quadratures. The symmetric and nonsymmetric cases are described separately. The text is organized in both cases analogously to show the differences in the understanding of the corresponding questions. We present several new results and discuss problems which remain open.

1. Introduction

Consider the problem of solving a linear algebraic system

$$Ax = b, \tag{1}$$

where A is a real n by n matrix, which is assumed to be nonsingular, and b is a real vector of length n. If the matrix A is large and sparse, then the methods based on Krylov subspaces, combined with a proper preconditioning, represent a competitive tool for computing the solution. As with other methods, the key questions are how precise approximation to the solution can be expected and at what expense it can be found. In a more specific language, we can ask what is the theoretical rate of convergence (assuming exact arithmetic) of a given method (on which problem parameters it depends and how it can be described) and how is the behaviour of the method affected by rounding errors. In our contribution, we restrict ourselves to real data (matrix A and the right hand side b). This is an obvious restriction

G. Winter Althaus and E. Spedicato (eds.), Algorithms for Large Scale Linear Algebraic Systems, 175–196.

and its only purpose is to simplify the exposition. Reformulation of all the results for the complex problem is straightforward.

Given A, b and the initial guess x^0, we can form the sequence $r^0, A^1r^0, A^2r^0...$ and consider Krylov spaces $K_m(A, r^0) = span\{r^0, Ar^0, ..., A^{m-1}r^0\}$, $m = 1, 2, ..., n$. Krylov space method seeks for the approximations to the solution from the varieties

$$x^m \in x^0 + K_m(A, r^0). \tag{2}$$

The spaces $K_m(A, r^0)$ are constructed via some orthogonal (or bi-orthogonal) basis and the approximation x^m is determined using some additional conditions (usually some minimization and/or orthogonality property). Note that for any Krylov space method, the error $x - x^m$ and the residual $r^m = b - Ax^m$ can be expressed as

$$\begin{aligned} x - x^m &= p(A)(x - x^0) \\ r^m &= p(A)\, r^0, \end{aligned} \tag{3}$$

where p is a polynomial of degree at most m satisfying $p(0) = 1$. A class of such polynomials will be denoted Π_m. More details about various aspects of the Krylov space methods can be found in this volume in the sections written by Arioli, Brezinski, Broyden, Saad and van der Vorst.

First, we assume exact arithmetic and attempt to analyze the theoretical rate of convergence of the Krylov space methods. It should be noted that the n by n linear algebraic system came in most cases from a discretization of the original, mostly infinite dimensional problem. Our primary goal is to solve effectively the original problem. It can hardly be done, however, without understanding of the linear algebraic solvers. It is worthless to develop a mathematical model of the real world problem, discretize it and end with a linear algebraic system which is almost or completely numerically unsolvable. Here we will discuss the rate of convergence of the Krylov space methods based on some properties of the linear algebraic system. Though we will not go beyond that, we underline that these properties are determined by the original problem and the method of discretization. This relation is not always transparent and it may be difficult to analyze (for some examples, see [58], [59]). It should be, however, exploited whenever possible.

Second, we turn to the effects of rounding errors. Basically, they deteriorate the rate of convergence by causing a loss of orthogonality among the computed (in exact arithmetic orthogonal) vectors. Fortunately, the loss of orthogonality is highly structured; this structure has been successfully studied (for the very competent review of the symmetric case, see [48]), and it has lead to fundamental conclusions about the implementation and applicability of the methods. Another unpleasant effect of rounding errors represents the fact that the error in the computed approximation does not decrease below some level called the ultimate attainable accuracy. Even if, in practical problems, we rarely iterate until this level is reached, it is important to determine (or bound) this level for particular methods. It completes the comparison of direct and iterative approaches for solving a given problem.

Our picture of both characterization of convergence and analysis of rounding errors will substantially change if we go from symmetric to general nonsymmetric systems. If A is symmetric, then we can form an orthonormal basis of R^n from the eigenvectors of A and the convergence can be described in terms of the simple quadratic forms and orthogonal polynomials. If A is nonsymmetric, then no such analogy is, in general, available. If A is not normal, things may become really complicated.

In our discussion of convergence and numerical properties, we will concentrate on basic unpreconditioned methods. In exact arithmetic, this does not mean a loss of generality because we can always consider a preconditioned method as the basic one applied to the preconditioned system. This is, of course, not true in finite precision arithmetic. The rounding error analysis of basic methods represents the first and fundamental step, which can be later modified to include the methods with preconditioning.

2. Characterization of convergence of the Krylov space methods

Given a Krylov space method, we wish to know how fast will the sequence of its approximate solutions $x^0, x^1, \ldots$ converge to the solution x. Our description of convergence should be simple and as close to the true behaviour as possible. It means that we must extract some essential information about the system which determines the convergence, develop convergence bounds, and, possibly, prove that the developed bounds are sharp, i.e., that no better bounds based on the extracted information can be achieved. Our characterization of convergence must lead into some useful insight into the behaviour of the method (it seems of little use to develop formal bounds containing potentially large or even unbounded terms and then ignore this fact in making conclusions about the given method).

In this part, we will not consider the effects of rounding errors, i. e., exact arithmetic is assumed. For simplicity of exposition, we assume that $K_m(A, r^0)$ has dimension m, $m = 1, 2, \ldots, n$. This assumption does not mean any loss of generality. If for some m $dim(K_m(A, r^0)) = m$ and $dim(K_{m+1}(A, r^0)) = m$, it represents the happy breakdown - the exact solution can be found in m steps. For this case, our considerations can be modified accordingly.

2.1. SYMMETRIC SYSTEMS

2.1.1. *Basic relations*

For A symmetric, we have a natural and very elegant tool in our hands. Any symmetric matrix can be decomposed as

$$A = U\Lambda U^T, \quad \Lambda = diag\{\lambda_1, \ldots, \lambda_n\}, \quad UU^T = U^TU = I, \tag{4}$$

where $\{\lambda_1, \lambda_2, \ldots\}$ represent eigenvalues and $\{u_1, u_2, \ldots\}$ the corresponding normalized eigenvectors of the matrix A. The matrix Λ is diagonal and the matrix $U = (u_1, u_2, \ldots, u_n)$ having orthonormal eigenvectors of A as its columns is orthonormal. It seems convenient to change the basis; if the basis $\{u_1, u_2, \ldots, u_n\}$ of R^n is used instead of the standard Euclidean basis $\{e_1, e_2, \ldots, e_n\}$, the information about the matrix A is reduced to the information about its eigenvalues.

In addition, Krylov space methods can be described using orthonormal basis $\{q^1, \ldots, q^m\}$ of the Krylov space $K_m(A, r^0)$. If $q_1 = r^0/\| r^0 \|$, then $\{q^1, \ldots, q^m\}$, called the Lanczos basis, is computed by the symmetric Lanczos process

$$AQ_m = Q_mT_m + \beta_{m+1}q^{m+1}e_m^T = Q_{m+1}T_{m+1,m}, \quad m = 1, 2, \ldots, \tag{5}$$

where $Q_m = (q^1, \ldots, q^m)$ and T_m is a symmetric tridiagonal matrix with positive subdiagonals, so called Jacobi matrix,

$$T_m = \begin{pmatrix} \alpha_1 & \beta_2 & & \\ \beta_2 & \alpha_2 & & \\ & & \ddots & \\ & & & \beta_m \\ & & \beta_m & \alpha_m \end{pmatrix}, T_{m+1,m} = \begin{pmatrix} \alpha_1 & \beta_2 & & \\ \beta_2 & \alpha_2 & & \\ & & \ddots & \\ & & & \beta_m \\ & & \beta_m & \alpha_m \\ & & & \beta_{m+1} \end{pmatrix}. \tag{6}$$

The eigendecomposition of the matrix T_m will be denoted as

$$T_m = S_m\Theta_mS_m^T, \tag{7}$$

where

$$\Theta_m = diag(\theta_1^{(m)}, \ldots, \theta_m^{(m)}), \quad S_mS_m^T = S_m^TS_m = I,$$
$$S_m = (s_1^{(m)}, \ldots, s_m^{(m)}).$$

It is worth to note that the Lanczos vectors can be expressed in the form of matrix polynomial applied to the initial vector q^1,

$$q^{m+1} = \psi_m(A)q^1/(\beta_2\beta_3 \ldots, \beta_{m+1}), \tag{8}$$

where ψ_m is the monic polynomial of degree m. Let Ψ_m be the class of all monic polynomials of degree m. From the orthogonality,

$$\| \psi_m(A)q^1 \| = \min_{\psi \in \Psi_m} \| \psi(A)q^1 \|, \tag{9}$$

or, using (4),

$$\sum_{i=1}^{n} (u_i, q^1)^2 \psi_m^2(\lambda_i) = \min_{\psi \in \Psi_m} \sum_{i=1}^{n} (u_i, q^1)^2 \psi^2(\lambda_i). \tag{10}$$

It is easy to see that $\{1, \psi_1, \psi_2, \ldots, \psi_n\}$ is the sequence of monic polynomials orthogonal with respect to the innerproduct

$$(\varphi, \psi) = \sum_{j=1}^{n} \omega_j \varphi(\lambda_j)\psi(\lambda_j), \qquad \omega_j = (u_j, q^1)^2 = (s_j^{(n)}, e_1)^2, \tag{11}$$

and ψ_m is the characteristic polynomial $det(\lambda I - T_m)$ of the Jacobi matrix T_m. Summarizing, we can look at the Lanczos process from the point of view of:

– the sequence of basis vectors $q^1, q^2, \ldots, q^n$,
– the sequence of Jacobi matrices $T_1, T_2, \ldots, T_n$,
– the sequence of monic orthogonal polynomials $\psi_1, \psi_2, \ldots \psi_n$.

This correspondence is fundamental and our feeling is that it has not been fully exploited yet. For details we refer to [21],[22], [49].

2.1.2. *Characterization of convergence*

We consider two examples: The conjugate gradient method (CG) and the minimal residual method (MINRES).

The CG approximation x_{CG}^m minimizes the A-norm of the error $\| x - x^m \|_A$, i.e., the A^{-1}-norm of the residual $\| b - Ax^m \|_{A^{-1}}$, at every step. Using the Lanczos basis Q_m, it is given by the formula

$$x^m = x^0 + Q_m y^m, \quad T_m y^m = \|r^0\| e^1. \tag{12}$$

For a symmetric positive definite matrix A, the approximation is well defined (T_m is nonsingular because the eigenvalues of the consecutive Jacobi matrices in the sequence $T_1, T_2, \ldots, T_n$ must interlace and the eigenvalues of T_n and A are identical). For A indefinite, however, T_m may become singular, or close to singular, and the CG approximation may not exist at every step. For details see [43]. From the minimizing property,

$$\| r_{CG}^m \|_{A^{-1}}^2 = \| \varphi_m^{CG}(A) r^0 \|_{A^{-1}}^2 = \min_{\varphi \in \Pi_m} \sum_{i=1}^{n} \{(u_i, r^0)^2 / \lambda_i\} \varphi^2(\lambda_i)$$

$$\leq \| r^0 \|_{A^{-1}}^2 \min_{\varphi \in \Pi_m} \max_j \varphi^2(\lambda_j)$$
$$= \| r^0 \|_{A^{-1}}^2 \min_{\varphi \in \Pi_m} \| \varphi(A) \|^2 \ . \quad (13)$$

The MINRES method minimizes the norm of the residual. Its approximations always exist and are given by the relation

$$x^m = x^0 + Q_m y^m, \quad \| \, \|r^0\| e^1 - T_{m+1,m} y^m \, \| = \min_y \| \, \|r^0\| e^1 - T_{m+1,m} y \, \| \ . \quad (14)$$

Clearly,

$$\| r_M^m \|^2 = \| \varphi_m^M(A) r^0 \|^2 = \min_{\varphi \in \Pi_m} \sum_{i=1}^n (u_i, r^0)^2 \varphi^2(\lambda_i)$$
$$\leq \| r^0 \|^2 \min_{\varphi \in \Pi_m} \max_i \varphi^2(\lambda_i)$$
$$= \| r^0 \|^2 \min_{\varphi \in \Pi_m} \| \varphi(A) \|^2 \ . \quad (15)$$

Relations (13) and (15) form a base for a convergence bound development. Taking a specific polynomial φ will lead to a specific bound. We are not going to review the existing bounds; they can be found elsewhere (see, e.g., [16]). We stress the fact that in (13) and (15) the norms of the residual are given by simple quadratic forms, based on the eigenvalues and the coordinates of the initial residual in the basis of eigenvectors. Assuming that the coordinates do not differ too much in magnitude, the convergence is essentially determined by the *distribution of eigenvalues.*

While $\| r_{CG}^m \|_{A^{-1}}$ and $\| r_M^m \|$ are monotonic, the norm of the CG residual $\| r_{CG}^m \|$ is not, in general. There is a simple but enlightening relation between the norms of the CG and MINRES residuals, namely,

$$\| r_{CG}^m \| = \frac{\| r_M^m \|}{\sqrt{1 - (\| r_M^m \| / \| r_M^{m-1} \|)^2}} \quad (16)$$

If the norm of the MINRES residual at step m is reduced by a large amount over its value at step $m-1$, then the norm of the CG residual will be close to the minimal possible value $\| r_M^m \|$. If the norm of the MINRES residual remains constant at a step (it must decrease every two steps, see e. g. [27]), then the corresponding CG iterate does not exist. For details see [11].

2.1.3. *Convergence of Ritz values*

From (13) and (15) it seems clear that if the roots of the polynomials φ_m^{CG} and φ_m^M approximate the eigenvalues λ_j of the system matrix, one may expect acceleration of convergence. For CG, φ_m^{CG} represents a properly normalized Lanczos polynomial

$$\varphi_m^{CG}(\mu) = \psi_m(\mu)/\psi_m(0), \tag{17}$$

and its roots are the eigenvalues of the Jacobi matrix T_m, i.e., the Ritz values. For nice quantitative description of the influence of the convergence of Ritz values to the acceleration of convergence of the *CG* method we refer to [51]. For MINRES the roots of φ_m^M are called the harmonic Ritz values. For a comprehensive description of the subject, see [46].

2.1.4. *Minimizing the matrix polynomial gives a sharp bound*
If we consider the so called ideal approximation problem

$$\min_{\varphi \in \Pi_m} \| \varphi(A) \|, \qquad m = 1, 2, \ldots, \tag{18}$$

then for any initial residual r^0 the MINRES convergence curve ($\| r_M^m \|$ as a function of m) lies below the ideal MINRES convergence curve, defined by (18). We may ask if the ideal MINRES convergence curve is a sharp envelope of all the possible MINRES convergence curves. In other words, given m, does there exist an initial residual vector r^0 such that

$$\| r_M^m \| \ / \ \| r^0 \| = \min_{\varphi \in \Pi_m} \| \varphi(A) \| \tag{19}$$

is satisfied? The answer is yes, see [23].

2.1.5. *Relation to Gauss quadrature and to continued fractions*
As the appearance of the orthogonal polynomials suggests, the question of convergence of the Krylov space methods is deeply connected with problems in the other (classical) areas. During the last decades, Golub has done and motivated a lot of work emphasizing these connections. We will briefly describe one example.

If we write the A-norm of the initial error in the CG method in the form

$$\| x - x^0 \|_A^2 = (r^0)^T A^{-1} r^0 = \| r^0 \|^2 \, e_1^T T_n^{-1} e_1 = \| r^0 \|^2 \, (T_n^{-1})_{11}$$

and consider the eigendecomposition (7), we obtain

$$(T_n^{-1})_{11} = \sum_{j=1}^{n} \frac{\omega_j}{\lambda_j} \tag{20}$$

where $\omega_j = (s_j^{(n)}, e_1)^2$. In addition, it is known (see [56], [34]) that (20) can be expanded into the continued fraction

$$(T_n^{-1})_{11} = \sum_{j=1}^{n} \frac{\omega_j}{\lambda_j} = C_n = \frac{G_n}{E_n} = \cfrac{1}{\alpha_1 - \cfrac{\beta_2^2}{\alpha_2 - \cfrac{\beta_3^2}{\ddots \, \overline{\alpha_{n-1} - \frac{\beta_n^2}{\alpha_n}}}}}, \tag{21}$$

where the denominator E_n and the numerator G_n are given by the three term recurrences

$$\begin{aligned} E_0 = 1, \quad E_1 = \alpha_1, \quad E_j &= -\alpha_j E_{j-1} - \beta_j^2 E_{j-2}, \\ G_0 = 0, \quad G_1 = 1, \quad G_j &= -\alpha_j G_{j-1} - \beta_j^2 G_{j-2}, \qquad j = 2, \ldots, n. \end{aligned} \tag{22}$$

Clearly, relations (20) - (21) hold true for any $m = 1, 2, \ldots, n$, i.e.,

$$(T_m^{-1})_{11} = \sum_{j=1}^{m} \frac{(s_j^{(m)}, e_1)^2}{\theta_j^{(m)}} = C_m, \tag{23}$$

where C_m is the m-th convergent of C_n. It seems therefore natural to look for a formula relating C_n, C_m, $\| x - x^0 \|_A$ and $\| x - x^m \|_A$.

The sum in the relation (20) can be considered as a Riemann-Stieltjes integral of the function $f(\lambda) = \lambda^{-1}$ with the distribution function having exactly n-points of increase. Exploiting the orthogonality of the Lanczos polynomials, the relation (23) can be considered as a result of the m-point Gauss quadrature applied to (20), i.e.,

$$(T_n^{-1})_{11} = (T_m^{-1})_{11} + R_m(1/\lambda). \tag{24}$$

Using the expression for the (truncation) error in the Gauss quadrature (e.g. [19], p.83), it can be shown that

$$R_m(1/\lambda) = \frac{1}{\| r^0 \|^2} \| x - x^m \|_A^2,$$

and, consequently,

$$\| x - x^m \|_A^2 = \| r^0 \|^2 \left[(T_n^{-1})_{11} - (T_m^{-1})_{11} \right], \tag{25}$$

or, in terms of continued fractions,

$$C_n - C_m = \| x - x^m \|_A^2 / \| r^0 \|^2 . \tag{26}$$

For details see [22], also [21]. For the original proof of (25), see [12].

2.2. NONSYMMETRIC SYSTEMS

2.2.1. *Basic relations*

For A nonsymmetric our principal tool is lost - we cannot, in general, use the basis of the space R^n composed of the matrix eigenvectors. If the matrix is defective, we simply do not have enough of them. One may suggest that this trouble is simply overcomed by the fact that the set of all diagonalizable matrices is dense in the class of all square matrices. However, we can easily see that the problem is not so simple.

If A is diagonalizable, then the decomposition

$$A = X\Lambda X^{-1}, \quad \Lambda = \text{diag}\,\{\lambda_1, \ldots, \lambda_n\} \tag{27}$$

looks similar to (4) with one but crucial difference: The matrix of eigenvectors is not unitary. It is unitary only if A is normal, i.e. $AA^T = A^T A$. If A is not diagonalizable, then for any $\delta > 0$ we may consider a diagonalizable matrix $\tilde{A}$, $\| A - \tilde{A} \| < \delta$, but two new difficulties may occur. The distance between the spectrum of A and $\tilde{A}$ is not, in general, of order δ; it depends on the departure from normality of the matrix A (the perturbation of individual eigenvalues depends on their condition numbers, see, e.g. [52]). In addition, the matrix of eigenvectors of the matrix $\tilde{A}$ can be very ill-conditioned. Any approach based on the Jordan normal form of the matrix A suffers from very similar difficulties.

The natural replacement for the spectral decomposition seems to be the Cauchy integral representation of the matrix function

$$f(A) = \frac{1}{2\pi i} \int_\Gamma (zI - A)^{-1} f(z)dz, \tag{28}$$

where $(zI - A)^{-1}$ is the resolvent and Γ is any simple closed curve (or union of simple closed curves) containing the spectrum of A (e.g., [20]). Using (28), the bound for the norm of the matrix polynomial $\| p(A) \|$ will be determined by the growth of the resolvent in the neighbourhood of the matrix spectrum. Though (28) looks as a universal tool, its use for the analysis of convergence of the Krylov space methods in the finite dimensional spaces is not free of difficulties.

In our discussion we will concentrate on the generalized minimal residual method (GMRES). Its approximations x_G^m minimize the residual norms among all the Krylov space methods (GMRES is a straightforward generalization of MINRES):

$$x^m \in x^0 + K_m(A, r^0), \quad \| r^m \| = \min_{u \in AK_m(A, r^0)} \| r^0 - u \|. \tag{29}$$

The orthonormal basis $\{v^1 = r^0/ \parallel r^0 \parallel, v^2, \ldots, v^m\}$ of $K_m(A, r^0)$ is computed by the Arnoldi process

$$AV_m = V_{m+1}H_{m+1,m}, \qquad m = 1, 2, \ldots, \tag{30}$$

where $V_m = (v^1, \ldots, v^m)$, and $H_{m+1,m}$ is an upper Hessenberg matrix. Then

$$x^m = x^0 + V_m y^m, \quad \parallel\parallel r^0 \parallel e_1 - H_{m+1,m} y^m \parallel = \min_y \parallel\parallel r^0 \parallel e_1 - H_{m+1,m} y \parallel . \tag{31}$$

The choice of GMRES is well justified. Its analysis is relatively simple (in comparison to the analysis of other methods) and it provides the information how accurate an approximation, in the sense of the size of the residual, can be found in the variety $x^0 + K_m(A, r^0)$.

2.2.2. *Characterization of convergence*

If A has a full set of eigenvectors, then, similarly to (15), the GMRES residual r_G^m can be bounded as

$$\parallel r_G^m \parallel \;\leq\; \parallel r^0 \parallel\parallel X \parallel\parallel X^{-1} \parallel \min_{\varphi \in \Pi_m} \max_i |\varphi(\lambda_i)|. \tag{32}$$

If the matrix A is normal, then $\parallel X \parallel\parallel X^{-1} \parallel = 1$, and the bound (32) is sharp ([28], [36]). Then, the convergence is determined by the spectrum of A . If the matrix A is not normal, the factor $\kappa(X) = \parallel X \parallel\parallel X^{-1} \parallel$ may be large and (32) may become a large overestimate of the actual residual.

In [27], the behaviour of GMRES applied to a nonnormal matrix and a given initial residual r^0 is related to its behaviour when applied to certain normal matrices and r^0. Since the convergence rate of the method applied to normal matrices is determined by the eigenvalues of the matrix, if the eigenvalues of the normal matrix could be related to some properties of the original matrix, then (32) would give a bound on the convergence rate, in terms of these properties.

Unfortunately, it is only in special cases that we are able to relate the eigenvalues of the normal matrix to meaningful properties of the original matrix. It is shown that any behaviour that can be seen with the GMRES method can be seen with the method applied to a unitary matrix. While certain properties of the original matrix - e.g., positive definiteness - appear to guarantee large gaps (large nonuniformity) in the spectrum of this unitary matrix, we have not been able to prove this. It is shown that if zero is outside the field of values of the original matrix A, then the GMRES method behaves just as it would for a certain Hermitian positive definite matrix. If A is close to Hermitian, then the GMRES method behaves just as it would for a Hermitian matrix whose eigenvalues are close to those of A. In this case then, the eigenvalues of A essentially determine the behaviour of the GMRES algorithm. In general, however, eigenvalue information alone cannot be sufficient to ensure fast convergence of the GMRES algorithm. We demonstrate it by the following theorem ([32], Theorem 2).

Theorem 1 *Given a nonincreasing positive sequence* $f(0) \geq f(1) \geq \ldots \geq f(n-1) > 0$ *and a set of nonzero complex numbers* $\{\lambda_1, \ldots, \lambda_n\}$, *there exists a matrix* A *with eigenvalues* $\lambda_1, \lambda_2, \ldots, \lambda_n$ *and an initial residual* r^0 *with* $\| r^0 \| = f(0)$ *such that the residual vectors* r^k *at each step of the GMRES method applied to* A *and* r^0 *satisfy* $\| r^k \| = f(k)$, $k = 1, 2, \ldots, n-1$. *If* $\{\lambda_1, \ldots, \lambda_n\}$ *represent roots of a real polynomial, then* A *can be chosen real.*

In the other words, any nonincreasing convergence curve is possible for GMRES with matrix having any prescribed eigenvalues. The proof in [32] is simple and it has several additional consequences. It allows to describe, e.g., the set of all matrices and initial residuals for which GMRES generates the given convergence history.

Clearly, eigenvalues alone are not enough to describe the convergence in the general nonsymmetric case. Despite that, one can still use (28) to develop convergence bounds which depend on the spectrum, see, e.g., [40], [9]. While such bounds describe well the asymptotic behaviour of the operators in infinite dimensional spaces, their application to GMRES in finite dimensional space must consider the following *problem of constant*. Assume that, using any approach, we have developed the bound

$$\| r_G^m \| \leq \| r^0 \| \, C(A, r^0) \, F(sp(A), m), \tag{33}$$

where $F(sp(A), m)$ is a function of the eigenvalues of A and the iteration step m, and $C(A, r^0)$ depends on some properties of the system to be solved. In the symmetric case the value of $C(A, r^0)$ can be bounded by a constant independent of A, r^0. In (15), e.g., $C(A, r^0) \leq 1$. For general nonsymmetric matrices, however, this cannot be done. Assume, for a moment, that $C(A, r^0) \leq K$ and K is independent of A, r^0. Then (33) becomes

$$\| r_G^m \| / \| r^0 \| \leq K \, F(sp(A), m). \tag{34}$$

The last bound is, however, completely irrelevant, because, due to Theorem 1, $KF(sp(A), m) \geq 1$ for $m = 1, 2, \ldots, n-1$. Indeed, if for any nonzero eigenvalues and for some m

$$KF(sp(A), m) < 1,$$

it would mean that for *any matrix* having these eigenvalues the relative residual norm $\| r_G^m \| / \| r^0 \|$ is less than 1, which is in contradiction with Theorem 1. Clearly, the value of $C(A, r^0)$ must always be taken into account. If we ignore it and base our conclusion on $F(sp(A), m)$ only, such a conclusion does not apply to all nonsymmetric matrices A and initial residuals r^0.

Despite this rather pessimistic picture, the common feeling is that in most cases (cf. [27], Fig. 1a, 1b) the GMRES behaviour is essentially determined by the spectrum of A. In [40] it is suggested that the cases in which the spectrum gives a wrong information about the convergence are pathological and the conflict described above can be explained by the fact that Theorem 1 considers all the possible initial residuals, including not "*generic*" ones. To our opinion the situation is not completely understood yet and it needs further work. The next subsection shows another complication.

To deal with the problem of nonnormality, Trefethen has introduced bounds based on the ε-pseudoeigenvalues of the matrix [58], [59]. For a given matrix A, the ε-pseudospectrum is a set of points in the complex plane which are the eigenvalues of some matrix $A + E$, with $\| E \| \leq \varepsilon$. The resolvent $(zI - A)^{-1}$ has norm equal to $1/\varepsilon$ on the boundary Γ_ε of the ε-pseudospectrum. This fact can be used in computing bounds for $f(A)$ when Γ_ε is used as the integration curve in (28). While in many cases this can lead to much better estimates than the approaches based on (32), there are other cases in which it gives a large overestimate for any value of ε. For details we refer to [27].

2.2.3. *Minimizing the matrix polynomial does not give a sharp bound*
As mentioned above, for A symmetric the ideal approximation problem (16) generates the envelope of all the possible MINRES convergence curves. An analogous statement is true for GMRES applied to normal matrices (and also to some other special classes of matrices, see [30]). However, matrices have been found for which

$$\frac{\min_{\varphi \in \Pi_m} \| \varphi(A) \|}{\max_{\|r^0\|} \| r_G^m \| / \| r^0 \|} > 1, \tag{35}$$

for certain iteration steps m, and, in addition, this ratio can be made arbitrarily large [57], [37]. The consequence is alarming. Most of the convergence analysis, including the approaches based on (28), have assumed that convergence rate can be analyzed in terms of a matrix approximation problem. Result in [57] suggests that this assumption should be checked. There exist examples in which Krylov space methods perform *for any ini-*

tial residual much better than any analysis based only on the matrix can explain. This indicates that the problem of characterizing the convergence of the nonsymmetric Krylov space iterations is a difficult one.

It should be noted that the practical effect of the results described in the last two subsections remains unclear. We do not know how these results are important in real world applications. All the "negative" results have been proved for matrices without showing any connection to possible operators representing some mathematical model of a real world problem. Moreover, some bad properties of the matrix may be created by an improper discretization of the corresponding operator. On the other hand, there are well-known examples of important nonnormal operators (e.g. [59], [10]). The situation needs further effort.

2.3. OTHER COMMENTS

Most of the nonsymmetric Krylov space methods use either some orthogonality (Galerkin - type methods) or some minimizing property (norm-minimizing methods) for determining an approximate solution.

In the first case it is necessary to solve a linear system with the upper Hessenberg matrix as coefficient matrix at every iteration step. In the second, an extended Hessenberg least squares problem is solved. Based on this correspondence, we can describe pairs of Krylov space methods. Examples include CG and MINRES described above; in the nonsymmetric case the FOM (Arnoldi) and the GMRES algorithms, the Bi-CG and the QMR algorithms. Their description can be found in other parts of this volume. The CG and MINRES residuals are related by (16) - similar relations hold true for the pairs of nonsymmetric Krylov space methods as well. This implies that if the Galerkin iterates are well defined and if one member of the pair performs very well, then the other member of the pair will also perform very well. If one member performs very poorly then the other member will also perform poorly. Details can be found in [8], [11].

In the symmetric case the acceleration of convergence can be explained by the convergence of the Ritz or the harmonic Ritz values. In the general case the situation is much less transparent, cf. [60], [4]. Its understanding is of great importance for the construction of hybrid or adaptive methods (e.g. [38], [39], [3]).

The relation of Krylov space methods to orthogonal polynomials and Gauss quadrature can be in some sense generalized to the nonsymmetric case, but it can hardly lead to conclusions of the same strength as for A symmetric [18], [17], [62]. Similarly, a unified description of the convergence of the Krylov space methods will be in the nonsymmetric case substantially different from the symmetric one [35].

3. Numerical behaviour of the Krylov space methods

In the presence of rounding errors the fundamental principle is shaken - the computed basic vectors will lose their orthogonality (or bi-orthogonality) property. As a consequence, the convergence may be delayed. In addition, the precision of the computed approximation is limited. Any iteration process will reach, possibly after a large number of iterations, its so called *ultimate attainable accuracy* level and, if not stopped, it will stagnate on this precision level forever. We will not describe the attainable accuracy of the Krylov space methods here (this subject is studied in [31] and in the papers referred to there). We will focus on the loss of orthogonality and on its effects to the rate of convergence. Similarly as in the previous section, we will study the symmetric and nonsymmetric cases separately.

3.1. SYMMETRIC SYSTEMS (NUMERICAL BEHAVIOUR)

3.1.1. *Loss of orthogonality*

Our first opinion is that the effect of rounding errors to the Lanczos process (and to all the methods mentioned in Section 2.1) is devastating. The orthogonality and even the linear independence of the computed basic vectors is usually lost very quickly. It was Paige who noticed, and then theoretically explained in a truly clever way, that the loss of orthogonality is closely related to the convergence of the computed Ritz values $\theta_j^{(m)}$ to the eigenvalues of the matrix A [42], [44], [45]. The three term recurrence relation (5) is replaced by the relation for the actually computed quantities

$$AQ_m = Q_m T_m + \beta_{m+1} q^{m+1} e_m^T + F_m, \tag{36}$$

where F_m represents the error term, $\| F_m \| \leq m^{1/2} \| A \| \varepsilon_1$, and ε_1 is a multiple of the machine precision ε. Let $z_j^{(m)} = Q_m s_j^{(m)}$ denote the computed Ritz vector and $\delta_{mj} = \beta_{m+1} |s_j^{(m)} e_m^T|$. Then the residual for the approximated eigenpair $\theta_j^{(m)}$, $z_j^{(m)}$ is bounded by

$$\| Az_j^{(m)} - \theta_j^{(m)} z_j^{(m)} \| \leq \delta_{mj} + m^{1/2} \| A \| \varepsilon_1. \tag{37}$$

Consequently, small δ_{mj} implies convergence of the Ritz pair in the residual sense. However, Paige proved much more.

Theorem 2 (Paige) *For any Ritz pair $\theta_j^{(m)}$, $z_j^{(m)}$ determined at the step m of the Lanczos process it holds*

$$\min_i |\lambda_i - \theta_j^{(m)}| \leq \max \{2.5(\delta_{mj} + m^{1/2} \| A \| \varepsilon_1), (m+1)^3 \| A \| \varepsilon_2\}, \tag{38}$$

$$|(z_j^{(m)}, q^{m+1})| = |\varepsilon_{jj}^{(m)}|/\delta_{mj}, \tag{39}$$

where $|\varepsilon_{jj}^{(m)}| \leq \| A \| \varepsilon_2$, ε_2 *is the multiple of* ε.

Summarizing, small δ_{mj} implies convergence of $\theta_j^{(m)}$ to some eigenvalue of A, and the orthogonality of the newly computed Lanczos vector q^{m+1} can be lost only in the directions of the converged Ritz vectors. If none of the Ritz values $\theta_j^{(m)}$ is close to some eigenvalue of A, then no significant loss of orthogonality occurs in steps one through m (see also [53]). We cannot describe the other results presented in [45] and all their numerous consequences, for more details we refer to [48], [47]. However, we wish to acknowledge the substantial role of Parlett, Kahan and the large group of their students in the subsequent developments.

3.1.2. *Backward error-like analysis*

A. Greenbaum proposed a new idea [24]. She studied the Lanczos algorithm for computing eigenvalues and eigenvectors of a symmetric matrix and the CG algorithm for solving symmetric positive definite linear systems. It was shown in [24] and in [25] that if the algorithms are used to compute eigenvalues of a matrix A or to solve a linear system $Ax = b$, then the eigenvalue approximations generated at each step and the errors in the linear system iterates generated at each step will be essentially the same as those of the exact algorithms applied to a larger matrix $\overline{A}$ whose eigenvalues lie in tiny intervals about the eigenvalues of A. There must be at least one eigenvalue of $\overline{A}$ close to every eigenvalue of A [54], but each interval around λ_i may contain many eigenvalues of $\overline{A}$. Note that this result does not represent the backward error analysis in the classical sense, because the dimensions of the original and the "perturbed" problem are different. However, it is a powerful tool for understanding the finite precision Lanczos and CG behaviour. For a detailed description we refer to [29].

3.1.3. *Consequences*

A small perturbation of the symmetric problem cause only a small change of the eigenvalues of the matrix A. Therefore if we apply the Lanczos method to the matrix A', where $\| A - A' \|$ is small and A' is symmetric, we may expect very similar behaviour as for the method applied to A. An analogous statement is true for the CG method applied to $A'x' = b'$, where $\|A' - A\|$ and $\|b' - b\|$ are small and A' is symmetric positive dfinite.

Clearly, the finite precision Lanczos and CG behaviour can again be described in terms of orthogonal polynomials. The computed polynomials are no longer orthogonal with respect to the measure defined by $\{\lambda_i, \omega_i\}_{i=1}^n$, see (11). However, they are orthogonal with respect to the measure that spreads the original weights over tiny intervals about the eigenvalues of the matrix A. Using this correspondence, one can apply the exact precision theory to matrix $\overline{A}$ or to the system $\overline{A}\overline{x} = \overline{b}$ and explain in this way the finite precision behaviour of the methods applied to the original problems. In particular, one can receive quantitative convergence bounds for the finite precision convergence rate [29], [15], and a quantitative description of the relation between the convergence of Ritz values and the acceleration of convergence of the CG process [41].

As it was pointed out in Section 2.1, the convergence of the continued fraction C_m to its final value C_n is in exact arithmetic determined by the A-norm of the error of the corresponding CG process. Due to rounding errors, the computed coefficients of the 3-term recurrence (elements of the tridiagonal matrices) may differ substantially from their theoretical counterparts. Despite that, the computed continued fraction C_m frequently gives very precise approximation of the final theoretical value $C = \| x - x^0 \|_A^2 / \| r^0 \|^2$. Similar behaviour was observed in Lanczos process based Gauss quadrature calculations [22]. Both these observations have the same background and can be explained using the relation between the Lanczos process and the Gauss quadrature. Using the analogy developed by Greenbaum, the problem of approximating the original sum

$$\sum_{j=1}^{n} \omega_j f(\lambda_j), \quad f(\lambda) = \lambda^{-1} \tag{40}$$

by the m-point Gauss quadrature computed in a *finite precision arithmetic* may be considered as a problem of computing the *exact* m-point Gauss quadrature approximation of the modified sum having the spread weight function. In this way, the *total error* (including the roundoff error) of the Gauss quadrature computed in finite precision arithmetic is expressed as the *truncation error* of the Gauss quadrature for the modified problem of larger dimension [22]. For the particular case $f(\lambda) = \lambda^{-1}$, this approach results in the relation between the actually computed quantities (the finite precision analogy of (26)),

$$C - C_m = \| x - x^m \|_A^2 / \| r^0 \|^2 + \Omega, \tag{41}$$

where Ω counts for small terms described in details in [22]. As a consequence, the convergence of C_m computed in a finite precision arithmetic to C is determined by the A-norm of the error of the corresponding finite precision CG run.

Some interesting questions remain open [55], [15]. However, to our opinion, though their solution will simplify and clarify the developed theory, it would not substantially change the whole picture.

3.2. NONSYMMETRIC SYSTEMS (NUMERICAL BEHAVIOUR)

A small perturbation of the nonsymmetric matrix may cause, if the matrix is nonnormal, a large perturbation of the eigenvalues. This again indicates that the role of the eigenvalues will not be in the nonsymmetric case the same as in the symmetric one.

3.2.1. *Finite precision analysis of the nonsymmetric Lanczos process*

All the practically used methods which preserve three term recurrences are related to the nonsymmetric Lanczos process (e.g. [20]). Some results of Paige were extended to the nonsymmetric case by Bai [2]. Though for both symmetric and nonsymmetric Lanczos process the results look formally similar, their implications are quite different. For example, the three term recurrences for the computed basis vectors contain in the nonsymmetric case error terms which cannot be bounded a-priori due to potential breakdown or nearbreakdown of the process. Based on the computed quantities, one can guarantee the convergence of a Ritz triplet in the residual sense - nothing more. There is no unsymmetric analogy of (38). There is an analogy of (39). However, it contains terms which can not be bounded a-priori. Consequently, the backward error-like analysis in [24] can hardly be extended in a simple way to the nonsymmetric case.

3.2.2. *Finite precision analysis of the GMRES method*

The loss of orthogonality in the Arnoldi process is not as dramatic as in the methods based on three term recurrences [6], [7] and preserving orthogonality has been emphasized in the GMRES implementations.

The orthogonality of the computed basis can be kept as close to the machine precision as possible using proper orthogonal transformations, e.g. Householder reflections. At the same time, under certain assumptions on the numerical nonsingularity of the system matrix, the GMRES implementation based on the Householder orthogonalization is backward stable (see [14], [1]) and its final accuracy is essentially the same as that guaranteed by direct solving of the system $Ax = b$ via the Householder or Givens QR decomposition.

The price for the Householder implementation is, however, too high for most of applications [61]. The Gram-Schmidt process is a cheaper alternative, and its modified version represents the most frequently used compromise. It is well known that the modified Gram-Schmidt GMRES performs surprisingly well, despite the fact that the loss of orthogonality is much worse than in the Householder implementation. We believe that we are able to offer an explanation [33]. We have found that there is a very important relation between the loss of orthogonality among the Arnoldi vectors and the decrease of the GMRES residual. Such relation has not been, to our knowledge, described and considered before. It can be proved that, for the modified Gram-Schmidt GMRES, the Arnoldi vectors will loose their orthogonality completely *only after* the residual of the computed approximation has dropped close to its final level of accuracy (which is proportional to the machine precision multiplied by the condition number of the system matrix). Until the orthogonality is completely lost, the modified Gram-Schmidt GMRES performs almost exactly as well as the Householder implementation. This suggests that unless the system matrix is extremely ill-conditioned, the use of the modified Gram-Schmidt GMRES is theoretically well justified.

The following question seems to be even more interesting. It can be demonstrated numerically, see also the next paragraph, that similar relation between the loss of orthogonality among the Arnoldi vectors and the decrease of the residual holds true even for the classical Gram-Schmidt GMRES implementation. The final level of accuracy is, however, much worse than in the modified Gram-Schmidt or Householder GMRES (it seems to be proportional to the square root of the machine precision).

Finite precision behaviour of the GMRES algorithm is illustrated by the numerical experiment with the matrix ADD1 of dimension 4960 used in the circuit simulation [13]. We used three implementations of the GMRES method preconditioned by the factorized approximate inverse preconditioner AIBC obtained by the incomplete biconjugation process (see [5], we used the drop tolerance 0.6). In the Figure 1, the norms of the true residual $\|b - Ax^m\|$ and of the Arnoldi residual $\|\|r^0\|e_1 - H_{m+1,m}y^m\|$ are plotted. The solid lines correspond to the Householder GMRES, the dotted lines to the modified Gram-Schmidt GMRES and the dashed lines to the classical GMRES. Figure 2 shows the loss of orthogonality among the computed basis vectors measured by the Frobenius norm of the matrix $(I - Q^TQ)$. We emphasize that this example represents the typical behaviour. It supports well our previous considerations. The computations were performed on the SGI Crimson workstation with the machine precision $\sim 2.2\ 10^{-16}$. It should be noted that no extra precision has been used for computing the showed values of the residual norms and they were therefore computed with the uncertainty caused by the corresponding rounding errors. Unless the value of the residual norm is close to the final accuracy level for the Householder or modified Gram-Schmidt GMRES, this uncertainty is insignificant.

What makes the observations described above attractive is the fact that the classical Gram-Schmidt is fully parallelizable. Further theoretical work on this problem may lead to reinstating of the classical Gram-Schmidt as a suitable alternative for the parallel implementation the preconditioned GMRES method. There will always be a difference between the classical and the modified Gram-Schmidt implementations in the level of the final

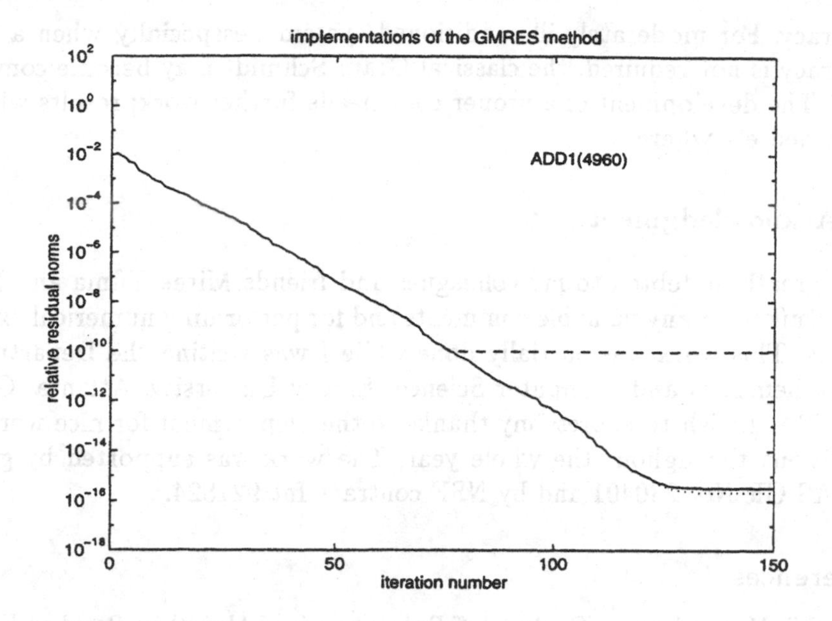

Figure 1. **Different implementations of the GMRES method**

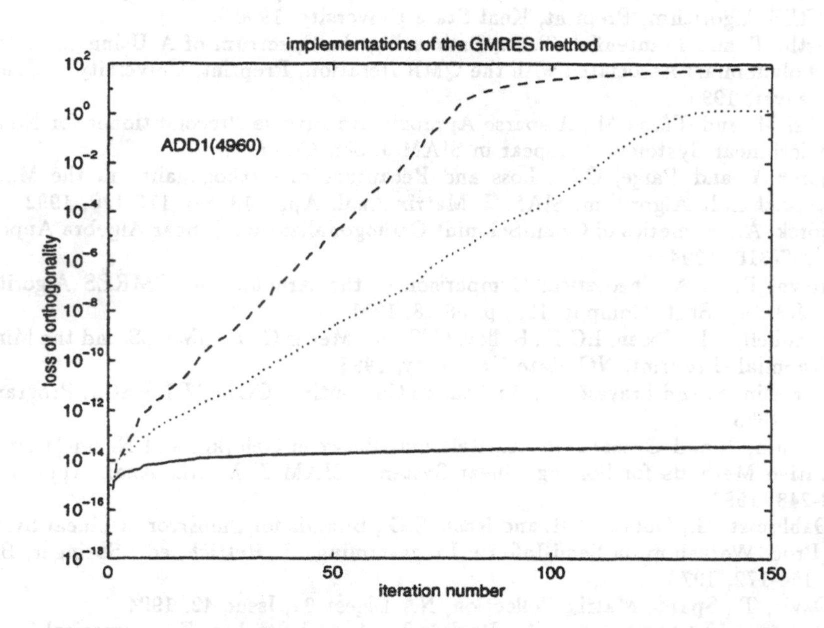

Figure 2. **The loss of orthogonality for different orthogonalization schemes**

accuracy. For moderately ill-conditioned systems, esspecially when a high accuracy is not required, the classical Gram-Schmidt may become competitive. The development of a proper code needs further work; results will be published elsewhere.

4. Acknowledgment

I am greatly indebted to my colleagues and friends Mirek Tůma and Miro Rozložník for many valuable comments and for performing numerical experiments. This work was partially done while I was visiting the Department of Mathematics and Computer Science, Emory University, Atlanta, Georgia, USA. I wish to express my thanks to the Department for nice working conditions throughout the whole year. The work was supported by grant GA AS CR No. 230401 and by NSF contract Int 921824.

References

1. Arioli, M. and Fassino, C., Roundoff Error Analysis of Algorithms Based on Krylov Subspace Methods, BIT 36, pp. 189-206, 1996
2. Bai, Z., Error Analysis of the Lanczos Algorithm for the Nonsymmetric Eigenvalue Problem, Math. Comp. 62, pp. 209-226, 1994
3. Baglama, J., Calvetti, D., Golub, G.H. and Reichel L., An Adaptively Preconditioned GMRES Algorithm, Preprint, Kent State University, 1996
4. Barth, T. and Manteuffel, T. A., Estimating the Spectrum of A Using the Roots of the Polynomials Associated with the QMR Iteration, Preprint, University of Colorado at Denver, 1995
5. Benzi M. and Tůma M., A sparse Approximate Inverse Preconditioner for Nonsymmetric Linear Systems, to appear in SIAM J. Sci. Comput.
6. Björck,Å. and Paige, C.C., Loss and Recapture of Orthogonality in the Modified Gram-Schmidt Algorithm, SIAM J. Matrix Anal. Appl. 13, pp. 176-190, 1992
7. Björck, Å., Numerics of Gram-Schmidt Orthogonalization, Linear Algebra Appl. 197, pp. 297-316, 1994
8. Brown, P.N., A Theoretical Comparison of the Arnoldi and GMRES Algorithms, SIAM J. Sci. Stat. Comput. 12, pp. 58-78, 1991
9. Campbell, S.L., Ipsen, I.C.F., Kelley, C.T. and Meyer C.D., GMRES and the Minimal Polynomial, Preprint, NC State University, 1995
10. Chatelin, F. and Frayssé, V., Qualitative Computing, COMETT Matari Programme, Orsay, 1993
11. Cullum, J. and Greenbaum, A., Relations Between Galerkin and Norm-Minimizing Iterative Methods for Solving Linear Systems, SIAM J. Matrix Anal. Appl. 17, pp. 223-248, 1996
12. Dahlquist, G., Golub, G.H. and Nash, S.G., Bounds for the Error in Linear Systems, in: Proc. Workshop on Semi-Infinite Programming, R. Hettich, ed., Springer, Berlin, pp. 154-172, 1978
13. Davis, T., Sparse Matrix Collection, NA Digest 94, Issue 42, 1994
14. Drkošová, J., Greenbaum, A., Rozložník, M. and Strakoš, Z., Numerical Stability of the GMRES Method, BIT 35, pp. 309-330, 1995
15. Druskin, V., Greenbaum, A. and Knizhnerman, L., Using Nonorthogonal Lanczos Vectors in the Computation of Matrix Functions, submitted to SIAM J. Num. Anal., 1996

16. Freund, R.W., Golub, G.H. and Nachtigal, N.N., Iterative Solution of Linear Systems, Acta Numerica 1, pp. 1-44, 1992
17. Freund. R., W. and Hochbruck, M., Gauss Quadratures Associated With The Arnoldi Process and The Lanczos Algorithm, in: Linear Algebra for Large Scale and Real-Time Applications, M.S. Moonen Ed., Kluwer Academic, Doordrecht, pp.377-380, 1993
18. Fischer, B. and Freund, R., On Adaptive Weighted Polynomial Preconditioning for Hermitian Positive Definite Matrices, submitted to SIAM J. Sci. Stat. Comput., 1994
19. Gautschi, W., A Survey of Gauss-Christoffel Quadrature Formulae, in: E.B. Christoffel- The Influence of His Work on Mathematics and the Physical Sciences, P.L. Bultzer and F. Fehér, Eds., Birkhauser, Boston, pp. 73-157, 1981
20. Golub, G.H. and van Loan, Ch., Matrix Computation (Second Edition), The Johns Hopkins Univ. Press, Baltimore, 1989
21. Golub, G.H. and Meurant, G., Matrices, Moments and Quadrature, Proceedings of the 15-th Dundee Conference, June 1993, D.F. Sciffeths and G.A. Watson, Eds., Longman Sci. Tech. Publ., 1994
22. Golub, G.H. and Strakoš, Z., Estimates in Quadratic Formulas, Numerical Algorithms 8, pp. 241-268, 1994
23. Greenbaum, A., Comparison of Splittings Used with the Conjugate Gradient Algorithm, Numer. Math. 33, pp. 181-194, 1979
24. Greenbaum, A. Behavior of Slightly Perturbed Lanczos and Conjugate Gradient Recurrences, Lin. Alg. Appl. 113, pp. 7-63, 1989
25. Greenbaum, A. and Strakoš, Z., Predicting the Behavior of Finite Precision Lanczos and Conjugate Gradient Computations, SIAM J. Matrix Anal. Appl. 13, pp. 121–137, 1992
26. Greenbaum, A.: A New Measure of Nonnormality, Preprint, Courant Institute of Mathematical Sciences, New York University, 1993.
27. Greenbaum, A. and Strakoš, Z., Matrices that Generate the Same Krylov Varieties, in: Recent Advances in Iterative Methods, G.H.Golub et al. (eds.), IMA Volumes in Maths and Its Applications, Springer, pp. 95–119, 1994
28. Greenbaum, A. and Gurvits, L., Max-Min Properties of Matrix Factor Norms, SIAM J. Sci. Comput., 15, pp. 348–358, 1994
29. Greenbaum, A., The Lanczos and Conjugate Gradient Algorithms in Finite Precision Arithmetic, In: Proceedins of the Lanczos Centennary Conference, SIAM, Philadelphia, 1994
30. Greenbaum, A., Trefethen, L.N., GMRES-CR and Arnoldi-Lanczos as Matrix Approximation Problems, SIAM J. Sci. Comput., 15, pp. 359–368, 1994
31. Greenbaum, A., Estimating the Attainable Accuracy of Recursively Computed Residual Methods, Tech. rep. TR95-1515, Cornell University, 1995
32. Greenbaum, A., Pták, V. and Strakoš, Z., Any Convergence Curve is Possible for GMRES, to appear in SIAM Matrix Anal. Appl. 17, 1996
33. Greenbaum, A., Rozložník, M. and Strakoš, Z., Numerical Behaviour of the Modified Gram-Schmidt GMRES Implementation (in preparation)
34. Hestenes, M.R. and Stiefel, E., Methods of Conjugate Gradients for Solving Linear Systems, J. Res. Nat. Bur. Stand. 49, pp. 409-436, 1952
35. Hochbruck, M., Lubich Ch., Error Analysis of the Krylov Methods in a Nutshell, submitted to SIAM J. Sci. Stat. Computing, 1995
36. Joubert, W., A Robust GMRES-Based Adaptive Polynomial Preconditioning Algorithm for Nonsymmetric Linear Systems, Preprint, University of Texas at Austin, 1993
37. Joubert, W., Faber, V., Knill, E. and Manteuffel,T., Minimal Residual Method Stronger than Polynomial Preconditioning, Proceedings of the Colorado Conf. on Iterative Method, Breckenridge, Colorado, 1994
38. Manteuffel, T. A. and Starke, G., On Hybrid Iterative Methods for Nonsymmetric Systems of Linear Equations, Preprint, University of Colorado at Denver, 1993

39. Morgan, R.B., A Restarted GMRES Method Augmented with Eigenvectors, SIAM J. Matrix Anal. Appl. 16, pp. 1154-1171, 1995
40. Nievanlinna, O., How Fast Can Iterative Methods Be?, in: Recent Advances in Iterative Methods, G.H.Golub et al. (eds.), IMA Volumes in Maths and Its Applications, Springer, pp. 135-149, 1994
41. Notay, Y., On the Convergence Rate of the Conjugate Gradients in the Presence of Rounding Errors, Numer Math. 65, pp. 301-317, 1993
42. Paige, C.C., Computational Variants of the Lanczos Method for the Eigenproblem, J. Inst. Maths. Applics 10, pp.373-381, 1972
43. Paige, C.C. and Saunders, M.A., Solution of Sparse Indefinite Systems of Linear Equations, SIAM J. Numer. Anal. 12, pp. 617-629, 1975
44. Paige, C.C., Error Analysis of the Lanczos Algorithm for Tridiagonalizing a Symmetric Matrix, J. Inst. Maths. Applics 18, pp. 341-349, 1976
45. Paige, C.C., Accuracy and Effectiveness of the Lanczos Algorithm for the Symmetric Eigenproblem, Linear Algebra Appl. 34, pp. 235-258, 1980
46. Paige, C.C., Parlett, B.N. and van der Vorst, H.A., Approximate Solutions and Eigenvalue Bounds from Krylov Subspaces, Numerical Lin. Alg. with Appl. 2, pp. 115-133, 1995
47. Parlett, B.N., The Symmetric Eigenvalue Problem, Prentice Hall, Englewood Cliffs, 1980
48. Parlett, B.N., Do We Fully Understand the Symmetric Lanczos Algorithm Yet?, in: Proceedins of the Lanczos Centennary Conference, SIAM, Philadelphia, pp.93-107, 1994
49. Pták, V., Krylov Sequences and Orthogonal Polynomials, Preprint, Institute of Computer Science, Academy of Sciences, Prague, 1996
50. Saad, Y., Schultz, M.H., GMRES: A Generalized Minimal Residual Algorithm for Solving Nonsymmetric Linear Systems, SIAM J. Sci. Stat. Comput. 7, pp. 856-869, 1986
51. van der Sluis, A. and van der Vorst, H.A., The Rate of Convergence of Conjugate Gradients, Numer. Math. 48, pp. 543-560, 1986
52. Stewart, G.W. and Sun, J., Matrix Perturbation Theory, Academic Press, Boston, 1990
53. Scott, D.S., How to Make the Lanczos Algorithm to Converge Slowly, Math. of Comput. 33, pp. 239-347, 1979
54. Strakos, Z., On the real convergence rate of the conjugate gradient method, Lin. Alg. Appl. 154-156, pp. 535-549, 1991
55. Strakos, Z. and Greenbaum, A., Open Questions in the Convergence Analysis of the Lanczos Process for the Real Symmetric Eigenvalue Problem, IMA Preprint Series 934, Univ. of Minnesota, 1992
56. Szego, G., Orthogonal Polynomials, AMS Colloq. Publ. 23, AMS, Providence, 1939
57. Toh K. Ch., GMRES vs. ideal GMRES, to appear in SIAM J. Matrix Analysis Appl.
58. Trefethen, L.N., Pseudospectra of matrices, in: Numerical Analysis 1991, D.F. Griffiths and G.A. Watson, Eds., Longman Sci. Tech. Publ., pp. 243-266, 1992
59. Trefethen, L.N., Pseudospectra of operators, ICIAM '95: Proceedings of the Third International Congress on Industrial and Applied Math., Akademie-Verlag, Berlin, 1996
60. van der Vorst, H.A. and Vuik, C., The Superlinear Convergence Behaviour of GMRES, J. Comp. Appl. Math. 48, pp. 327-341, 1993
61. Walker, H., Implementations of the GMRES Using Householder Transformations, SIAM J. Sci. Stat. Comp. 9, pp. 152-163, 1989
62. Watkins, D.S., Some Perspectives of The Eigenvalue Problem, SIAM Review 35, pp. 430-471, 1993

LOOK-AHEAD BLOCK-CG ALGORITHMS

C. G. BROYDEN
Università di Bologna,
Facoltà di Scienze MM.FF.NN.,
via Sacchi N.3,
47023 Cesena (Fo),
Italy

Abstract. In this paper necessary and sufficient conditions are obtained for look-ahead versions of the block-CG algorithm to be free from both serious and incurable breakdown. Following from this, unstable versions of the algorithm are identified and stable ones proposed.

1. Introduction

The block CG method [12] is a method of conjugate-gradient type for solving, in some sense, the equations

$$\mathbf{GX} = \mathbf{H} \tag{1}$$

where $\mathbf{G} \in \mathbf{R}^{n \times n}$ is a real symmetric (but not necessarily positive-definite or even nonsingular) matrix, $\mathbf{H} \in \mathbf{R}^{n \times r}$ is a real matrix of right-hand sides and $\mathbf{X} \in \mathbf{R}^{n \times r}$ is the solution to be determined. Although the original motivation for the algorithm was the desire to solve simultaneously multiple linear systems, or to solve a single linear system in which the matrix has several separated eigenvalues or is not easily accessed on a computer, a major reason for the algorithm's importance is that it includes so many apparently unrelated methods as special cases [4], [2]. Thus any theoretical results obtained, or computational variations derived, for the algorithm can be applied immediately to to a wide variety of other methods, simplifying and unifying the theory or simultaneously modifying a wide range of algorithms.

The key to the algorithm is the construction of a sequence of matrices $\{\mathbf{P}_i\}$ having the same dimensions as $\mathbf{X}$ and $\mathbf{H}$ such that $\mathbf{P}_j^T \mathbf{G} \mathbf{P}_k = \mathbf{0}$ for

G. Winter Althaus and E. Spedicato (eds.), Algorithms for Large Scale Linear Algebraic Systems, 197–215.

$j \neq k$. It does this, *formally*, by computing a sequence of oblique projection matrices $\mathbf{Q}_i$, $i = 0, 1, ...$, and computing the next conjugate matrix, $\mathbf{P}_{i+1}$, by

$$\mathbf{P}_{i+1} = \mathbf{Q}_i^T \mathbf{W}_{i+1} \tag{2}$$

where $\mathbf{Q}_i$ is chosen so that $\mathbf{P}_{i+1}$ is conjugate to all the matrices $\mathbf{P}_j$ already computed and $\mathbf{W}_{i+1}$ is a matrix which may be chosen arbitrarily. In fact if $\mathbf{Q}_i$ is defined by

$$\mathbf{Q}_i = \mathbf{I} - \mathbf{G}\overline{\mathbf{P}}_i \overline{\mathbf{D}}_i^{-1} \overline{\mathbf{P}}_i^T \tag{3}$$

where

$$\overline{\mathbf{P}}_i = [\ \mathbf{P}_1 \quad \mathbf{P}_2 \quad ..., \quad \mathbf{P}_i\ ] \tag{4}$$

and

$$\overline{\mathbf{D}}_i = \overline{\mathbf{P}}_i^T \mathbf{G} \overline{\mathbf{P}}_i \tag{5}$$

it is evident that if $\mathbf{P}_{i+1}$ is computed by equation (2) then $\overline{\mathbf{P}}_i^T \mathbf{G}\mathbf{P}_{i+1} = \mathbf{0}$ for any choice of $\mathbf{W}_{i+1}$ since, trivially, $\mathbf{Q}_i \mathbf{G} \overline{\mathbf{P}}_i = \mathbf{0}$. Now it follows from equation (4) and the conjugacy of the matrices $\mathbf{P}_j$ that $\overline{\mathbf{D}}_i$ is a block diagonal matrix whose diagonal blocks $\mathbf{D}_j$ are given by

$$\mathbf{D}_j = \mathbf{P}_j^T \mathbf{G} \mathbf{P}_j \tag{6}$$

Thus $\overline{\mathbf{D}}_i$ is nonsingular and the algorithm properly defined if and only if the matrices $\mathbf{D}_j$ are all nonsingular. Moreover, the block-diagonality of $\overline{\mathbf{D}}_i$ permits $\mathbf{Q}_i$ to be expressed alternatively by

$$\mathbf{Q}_i = \mathbf{I} - \mathbf{G} \sum_{j=1}^{i} \mathbf{P}_j \mathbf{D}_j^{-1} \mathbf{P}_j^T \tag{7}$$

and it is this form that is implicitly used in the following formal description.

Algorithm 1

1. Set $i = 1$, $\mathbf{Q}_0 = \mathbf{I}$ and select $\mathbf{X}_1$. Compute $\mathbf{F}_1 = \mathbf{G}\mathbf{X}_1 - \mathbf{H}$ and determine $\mathbf{P}_1$.
2. compute $\mathbf{X}_{i+1} = \mathbf{X}_i - \mathbf{P}_i \mathbf{D}_i^{-1} \mathbf{P}_i^T \mathbf{F}_i$
3. compute $\mathbf{F}_{i+1} = \mathbf{G}\mathbf{X}_{i+1} - \mathbf{H}$
4. compute $\mathbf{Q}_i = \mathbf{Q}_{i-1} - \mathbf{G}\mathbf{P}_i \mathbf{D}_i^{-1} \mathbf{P}_i^T$
5. compute $\mathbf{P}_{i+1} = \mathbf{Q}_i^T \mathbf{W}_{i+1}$ for some $\mathbf{W}_{i+1}$
6. set $i = i + 1$,
7. return to step (2).

Now this algorithm suffices as long as the matrices $\mathbf{D}_j$ are nonsingular. If, though, $\mathbf{D}_j$ is singular for some j then, depending on the cause of the singularity, different remedial actions have to be taken.

The two basic causes of singularity of the matrices $\mathbf{D}_j$ are either that $\mathbf{P}_j$ suffers a loss of rank or that $\mathbf{G}$ is indefinite, when $\mathbf{D}_j$ may be singular even though the columns of $\mathbf{P}_j$ are linearly independent. The loss of rank of $\mathbf{P}_j$ is not particularly disastrous (but see below) and can happen even if $\mathbf{G}$ is definite. In the context of the QMR algorithm Freund and Nachtigal [6] refer to its occurrence as *regular termination* and the problems that it poses have been discussed by O'Leary [12] and, more recently, by Nikishin and Yeremin [11]. The last two authors describe a scheme by which, if $\mathbf{P}_j$ becomes rank-deficient, its size is reduced until it is equal to its rank. This rather neat inversion of the expected does, apparently, give good results. We may therefore assume that the problems due to the rank-deficiency of $\mathbf{P}_j$ are well on the way to being solved. We shall also assume in the remainder of the discussion that $\mathbf{P}_j$ always has full rank.

The other cause of singularity of $\mathbf{D}_j$ occurs when $\mathbf{P}_j$ has full rank and $\mathbf{G}$ is indefinite, and is referred to as *serious breakdown.* There is no guarantee in this case that reducing the number of columns of $\mathbf{P}_j$ would solve the problem since it is quite possible that if $\mathbf{G}$ is indefinite then $\mathbf{p}^T\mathbf{G}\mathbf{p} = 0$ where $\mathbf{p}$ is a vector, i.e. a single column. Another strategy therefore has to be found and the following rather obvious one has been suggested. If $\mathbf{Q}_i$ cannot be computed because $\mathbf{P}_i^T\mathbf{G}\mathbf{P}_i$ is singular then the matrices $\mathbf{P}_{i+j}$, $j = 1, 2, ...$, should be computed using the *old* oblique projector $\mathbf{Q}_{i-1}$ instead of an updated one. This at least ensures that the matrices $\mathbf{P}_{i+j}$, $j = 0, 1, ...$, are conjugate to the matrices $\mathbf{P}_k$, $k = 1, 2, ..., i-1$, even though they will not be conjugate to each other. However this process cannot be allowed to continue indefinitely since the algorithm functions by constructing a sequence of conjugate matrices, the linear-algebraic costs of not doing so being prohibitive. The obvious way forward is to keep adjoining the matrices $\mathbf{P}_{i+j}$ to the matrix $\overline{\mathbf{P}}_i$ just as in Algorithm 1 in the hope that, sooner or later, $\mathbf{Q}_i$ as defined by equation (3) will become computable. If it does then it may be calculated and the algorithm can proceed happily on its way. If it does not then we have the problem of *incurable breakdown* for which, as the name implies, there is no known solution.

From the above description of the algorithm it might be thought that the name *deferred* (or *delayed*) *update method* might be appropriate since the updating of $\mathbf{Q}_{i-1}$ is deferred until it is possible to do so. It is, though, known as the *look-ahead Lanczos method* [6], [13] for some versions of the algorithm while for others it has been referred to as the *method of hyperbolic pairs* [10]. Other authors again [1] refer to their methods as being *breakdown-free.* However not all these versions or near-versions of the block

CG algorithm exhibit the same behaviour. The method of hyperbolic pairs, for example, does not suffer from incurable breakdown, neither does the method of Hegedüs [8]. This fate is reserved for the Lanczos-type methods alone [6], [7], and [13].

In this paper we give a general deferred update (or look-ahead) algorithm for the block CG algorithm, of which the others cited above are special cases. We relate the crucial matrices $\mathbf{D}_j$ to certain matrices derived from the underlying Krylov sequence generated implicitly by the algorithm and obtain necessary and sufficient conditions for the $\mathbf{D}_j$'s to be non-singular. We show that the behaviour of these methods is determined by two matrices which effectively define the algorithm and finally explain why some versions can experience incurable breakdown while others are immune to it.

2. Notation

In this section we define certain matrices that are used in the course of the discussion. We also explain the notation, which is dictated by the need to accomodate matrices that on one hand are best viewed as elements in a single sequence $\{\mathbf{P}_h\}$ and on the other are more appropriately regarded as submatrices of a single sequence of larger matrices $\left\{\widetilde{\mathbf{P}}_i\right\}$. This is because, in Algorithm 1, the matrices $\mathbf{P}_i$ computed in Step 3 are the same as those used in Step 2 to compute $\mathbf{Q}_i$ whereas in the look-ahead version the computed matrices $\mathbf{P}_h$ may only be submatrices of the matrices $\widetilde{\mathbf{P}}_i$ used in the calculation of $\widetilde{\mathbf{Q}}_i$. The following formal description should make matters clear.

Algorithm 2

1. Set $i = 1$, $h = 1$, $\widetilde{\mathbf{Q}}_0 = \mathbf{I}$ and select $\mathbf{X}_1$. Compute $\mathbf{F}_1 = \mathbf{G}\mathbf{X}_1 - \mathbf{H}$ and determine $\mathbf{P}_1$
2. Set $\widetilde{\mathbf{P}}_i = \mathbf{P}_h$
3. while $\widetilde{\mathbf{P}}_i^T\mathbf{G}\widetilde{\mathbf{P}}_i$ is singular
   (a) compute $\mathbf{P}_{h+1} = \widetilde{\mathbf{Q}}_{i-1}^T\mathbf{W}_{h+1}$ for some $\mathbf{W}_{h+1}$
   (b) set $\widetilde{\mathbf{P}}_i = \begin{bmatrix} \widetilde{\mathbf{P}}_i & \mathbf{P}_{h+1} \end{bmatrix}$
   (c) set $h = h + 1$
4. endwhile
5. compute $\mathbf{X}_{i+1} = \mathbf{X}_i - \widetilde{\mathbf{P}}_i(\widetilde{\mathbf{P}}_i^T\mathbf{G}\widetilde{\mathbf{P}}_i)^{-1}\widetilde{\mathbf{P}}_i^T\mathbf{F}_i$
6. compute $\mathbf{F}_{i+1} = \mathbf{G}\mathbf{X}_{i+1} - \mathbf{H}$
7. compute $\widetilde{\mathbf{Q}}_i = \widetilde{\mathbf{Q}}_{i-1} - \mathbf{G}\widetilde{\mathbf{P}}_i(\widetilde{\mathbf{P}}_i^T\mathbf{G}\widetilde{\mathbf{P}}_i)^{-1}\widetilde{\mathbf{P}}_i^T$

8. compute $\mathbf{P}_{h+1} = \tilde{\mathbf{Q}}_i^T \mathbf{W}_{h+1}$
9. set $i = i + 1$, $h = h + 1$
10. return to step (2).

It follows from the above description that if $\mathbf{P}_h^T \mathbf{G} \mathbf{P}_h$ is nonsingular for $\forall h$ the while-loop is never entered and the algorithm reduces to Algorithm 1. If, on the other hand, $\mathbf{P}_h^T \mathbf{G} \mathbf{P}_h$ is singular for some h then the while-loop will be activated and the resulting matrix computed, $\tilde{\mathbf{P}}_i$, will have dimensions determined by the number of times the while-loop is traversed. Suppose that the loop is traversed $\lambda_i - 1$ times. Then $\tilde{\mathbf{P}}_i$ will consist of λ_i blocks each consisting of one of the matrices $\mathbf{P}_h$ for appropriate values of h. In fact, if we define $\sigma_0 = 0$ and σ_i, $i = 1, 2, ...,$ by

$$\sigma_i = \sum_{j=1}^{i} \lambda_j \tag{8}$$

it is readily seen that

$$\tilde{\mathbf{P}}_i = [\ \mathbf{P}_{\sigma_{i-1}+1} \quad \mathbf{P}_{\sigma_{i-1}+2} \quad ... \quad \mathbf{P}_{\sigma_i}\ ] \tag{9}$$

By analogy with Algorithm 1 the algorithm generates a sequence of matrices $\tilde{\mathbf{P}}_i$ that are mutually conjugate and for which the the matrices $\tilde{\mathbf{P}}_j^T \mathbf{G} \tilde{\mathbf{P}}_j$ are nonsingular for $\forall j$, and it does this by implicitly computing the projectors $\tilde{\mathbf{Q}}_j$, $1 \le j \le i$, by

$$\tilde{\mathbf{Q}}_j = \mathbf{I} - \mathbf{G} \overline{\mathbf{P}}_{\sigma_j} \overline{\mathbf{D}}_{\sigma_j}^{-1} \overline{\mathbf{P}}_{\sigma_j}^T \tag{10}$$

Now if $\sigma_i < h \le \sigma_{i+1}$ we may define $\tilde{\mathbf{P}}_{R_{i+1}}$ and $\tilde{\mathbf{D}}_{R_{i+1}}$ by

$$\overline{\mathbf{P}}_h = \left[\ \overline{\mathbf{P}}_{\sigma_i} \quad \tilde{\mathbf{P}}_{R_{i+1}}\ \right] \tag{11}$$

and

$$\tilde{\mathbf{D}}_{R_{i+1}} = \tilde{\mathbf{P}}_{R_{i+1}}^T \mathbf{G} \tilde{\mathbf{P}}_{R_{i+1}} \tag{12}$$

If $\tilde{\mathbf{D}}_{R_{i+1}}$ is nonsingular then $\tilde{\mathbf{P}}_{R_{i+1}} = \tilde{\mathbf{P}}_{i+1}$ and $h = \sigma_{i+1}$. If $\tilde{\mathbf{D}}_{R_{i+1}}$ is singular then $\tilde{\mathbf{P}}_{R_{i+1}}$ is a submatrix of $\tilde{\mathbf{P}}_{i+1}$ and $h < \sigma_{i+1}$. It follows from Steps (3a) and (8) of Algorithm 2 and the above definitions that, for $\sigma_i < h \le \sigma_{i+1}$, $\mathbf{P}_h$ is computed by

$$\mathbf{P}_h = \tilde{\mathbf{Q}}_i^T \mathbf{W}_h \tag{13}$$

so that

$$\tilde{\mathbf{P}}_{R_{i+1}} = \tilde{\mathbf{Q}}_i^T\ [\ \mathbf{W}_{\sigma_i+1} \quad \mathbf{W}_{\sigma_i+2} \quad ... \quad \mathbf{W}_h\ ] \tag{14}$$

for any h satisfying the above inequality. Clearly, since $\tilde{\mathbf{Q}}_i$ is idempotent,

$$\tilde{\mathbf{P}}_{R_{i+1}} = \tilde{\mathbf{Q}}_i^T \tilde{\mathbf{P}}_{R_{i+1}} \tag{15}$$

As soon as $\tilde{\mathbf{D}}_{R_{i+1}}$ is found to be nonsingular, $\tilde{\mathbf{P}}_{i+1}$ is set equal to $\tilde{\mathbf{P}}_{R_{i+1}}$ and $\tilde{\mathbf{Q}}_i$ is updated. The identification of necessary and sufficient conditions for the nonsingularity of $\tilde{\mathbf{D}}_{R_{i+1}}$ forms one of the principal objectives of this paper since they determine the feasibility or otherwise of the look-ahead algorithm.

In addition to the matrices defined above we shall also need matrices based on the Krylov sequence $\left\{(\mathbf{KG})^{i-1}\mathbf{P}_1\right\}$, $i = 1, 2, ..., h$, where $\mathbf{K}$ is a symmetric matrix (the *preconditioning matrix*) to be discussed below. By analogy with equation (4) (with $i = h$) we define

$$\overline{\mathbf{V}}_h = \left[\begin{array}{ccccc} \mathbf{P}_1 & \mathbf{KGP}_1 & (\mathbf{KG})^2\mathbf{P}_1 & \ldots & (\mathbf{KG})^{h-1}\mathbf{P}_1 \end{array}\right] \tag{16}$$

and, for $\sigma_i < h \leq \sigma_{i+1}$, equation (11) suggests

$$\overline{\mathbf{V}}_h = \left[\begin{array}{cc} \overline{\mathbf{V}}_{\sigma_i} & \tilde{\mathbf{V}}_{R_{i+1}} \end{array}\right] \tag{17}$$

where $\tilde{\mathbf{V}}_{R_{i+1}}$ has the same dimensions as $\tilde{\mathbf{P}}_{R_{i+1}}$.

We finally anticipate a result that will be proved in the next section. For appropriate choices of $\mathbf{W}_h$ we will see that

$$\overline{\mathbf{P}}_h = \overline{\mathbf{V}}_h \overline{\mathbf{U}}_h \tag{18}$$

where $\overline{\mathbf{U}}_h = [\mathbf{U}_{jk}]$ is a block upper-triangular matrix and where $\mathbf{U}_{jk} \in \mathbf{R}^{r \times r}$, $1 \leq j, k \leq h$. The integer r is the number of columns of the matrices $\mathbf{P}_j$. This result, somewhat surprisingly, is independent of the values of λ_i. An alternative form of the equation, derived from equations (11), (17) and (18), is

$$\left[\begin{array}{cc} \overline{\mathbf{P}}_{\sigma_i} & \tilde{\mathbf{P}}_{R_{i+1}} \end{array}\right] = \left[\begin{array}{cc} \overline{\mathbf{V}}_{\sigma_i} & \tilde{\mathbf{V}}_{R_{i+1}} \end{array}\right] \left[\begin{array}{cc} \overline{\mathbf{U}}_{\sigma_i} & \mathbf{Y}_{i+1} \\ \mathbf{0} & \mathbf{Z}_{i+1} \end{array}\right] \tag{19}$$

It is these last two equations that justify the term *Krylov Sequence* when applied to conjugate gradient methods.

3. The Lanczos version

The Lanczos version of the block-CG algorithm is defined by choosing $\mathbf{P}_1$ arbitrarily (but of full rank) and $\mathbf{W}_h$, $h = 2, 3, ...,$ by

$$\mathbf{W}_h = \mathbf{KGP}_{h-1} \tag{20}$$

so that from equation (13) we have, for $\sigma_i \leq h < \sigma_{i+1}$,

$$\mathbf{P}_{h+1} = \tilde{\mathbf{Q}}_i^T \mathbf{KGP}_h. \tag{21}$$

The arbitrary symmetric matrix $\mathbf{K}$ (the preconditioning matrix) is usually chosen to improve the convergence of the algorithm. It also has a powerful effect on the stability of the algorithm and it is this aspect of its influence that we consider here. We begin this section with two lemmas, the first a technical one that permits alternative expressions for the matrices $\tilde{\mathbf{Q}}_j$ while the second establishes the fundamental equation (18).

Lemma 1

The replacement of $\overline{\mathbf{P}}_{\sigma_j}$ by $\overline{\mathbf{P}}_{\sigma_j}\mathbf{M}_j$ in equation (10), where $\mathbf{M}_j$ is *any* nonsingular matrix, leaves $\tilde{\mathbf{Q}}_j$ unaltered.

Proof

Straightforward, by substitution ∎

Lemma 2

Let the sequence $\{\mathbf{P}_h\}$ be computed by the Lanczos version of Algorithm 2. Then

$$\mathbf{P}_h = (\mathbf{KG})^{h-1}\mathbf{P}_1 + \sum_{k=1}^{h-1} (\mathbf{KG})^{k-1}\mathbf{P}_1\mathbf{U}_{kh} \tag{22}$$

for some matrices $\mathbf{U}_{kh}$.

Proof

By induction. The Lemma is clearly true for $h = 1$. We show that if it is true for $1 \leq h \leq s$ then it is also true for $h = s + 1$.

Assume that equation (22) holds for $1 \leq h \leq s$ and that $\sigma_i \leq s < \sigma_{i+1}$ (this is always true for some i even if $h = 1$ and $\sigma_1 > 1$ since $\sigma_0 = 0$). Then $\mathbf{P}_{s+1}$ is computed from equation (21) so that, from equation (10) with $j = i$ and equation (4),

$$\mathbf{P}_{s+1} = \mathbf{KGP}_s + \sum_{k=1}^{\sigma_i} \mathbf{P}_k\mathbf{Y}_{ks}$$

for some matrices $\mathbf{Y}_{ks}$. Since $\sigma_i \leq s$ it follows from the induction hypothesis that we can substitute for both $\mathbf{P}_s$ and the matrices $\mathbf{P}_k$ the values given by equation (22). This yields, on collecting terms, equation (22) with h equal to $s + 1$, completing the proof ∎

Not only does Lemma 2 validate equation (18) but it also shows that the diagonal blocks of $\mathbf{U}_{jj}$ of $\overline{\mathbf{U}}_h$ are unit matrices. $\overline{\mathbf{U}}_h$ is thus nonsingular for $\forall h$, an essential result for what follows.

Now the aim of this section of the paper is to obtain an expression for $\widetilde{\mathbf{D}}_{R_{i+1}}$ so our next step is to derive an expression for $\widetilde{\mathbf{P}}_{R_{i+1}}$.

Lemma 3

Let $\widetilde{\mathbf{P}}_{R_{i+1}}$ be computed by the Lanczos version of Algorithm 2. Then

$$\widetilde{\mathbf{P}}_{R_{i+1}} = \widetilde{\mathbf{Q}}_i^T \widetilde{\mathbf{V}}_{R_{i+1}} \mathbf{Z}_{i+1} \tag{23}$$

for some nonsingular matrix $\mathbf{Z}_{i+1}$.

Proof

From Lemma 1, equation (18) with $h = \sigma_i$ and equation (10) we have

$$\widetilde{\mathbf{Q}}_i = \mathbf{I} - \mathbf{G}\overline{\mathbf{V}}_{\sigma_i}\overline{\mathbf{D}}_{\sigma_i}^{-1}\overline{\mathbf{V}}_{\sigma_i}^T \tag{24}$$

so that $\widetilde{\mathbf{Q}}_i^T \overline{\mathbf{V}}_{\sigma_i} = \mathbf{0}$ while from equation (19) we have

$$\widetilde{\mathbf{P}}_{R_{i+1}} = \overline{\mathbf{V}}_{\sigma_i}\mathbf{Y}_{i+1} + \widetilde{\mathbf{V}}_{R_{i+1}}\mathbf{Z}_{i+1} \tag{25}$$

The result follows immediately from equations (15) and (25) and the nonsingularity of $\overline{\mathbf{U}}_h$, which guarantees the nonsingularity of $\mathbf{Z}_{i+1}$∎

We now derive the necessary and sufficient condition for $\widetilde{\mathbf{D}}_{R_{i+1}}$ to be nonsingular.

Theorem 1

Let $\widetilde{\mathbf{P}}_{R_{i+1}}$ be calculated by the Lanczos version of Algorithm 2 and let $\mathbf{S}_h$ be defined by

$$\mathbf{S}_h = \overline{\mathbf{V}}_h^T \mathbf{G} \overline{\mathbf{V}}_h \tag{26}$$

Then, for $\sigma_i < h \leq \sigma_{i+1}$, $\widetilde{\mathbf{D}}_{R_{i+1}}$ is nonsingular if and only if $\mathbf{S}_h$ is nonsingular.

Proof

From Lemma 3, $\widetilde{\mathbf{D}}_{R_{i+1}} = (\widetilde{\mathbf{Q}}_i^T \widetilde{\mathbf{V}}_{R_{i+1}} \mathbf{Z}_{i+1})^T \mathbf{G} \widetilde{\mathbf{Q}}_i^T \widetilde{\mathbf{V}}_{R_{i+1}} \mathbf{Z}_{i+1}$ so from equation (10), the idempotency of $\widetilde{\mathbf{Q}}_i$ and the nonsingularity of $\mathbf{Z}_{i+1}$, $\widetilde{\mathbf{D}}_{R_{i+1}}$ is nonsingular if and only if $\widetilde{\mathbf{V}}_{R_{i+1}}^T \widetilde{\mathbf{Q}}_i \mathbf{G} \widetilde{\mathbf{V}}_{R_{i+1}}$ is nonsingular. Now from equation (24)

$$\widetilde{\mathbf{V}}_{R_{i+1}}^T \widetilde{\mathbf{Q}}_i \mathbf{G} \widetilde{\mathbf{V}}_{R_{i+1}} = \widetilde{\mathbf{V}}_{R_{i+1}}^T \mathbf{G} \widetilde{\mathbf{V}}_{R_{i+1}} - \widetilde{\mathbf{V}}_{R_{i+1}}^T \mathbf{G} \overline{\mathbf{V}}_{\sigma_i} \overline{\mathbf{D}}_{\sigma_i}^{-1} \overline{\mathbf{V}}_{\sigma_i}^T \mathbf{G} \widetilde{\mathbf{V}}_{R_{i+1}} \tag{27}$$

and from equations (17) and (26),

$$\mathbf{S}_h = \begin{bmatrix} \overline{\mathbf{V}}_{\sigma_i}^T \\ \tilde{\mathbf{V}}_{R_{i+1}}^T \end{bmatrix} \mathbf{G} \begin{bmatrix} \overline{\mathbf{V}}_{\sigma_i} & \tilde{\mathbf{V}}_{R_{i+1}} \end{bmatrix}$$

It is now readily verified that the expression on the right-hand side of equation (27) is just the Schur complement of $\overline{\mathbf{V}}_{\sigma_i}^T \mathbf{G} \overline{\mathbf{V}}_{\sigma_i}$ in $\mathbf{S}_h$. This is nonsingular, from the properties of Schur complements (see e.g. [9]), if and only if $\mathbf{S}_h$ is nonsingular, and the result follows ∎

4. The Hestenes-Stiefel version

In the Lanczos version of the algorithm the generators $\{\mathbf{W}_h\}$ are given by equation (20) but for the Hestenes-Stiefel (HS) version they are given by $\mathbf{W}_h = \mathbf{K}\mathbf{F}_h$. The HS choice of generators removes a degree of freedom since $\mathbf{P}_1$ may no longer be chosen independently of $\mathbf{X}_1$ but must now be set equal to $\mathbf{K}\mathbf{F}_1$. Another difference is that since a new value of $\mathbf{F}$ is computed only after emerging from the while-loop the above generators can only be used in Step (8). In Step (3a) the old generators given by equation (20) must still be used.

From Step (5) of the algorithm we see that

$$\mathbf{X}_{i+1} = \mathbf{X}_i + \widetilde{\mathbf{P}}_i \widetilde{\mathbf{M}}_i \tag{28}$$

where

$$\widetilde{\mathbf{M}}_i = -(\widetilde{\mathbf{P}}_i^T \mathbf{G} \widetilde{\mathbf{P}}_i)^{-1} \widetilde{\mathbf{P}}_i^T \mathbf{F}_i \tag{29}$$

so that from Step (6) and equation (28) we obtain

$$\mathbf{F}_{i+1} = \mathbf{F}_1 + \mathbf{G} \sum_{j=1}^{i} \widetilde{\mathbf{P}}_j \widetilde{\mathbf{M}}_j$$

It is, however, convenient to re-write this equation. Since $\widetilde{\mathbf{M}}_i \in \mathbf{R}^{r\lambda_i \times r}$ it may, by analogy with equation (9), be written

$$\widetilde{\mathbf{M}}_i = \begin{bmatrix} \mathbf{M}_{\sigma_{i-1}+1}^T & \mathbf{M}_{\sigma_{i-1}+2}^T & \dots & \mathbf{M}_{\sigma_i}^T \end{bmatrix}^T \tag{30}$$

and from this, equation (9) and the previous equation we have

$$\mathbf{F}_{i+1} = \mathbf{F}_1 + \mathbf{G} \sum_{k=1}^{\sigma_i} \mathbf{P}_k \mathbf{M}_k \tag{31}$$

We can now prove Lemma 4, the HS version of Lemma 2.

Lemma 4

Let the sequence $\{\mathbf{P}_h\}$ be computed by the HS version of Algorithm 2. Then, for $\sigma_i + 1 \leq h \leq \sigma_{i+1}$,

$$\mathbf{P}_h = (\mathbf{KG})^{h-1}\mathbf{P}_1\mathbf{L}_i + \sum_{k=1}^{h-1}(\mathbf{KG})^{k-1}\mathbf{P}_1\mathbf{U}_{kh} \tag{32}$$

for some matrices $\mathbf{U}_{kh}$, where $\mathbf{L}_0 = \mathbf{I}$ and

$$\mathbf{L}_i = \mathbf{M}_{\sigma_1}\mathbf{M}_{\sigma_2}...\mathbf{M}_{\sigma_i} \tag{33}$$

Proof

By induction. The Lemma is clearly true for $h = 1$. We show that if it is true for $1 \leq h \leq s$ then it is also true for $h = s + 1$.

There are two cases to consider.

(a) $\quad h = \sigma_i + 1$

Assume that equation (32) holds for $1 \leq h \leq \sigma_i$. For this case $\mathbf{P}_{\sigma_i+1} = \tilde{\mathbf{Q}}_i^T\mathbf{KF}_{i+1}$ so that, from equation (31) and the fact that $\mathbf{P}_1 = \mathbf{KF}_1$ we get, with a slight rearrangement of the sum,

$$\mathbf{P}_{\sigma_i+1} = \tilde{\mathbf{Q}}_i^T\left[\mathbf{P}_1 + \mathbf{KGP}_{\sigma_i}\mathbf{M}_{\sigma_i} + \mathbf{KG}\sum_{k=1}^{\sigma_i-1}\mathbf{P}_k\mathbf{M}_k\right]$$

so that, from the induction hypothesis,

$$\mathbf{P}_{\sigma_i+1} = \tilde{\mathbf{Q}}_i^T\left[(\mathbf{KG})^{\sigma_i}\mathbf{P}_1\mathbf{L}_{i-1}\mathbf{M}_{\sigma_i} + \sum_{k=1}^{\sigma_i}(\mathbf{KG})^{k-1}\mathbf{P}_1\mathbf{Y}_{k,\sigma_i+1}\right]$$

for some matrices $\mathbf{Y}_{k,\sigma_i+1}$. Expanding $\tilde{\mathbf{Q}}_i^T$ by equation (10) with $j = i$ and again applying the induction hypothesis yields

$$\mathbf{P}_{\sigma_i+1} = (\mathbf{KG})^{\sigma_i}\mathbf{P}_1\mathbf{L}_i + \sum_{k=1}^{\sigma_i}(\mathbf{KG})^{k-1}\mathbf{P}_1\mathbf{U}_{k,\sigma_i+1}$$

for some matrices $\mathbf{U}_{k,\sigma_i+1}$ and where $\mathbf{L}_i = \mathbf{L}_{i-1}\mathbf{M}_{\sigma_i}$. The Lemma is thus true for $h = \sigma_i + 1$.

Case (b) $\quad \sigma_i + 1 < h \leq \sigma_{i+1}$

Assume now the Lemma to be true for $1 \leq h \leq s$ where $\sigma_i + 1 < s \leq \sigma_{i+1}$. It has just been shown to be true for $h = \sigma_i+1$ and since the generators for $\sigma_i + 1 < s \leq \sigma_{i+1}$ are $\mathbf{KGP}_{s-1}$ the identical argument used to prove Lemma 2 establishes the validity of equation (32) for $\sigma_i + 1 < h \leq \sigma_{i+1}$ with no further change to $\mathbf{L}_i$. This completes the proof ▪

5. Some implications

It is convenient now, before proceeding further, to examine the implications of the theory so far derived and we begin by considering Case (a) of Lemma 4 in which it was assumed that $h = \sigma_i + 1$ for some i. This can always be true even if $h = 1$ since then $i = 0$, $\sigma_i = 0$ and $\widetilde{\mathbf{Q}}_0 = \mathbf{I}$. If it is also true for some $i > 0$ it implies that $\widetilde{\mathbf{Q}}_i$ has been successfully computed so that $\overline{\mathbf{P}}_{\sigma_i}^T \mathbf{G} \overline{\mathbf{P}}_{\sigma_i}$ is nonsingular and $\overline{\mathbf{P}}_{\sigma_i}$ therefore has full rank. This further implies, from equation (18) with $h = \sigma_i$, that $\overline{\mathbf{V}}_{\sigma_i}$ has full rank and that $\overline{\mathbf{U}}_{\sigma_i}$ is nonsingular. These last two statements are manifestly true for the Lanczos version of the algorithm, from Theorem 1 and the definition of $\mathbf{S}_{\sigma_i}$ for the first one and from the fact that all the diagonal elements of $\overline{\mathbf{U}}_{\sigma_i}$ are equal to the identity matrix for the second.

For the HS version of the algorithm, though, matters are slightly different. The diagonal blocks $\mathbf{U}_{kk}$ of $\overline{\mathbf{U}}_{\sigma_i}$ are, from Lemma 4, equal to $\mathbf{L}_{j-1}$ for $\sigma_{j-1} < k \leq \sigma_j$, $1 \leq j \leq i$, so that the fact that $\overline{\mathbf{U}}_{\sigma_i}$ is nonsingular implies that $\mathbf{L}_{j-1}$ is nonsingular for $1 \leq j \leq i$. Now this is trivially true for $\mathbf{L}_0$ which, by definition, is equal to the identity but for $2 \leq j \leq i$ the nonsingularity of $\mathbf{L}_{j-1}$ implies in turn, from equation (33), that the matrices $\mathbf{M}_{\sigma_j}$ are nonsingular for $1 \leq j \leq i - 1$.

For $\sigma_i < h \leq \sigma_{i+1}$ we find, from Lemma 4, that $\widetilde{\mathbf{P}}_{R_{i+1}}$ satisfies equation (23) but where $\mathbf{Z}_{i+1}$ is block upper-triangular with diagonal blocks now equal to $\mathbf{L}_i$. Thus $\mathbf{Z}_{i+1}$ is nonsingular if and only if $\mathbf{M}_{\sigma_i}$ is nonsingular so that the HS version of the algorithm enjoys an extra dimension of uncertainty compared to the Lanczos one. Regardless though of the singularity or otherwise of $\mathbf{Z}_{i+1}$ an argument identical to that used to prove Lemma 3 shows that $\widetilde{\mathbf{P}}_{R_{i+1}}$ satisfies equation (23) also in the HS case, thus providing the same simple expression for this vital matrix.

Assume now that $\mathbf{Z}_{i+1}$ is nonsingular, i.e. that $\mathbf{M}_{\sigma_i}$ is nonsingular. An identical argument to that used to establish Theorem 1 now yields the same result for the HS version of the algorithm. We state this formally as

Theorem 2

Let $\widetilde{\mathbf{P}}_{R_{i+1}}$ be calculated by the HS version of Algorithm 2, let $\mathbf{S}_h$ be defined by equation (26) and let $\mathbf{M}_{\sigma_i}$ be nonsingular. Then, for $\sigma_i < h \leq \sigma_{i+1}$, $\widetilde{\mathbf{D}}_{R_{i+1}}$ is nonsingular if and only if $\mathbf{S}_h$ is nonsingular.

We see from the above theorems that the matrices that govern the behaviour of Algorithm 2 are $\mathbf{S}_h$ and, additionally for the HS version, $\mathbf{M}_{\sigma_i}$. From equation (26), $\mathbf{S}_h$ is singular if $\overline{\mathbf{V}}_h$ is rank-deficient and can be singular, even if if $\overline{\mathbf{V}}_h$ is not rank-deficient, if $\mathbf{G}$ is indefinite. For the Lanczos case, since $\overline{\mathbf{U}}_h$ is always nonsingular, $\overline{\mathbf{P}}_h$ is rank-deficient if and only if $\overline{\mathbf{V}}_h$ is rank-deficient and it is the advent of rank-deficiency in $\overline{\mathbf{V}}_h$ that leads to the regular termination of the algorithm. This is also the circumstance that is dealt with by reducing r as described by Nikishin and Yeremin [11] and is not regarded as unduly serious. Clearly, if $\mathbf{S}_h$ is singular because $\overline{\mathbf{V}}_h$ is rank-deficient then $\mathbf{S}_{h+j}$ is singular for $\forall j \geq 0$ and Algorithm 2 will execute the while-loop for ever. Look-ahead is not designed to cope with this so we will assume either that it does not occur or that the algorithm may be suitably modified to deal with it. The problem dealt with by look-ahead occurs when $\overline{\mathbf{V}}_h$ has full rank and $\mathbf{S}_h$ is singular because $\mathbf{G}$ is indefinite. In this case it is entirely possible for $\mathbf{S}_h$ to be singular but for $\mathbf{S}_{h+j}$ to be nonsingular for some $j > 0$, and if this happens Theorem 1 guarantees that Algorithm 2 will eventually escape from the while-loop and get back on course.

The same considerations apply to the HS version of the algorithm although here the picture is complicated by the presence of the matrices $\mathbf{M}_{\sigma_i}$. Provided though that these are nonsingular, Theorem 2 guarantees the emergence from the while-loop if the same conditions hold. It is only if $\mathbf{S}_{h+j}$ is singular for $\forall j \geq 0$ when $\overline{\mathbf{V}}_h$ has full rank that, for both versions of the algorithm, incurable breakdown occurs.

Suppose now that $\mathbf{M}_{\sigma_i}$ is singular. Then $\mathbf{L}_i$ is singular and $\mathbf{Z}_{i+1}$ will be singular. Since (see above) $\widetilde{\mathbf{P}}_{R_{i+1}}$ satisfies equation (23) it will be rank-deficient so that, from equation (12), $\widetilde{\mathbf{D}}_{R_{i+1}}$ will be singular and will remain so as h increases. Again the algorithm will be trapped in the while-loop and an incurable breakdown will have occurred. Thus the HS version can suffer incurable breakdown under circumstances in which the Lanczos version would survive, showing the Lanczos version to be inherently more stable than the HS one.

In the next section we will examine the behaviour of some particular cases of the block-CG algorithm for just one step of look-ahead. These cases include all the look-ahead variants of the block-CG algorithm known to the author, and it will be shown that for these algorithms the ability to escape incurable breakdown is associated with the definiteness or otherwise of $\mathbf{K}$, the preconditioning matrix.

6. Some specific examples

In the examples considered here (see table) the block size r is either one or two. If it is one (for the CGi algorithm - i for indefinite) then $\mathbf{A} \in^{m\times m}$ is assumed to be symmetric, indefinite but nonsingular. If it is two (all other cases) then $\mathbf{A} \in^{m\times m}$ is assumed to be nonsymmetric but nonsingular. Thus in every case both $\mathbf{G}$ and $\mathbf{K}$ are nonsingular and in every case for which $r = 2$, $\mathbf{G}$ is block skew diagonal and hence indefinite. Thus all algorithms to be considered are susceptible to serious breakdown since in every case $\mathbf{G}$ is indefinite.

Method	G	K	h
CGi	$\mathbf{A}$	$\mathbf{I}$	$\mathbf{b}$
BCG, MRZ	$\begin{bmatrix} \mathbf{0} & \mathbf{A}^T \\ \mathbf{A} & \mathbf{0} \end{bmatrix}$	$\begin{bmatrix} \mathbf{0} & \mathbf{I} \\ \mathbf{I} & \mathbf{0} \end{bmatrix}$	$\begin{bmatrix} \mathbf{0} & \mathbf{c} \\ \mathbf{b} & \mathbf{0} \end{bmatrix}$
HG	$\begin{bmatrix} \mathbf{0} & \mathbf{A}^T \\ \mathbf{A} & \mathbf{0} \end{bmatrix}$	$\begin{bmatrix} \mathbf{I} & \mathbf{0} \\ \mathbf{0} & \mathbf{I} \end{bmatrix}$	$\begin{bmatrix} \mathbf{0} & \mathbf{c} \\ \mathbf{b} & \mathbf{0} \end{bmatrix}$
QMR	$\begin{bmatrix} \mathbf{0} & \mathbf{I} \\ \mathbf{I} & \mathbf{0} \end{bmatrix}$	$\begin{bmatrix} \mathbf{0} & \mathbf{A} \\ \mathbf{A}^T & \mathbf{0} \end{bmatrix}$	$\begin{bmatrix} \mathbf{0} & \mathbf{A}^{-T}\mathbf{c} \\ \mathbf{A}^{-1}\mathbf{b} & \mathbf{0} \end{bmatrix}$

Table 1.

It will further be assumed that in all cases for which $r = 2$, the initial values of $\mathbf{X}$ and $\mathbf{P}$ have the forms $\mathbf{X}_1 = \begin{bmatrix} \mathbf{x}_1 & \mathbf{0} \\ \mathbf{0} & \mathbf{z}_1 \end{bmatrix}$ and $\mathbf{P}_1 = \begin{bmatrix} \mathbf{u}_1 & \mathbf{0} \\ \mathbf{0} & \mathbf{v}_1 \end{bmatrix}$, where $\mathbf{x}_1, \mathbf{z}_1, \mathbf{u}_1, \mathbf{v}_1 \in \mathbf{R}^m$. With these forms and the particular choices of $\mathbf{G}$ and $\mathbf{K}$ it is straightforward to show that, for both Lanczos and HS versions of the algorithm, $\mathbf{P}_h = \begin{bmatrix} \mathbf{u}_h & \mathbf{0} \\ \mathbf{0} & \mathbf{v}_h \end{bmatrix}$ or $\mathbf{P}_h = \begin{bmatrix} \mathbf{0} & \mathbf{u}_h \\ \mathbf{v}_h & \mathbf{0} \end{bmatrix}$, $\forall h$. Either way $\mathbf{P}_h^T\mathbf{G}\mathbf{P}_h$ is a 2×2 symmetric skew-diagonal matrix. Thus for all these algorithms $\mathbf{P}_h^T\mathbf{G}\mathbf{P}_h$ will either be nonsingular or null (this is trivial if $r = 1$) and this implies, from Theorems 1 and 2 and the properties of the Schur complements, that if $\mathbf{S}_{h-1}$ is nonsingular then $\mathbf{S}_h$ is either nonsingular or that its rank will be equal to that of $\mathbf{S}_{h-1}$. We now use this result to prove the following theorem.

Theorem 3

Let $\overline{\mathbf{V}}_h$ and $\mathbf{S}_h$ be as defined above, and let

1. $\mathbf{S}_{h-1}$ be nonsingular
2. $\overline{\mathbf{V}}_h$ have full rank
3. $rank(\mathbf{S}_h) = rank(\mathbf{S}_{h-1})$
4. $\mathbf{G}$ be nonsingular
5. $\mathbf{K}$ be definite

Then $\mathbf{S}_{h+1}$ is nonsingular.

Proof

From equations (16) and (26) we may, after some adjustment, write

$$\mathbf{S}_{h+1} = \begin{bmatrix} \mathbf{P}_1^T\mathbf{G}\overline{\mathbf{V}}_h & \mathbf{P}_1^T\mathbf{G}(\mathbf{KG})^h\mathbf{P}_1 \\ \overline{\mathbf{V}}_h^T\mathbf{GKG}\overline{\mathbf{V}}_h & \overline{\mathbf{V}}_h^T\mathbf{G}(\mathbf{KG})^{h+1}\mathbf{P}_1 \end{bmatrix}$$

From hypotheses (2), (4) and (5), $\overline{\mathbf{V}}_h^T\mathbf{GKG}\overline{\mathbf{V}}_h$ is nonsingular so that if $\mathbf{S}_{h+1}$ is singular every null vector $\mathbf{z}$ must have the form $\mathbf{z}^T = \begin{bmatrix} \mathbf{z}_1^T & \mathbf{z}_2^T \end{bmatrix}$ where $\mathbf{z}_2 \neq \mathbf{0}$. Let now

$$\mathbf{S}_{h+1} = \begin{bmatrix} \mathbf{S}_h & \mathbf{S}_2 \\ \mathbf{S}_2^T & \mathbf{S}_3 \end{bmatrix}$$

so that if $\mathbf{S}_{h+1}\mathbf{z} = \mathbf{0}$ then

$$\mathbf{S}_h\mathbf{z}_1 + \mathbf{S}_2\mathbf{z}_2 = \mathbf{0} \tag{34}$$

Now since $\mathbf{S}_h$ has rank $(\mathbf{S}_{h-1})$ and since $\mathbf{S}_h \in \mathbf{R}^{hr\times hr}$, $\mathbf{S}_h$ will have nullity r so that $\exists\mathbf{X} \in \mathbf{R}^{hr\times r}$ of rank r such that $\mathbf{S}_h\mathbf{X} = \mathbf{0}$. Premultiplying equation (34) by $\mathbf{X}^T$ then gives, since $\mathbf{S}_h$ is symmetric, $\mathbf{X}^T\mathbf{S}_2\mathbf{z}_2 = \mathbf{0}$ where $\mathbf{z}_2 \neq \mathbf{0}$. Now $\mathbf{X}^T\mathbf{S}_2$ is square $(r \times r)$ so that there are now two possibilities. If $\mathbf{X}^T\mathbf{S}_2$ is nonsingular then $\mathbf{z}_2 = \mathbf{0}$ and we have an immediate contradiction. If $\mathbf{X}^T\mathbf{S}_2$ is singular then $\exists\mathbf{y} \neq \mathbf{0}$ such that $\mathbf{S}_2^T\mathbf{X}\mathbf{y} = \mathbf{0}$ and since $\mathbf{X}$ has full rank this implies that $\mathbf{X}\mathbf{y} \neq \mathbf{0}$. Thus if $\mathbf{z}^T = \begin{bmatrix} (\mathbf{X}\mathbf{y})^T & \mathbf{0}^T \end{bmatrix}$ we have, since $\mathbf{S}_h\mathbf{X} = \mathbf{0}$, $\mathbf{S}_{h+1}\mathbf{z} = \mathbf{0}$ giving another contradiction. The supposition that $\exists\mathbf{z} \neq \mathbf{0}$ such that $\mathbf{S}_{h+1}\mathbf{z} = \mathbf{0}$ is thus false, proving the Theorem ∎

Consider now our seven algorithms (only QMR has no HS version) in the light of this theorem. There is no reason why all of them should not satisfy Hypotheses (1) - (4) but only CGi and HG satisfy Hypothesis (5). Only these two algorithms, therefore, are immune from incurable breakdown and this has indeed been proved for the individual algorithms by Luenberger [10] and Hegedüs [8]. All the other algorithms are known to be susceptible to incurable breakdown and we deduce from this that the final hypothesis, the definiteness of $\mathbf{K}$, is the crucial factor that guarantees recovery.

We can also now look at the problem of the singularity or otherwise of the matrices $\mathbf{M}_{\sigma_i}$. If $h = \sigma_i + 1$ for some value of i then $\widetilde{\mathbf{P}}_{R_{i+1}} = \mathbf{P}_h$ and $\mathbf{P}_h^T\mathbf{G}\mathbf{P}_h$ is either nonsingular or null. If it is nonsingular then, from equations (29) and (30), $\mathbf{M}_{\sigma_{i+1}} = -(\mathbf{P}_h^T\mathbf{G}\mathbf{P}_h)^{-1}\mathbf{P}_h^T\mathbf{F}_{i+1}$ so that $\mathbf{M}_{\sigma_{i+1}}$ will be singular if and only if (since $h = \sigma_i + 1$) $\mathbf{P}_{\sigma_i+1}^T\mathbf{F}_{i+1}$ is singular. If, on the other hand, $\mathbf{P}_h^T\mathbf{G}\mathbf{P}_h$ is null, $\widetilde{\mathbf{P}}_{R_{i+1}} = \left[\ \mathbf{P}_h \quad \mathbf{P}_{h+1}\ \right]$ and $\widetilde{\mathbf{D}}_{R_{i+1}}$ is nonsingular (either by hypothesis or because the algorithm satisfies the conditions of Theorem 3) then, by equation (29),

$$\widetilde{\mathbf{M}}_{i+1} = -\begin{bmatrix} \mathbf{0} & \mathbf{D}_0 \\ \mathbf{D}_0^T & \mathbf{D}_1 \end{bmatrix}^{-1} \begin{bmatrix} \mathbf{P}_h^T\mathbf{F}_{i+1} \\ \mathbf{P}_{h+1}^T\mathbf{F}_{i+1} \end{bmatrix}$$

where $\mathbf{D}_j = \mathbf{P}_{h+j}^T\mathbf{G}\mathbf{P}_{h+1}$. Thus, since $\mathbf{D}_0$ cannot be singular, we obtain from equation (30) and the supposition that $h = \sigma_i + 1$

$$\mathbf{M}_{\sigma_{i+1}} = -\mathbf{D}_0^{-1}\mathbf{P}_{\sigma_i+1}^T\mathbf{F}_{i+1}$$

Thus both when $\mathbf{P}_h^T\mathbf{G}\mathbf{P}_h$ is nonsingular and when it is null, $\mathbf{M}_{\sigma_{i+1}}$ is singular if and only if $\mathbf{P}_{\sigma_i+1}^T\mathbf{F}_{i+1}$ is singular so in order to determine the second factor influencing the stability of the HS versions we need only find an expression for $\mathbf{P}_{\sigma_i+1}^T\mathbf{F}_{i+1}$.

Now it was shown in [2] that $\mathbf{F}_{i+1} = \widetilde{\mathbf{Q}}_i\mathbf{F}_1$ and since for $h = \sigma_i + 1$ we know that $\mathbf{P}_{\sigma_i+1} = \widetilde{\mathbf{P}}_{R_{i+1}}, \widetilde{\mathbf{V}}_{R_{i+1}} = (\mathbf{K}\mathbf{G})^{\sigma_i}\mathbf{P}_1$ and equation (23) is valid we see that

$$\mathbf{P}_{\sigma_i+1}^T\mathbf{F}_{i+1} = \mathbf{Z}_{i+1}^T\mathbf{P}_1^T(\mathbf{G}\mathbf{K})^{\sigma_i}\widetilde{\mathbf{Q}}_i^2\mathbf{F}_1$$

Thus, since $\mathbf{Z}_{i+1}$ is nonsingular and $\widetilde{\mathbf{Q}}_i$ is idempotent, $\mathbf{P}_{\sigma_i+1}^T\mathbf{F}_{i+1}$ is singular if and only if $\mathbf{P}_1^T(\mathbf{G}\mathbf{K})^{\sigma_i}\widetilde{\mathbf{Q}}_i\mathbf{F}_1$ is singular. Define now $\mathbf{T}_h$ by

$$\mathbf{T}_h = \overline{\mathbf{V}}_h^T \left[\ \mathbf{F}_1 \quad \mathbf{G}\overline{\mathbf{V}}_{h-1}\ \right] \tag{35}$$

Expanding $\mathbf{P}_1^T(\mathbf{G}\mathbf{K})^{\sigma_i}\widetilde{\mathbf{Q}}_i\mathbf{F}_1$ by equation (24) and writing $\mathbf{T}_{\sigma_i+1}$ as

$$\mathbf{T}_{\sigma_i+1} = \begin{bmatrix} \overline{\mathbf{V}}_{\sigma_i}^T \\ \mathbf{P}_1^T(\mathbf{G}\mathbf{K})^{\sigma_i} \end{bmatrix} \left[\ \mathbf{F}_1 \quad \mathbf{G}\overline{\mathbf{V}}_{\sigma_i}\ \right]$$

leads straightforwardly to the result that $\mathbf{P}_1^T(\mathbf{G}\mathbf{K})^{\sigma_i}\widetilde{\mathbf{Q}}_i\mathbf{F}_1$ is merely the Schur complement of $\overline{\mathbf{V}}_{\sigma_i}^T\mathbf{G}\overline{\mathbf{V}}_{\sigma_i}$ in $\mathbf{T}_{\sigma_i+1}$. Thus $\mathbf{M}_{\sigma_{i+1}}$ is singular if and only if $\mathbf{T}_{\sigma_i+1}$ is singular.

Now for the HS versions, $\mathbf{P}_1 = \mathbf{K}\mathbf{F}_1$ and substituting $\mathbf{K}^{-1}\mathbf{P}_1$ for $\mathbf{F}_1$ in equation (35) yields, from equation (16),

$$\mathbf{T}_h = \overline{\mathbf{V}}_h^T \mathbf{K}^{-1} \overline{\mathbf{V}}_h$$

Thus again the definiteness of $\mathbf{K}$ has a positive influence on the HS versions of the algorithm since if $\mathbf{K}$ is definite and $\overline{\mathbf{V}}_h$ has full rank then $\mathbf{T}_h$ is nonsingular.

If $\mathbf{K}$ is indefinite the picture is very much less clear, but it is certainly possible in principle for $\mathbf{T}_h$ to become singular for some h. The conclusion is that the HS versions of Algorithm 2 are best avoided if $\mathbf{K}$ is indefinite.

7. Some computational aspects

In practice we do not, of course, compute the matrices $\mathbf{Q}_i$ and $\widetilde{\mathbf{Q}}_i$. Their use here is to provide a concise description of the algorithms uncluttered by computational detail to enable us to concentrate on the salient features of the theory. To obtain the computational form of Algorithm 2 (see [12] for Algorithm 1) we need just one more definition which expresses $\widetilde{\mathbf{P}}_i(\widetilde{\mathbf{P}}_i^T \mathbf{G} \widetilde{\mathbf{P}}_i)^{-1}$ analogously to $\widetilde{\mathbf{P}}_i$ and $\widetilde{\mathbf{M}}_i$, thus:

$$\widetilde{\mathbf{P}}_i(\widetilde{\mathbf{P}}_i^T \mathbf{G} \widetilde{\mathbf{P}}_i)^{-1} = [\ \mathbf{Y}_{\sigma_{i-1}+1} \quad \mathbf{Y}_{\sigma_{i-1}+2} \quad \dots \quad \mathbf{Y}_{\sigma_i}\ ] \tag{36}$$

Then a simple generalisation of the argument given in [3] shows that, for the Lanczos version,

$$\mathbf{P}_{\sigma_i+1} = (\mathbf{I} - \mathbf{Y}_{\sigma_{i-1}} \mathbf{P}_{\sigma_{i-1}}^T \mathbf{G} - \widetilde{\mathbf{P}}_i(\widetilde{\mathbf{P}}_i^T \mathbf{G} \widetilde{\mathbf{P}}_i)^{-1} \widetilde{\mathbf{P}}_i^T \mathbf{G}) \mathbf{K} \mathbf{G} \mathbf{P}_{\sigma_i} \tag{37}$$

and for the HS version

$$\mathbf{P}_{\sigma_i+1} = (\mathbf{I} - \mathbf{Y}_{\sigma_i} \mathbf{P}_{\sigma_i}^T \mathbf{G}) \mathbf{K} \mathbf{F}_{i+1} \tag{38}$$

while for both versions, for $\sigma_i + 2 \leq h \leq \sigma_{i+1}$,

$$\mathbf{P}_h = (\mathbf{I} - \mathbf{Y}_{\sigma_i} \mathbf{P}_{\sigma_i}^T \mathbf{G}) \mathbf{K} \mathbf{G} \mathbf{P}_{h-1} \tag{39}$$

Note that, if the while-loop is never activated, equations (37) and (38) reduce to the standard block-Lanczos and block-HS forms respectively.

Now it so happens that $\mathbf{P}_{\sigma_i+1}$ is computed by Step (8) while $\mathbf{P}_h$, $\sigma_i + 2 \leq h \leq \sigma_{i+1}$, is computed by Step (3a). It therefore suffices to replace Step (3a) by equation (39) and Steps (7) and (8) (since we now have no need of $\widetilde{\mathbf{Q}}_i$) by equation (37) or equation (38) as required. Substitution of the particular values from the table then leads to the particular computational forms, although in some cases these differ from those given by other authors.

In particular the equations for the look-ahead Lanczos algorithm given by Freund, Gutknecht and Nachtigal [5], [6] are somewhat more complicated than those derived by the above substitutions. What appears to happen is that the matrices $\mathbf{P}_h$ generated by the FGN version of the while-loop are permitted to stray from the path of true conjugacy, and are then jerked back into line by the FGN equivalent of our Step (8). Thus the above theory does not strictly apply to this version of the algorithm though we conjecture that it could be modified to do so. We further conjecture that Theorem 1 would be valid as it stands also for the FGN algorithm.

Another complication introduced by the need to avoid overflow in practical cases is the scaling of the matrices $\mathbf{P}_h$ by post-multiplication by a nonsingular scaling matrix $\mathbf{C}_h$. This is often chosen so that $\mathbf{C}_h^T\mathbf{P}_h^T\mathbf{G}\mathbf{P}_h\mathbf{C}_h = \mathbf{I}$ and its calculation can be used to detect the onset of linear dependence of the columns of $\mathbf{P}_h$. This scaling has no effect on the above theory (as long as every $\mathbf{C}_h$ is nonsingular) since all it does is to postmultiply $\overline{\mathbf{U}}_h$ by a nonsingular block-diagonal matrix. Nothing essential changes and the three theorems remain unscathed. No essential changes occur either if, in Step (6), $\mathbf{F}_{i+1}$ is computed recursively by

$$\mathbf{F}_{i+1} = \mathbf{F}_i - \mathbf{G}\widetilde{\mathbf{P}}_i(\widetilde{\mathbf{P}}_i^T\mathbf{G}\widetilde{\mathbf{P}}_i)^{-1}\widetilde{\mathbf{P}}_i^T\mathbf{F}_i$$

thereby eliminating the need to compute $\mathbf{G}\mathbf{X}_{i+1}$ and enabling the calculation to be further simplified.

Finally we note that since, in practice, we rarely if ever get a singular matrix but often get very ill-conditioned ones there is an element of real judgement in deciding whether or not to activate, and then when to exit from, the while-loop. We do not discuss this here but note that all practical look-ahead algorithms have checks, some quite sophisticated, to enable these decisions to be made with sufficient precision. Results are, however, not always entirely satisfactory and the need for such decisions often introduces unwelcome elements of complication and empiricism into the elegant CG structure.

8. Further implications and conclusions

Our analysis of a look-ahead version of the block-CG algorithm indicates that, just as in the case of no look-ahead [2], the process is determined by the Krylov matrix $\overline{\mathbf{V}}_h$ and the two subsidiary matrices $\mathbf{S}_h$ and $\mathbf{T}_h$ which in turn depend on the two matrices $\mathbf{G}$ and $\mathbf{K}$ that define the original algorithm. If $\mathbf{S}_h$ is singular for some h and $\mathbf{S}_{h+j}$ is nonsingular for some $j \geq 1$ then a look-ahead method of the type described above will bring the algorithm back to normality. If, however, $\mathbf{S}_{h+j}$ is singular for $\forall j \geq 0$ then we have an incurable breakdown on our hands, but we have shown that for a certain class of algorithms (that satisfy the conditions of Theorem 3) this only occurs if $\mathbf{K}$ is indefinite.

For all the seven algorithms quoted, where $\mathbf{A} \in \mathbf{R}^{m \times m}$, the maximum value that h can attain before $\overline{\mathbf{V}}_h$ becomes rank-deficient is m, when $\overline{\mathbf{V}}_h$ $(= \overline{\mathbf{V}}_m)$ becomes square. In this case, if it is not rank-deficient, $\mathbf{S}_h$ *must* be nonsingular if $\mathbf{G}$ is nonsingular. Thus if the algorithm runs its full course then look-ahead must work, albeit for some possibly very large value of j. Incurable breakdown therefore only occurs if $\overline{\mathbf{V}}_h$ becomes rank-deficient for some $h < m$. But $\overline{\mathbf{V}}_h$ becoming rank-deficient is referred to (perhaps optimistically) as regular termination (the term *premature termination* might be preferred if $\overline{\mathbf{V}}_h$ becomes rank-deficient for some $h < m$) so we have the slightly whimsical equation

serious breakdown + premature termination = incurable breakdown

Premature termination arises when there are fewer than the maximum possible number of distinct eigencomponents in $\mathbf{P}_1$ and can occur either because of an unfortunate choice of $\mathbf{P}_1$ or, more likely, because $\mathbf{KG}$ has multiple eigenvalues. If either of these two contingencies arises there will always be the possibility that premature termination will occur while the algorithm is in the throes of serious breakdown. That this sometimes cannot happen is the somewhat surprising consequence of Theorem 3, which shows that if $rank(\mathbf{S}_h) = rank(\mathbf{S}_{h-1})$ and $\overline{\mathbf{V}}_h$ has full rank then $\overline{\mathbf{V}}_{h+1}$ *also* has full rank. The theorem thus makes a statement about the next term of the Krylov sequence, a term yet to be computed.

Since incurable breakdown occurs only in the case of premature termination it would seem that multiple eigenvalues of $\mathbf{KG}$ predispose an algorithm towards incurable breakdown. Since, though, they predispose an algorithm to rapid termination if serious breakdown does not occur they are not regarded as necessarily undesirable. The lesson here seems to be that the algorithm should avoid serious breakdown if at all possible, and it can be guaranteed to do so only if $\mathbf{G}$ is definite. This conclusion is corroborated by the difficulty of deciding whether or not serious breakdown has occurred. In some practical algorithms no fewer than three independent checks are carried out to test for this, giving an indication of the difficulty of making this particular decision.

The author believes that, in view of the problems outlined above, it might be better to sidestep completely the threat of serious breakdown. This can be done by using the Lanczos versions of the conjugate residual algorithm (for symmetric indefinite problems) and the biconjugate residual algorithm (for nonsymmetric problems) which are, in theory, completely breakdown-free [2]. The Lanczos forms of these two algorithms do not appear to have been experimentally tested to date and testing might reveal hitherto unsuspected weaknesses. In view of their theoretical advantages, though, their comparative simplicity and their robustness, they should at least be tried.

Acknowledgments

The author thanks CNR (Italy) for partial financial support under Progetto Finalizzato Trasporti 2, Contract No. 93.01861.PF74 and Professors Aristide Mingozzi and Emilio Spedicato for their continuing interest.

References

1. C. Brezinski, M. R. Zaglia, and H. Sadok. A breakdown-free Lanczos-type algorithm for solving linear systems. *Numer. Math.*, 63:29–38, 1992.
2. C. G. Broyden. A breakdown of the block-cg method. *Optimization Methods and Software*. To be published.
3. C. G. Broyden. Block conjugate gradient methods. *Optimization Methods and Software*, 2:1–17, 1993.
4. C. G. Broyden. A note on the block conjugate gradient algorithm of O'Leary. *Optimization Methods and Software*, 5:347–350, 1995.
5. R. W. Freund, M. H. Gutknecht, and N. M. Nachtigal. An implementation of the look-ahead Lanczos algorithm for non-hermitian matrices. Technical Report 91-06, IPS, ETH-Zentrum, Zurich, 1991.
6. R. W. Freund and N. M. Nachtigal. QMR: A quasi-minimal residual method for non-hermitian linear systems. *Numer. Math.*, 60:315–339, 1991.
7. M. Gutknecht. A completed theory of the unsymmetric Lanczos process and related algorithms, part 1. *SIAM J. Matrix Anal. Appl.*, 13:594–639, 1992.
8. Cs. Hegedüs. Generating conjugate directions for arbitrary matrices by matrix equations. Technical Report KFKI-1990-36/M, Hungarian Academy of Sciences, Central Research Institute for Physics, Budapest, 1990.
9. R. A. Horn and C. A. Johnson. *Matrix Analysis*. Cambridge University Press, 1985.
10. D. G. Luenberger. Hyperbolic pairs in the method of conjugate gradients. *SIAM J. Appl. Math.*, 17:1263–1267, 1969.
11. A. A. Nikishin and A. Y. Yeremin. Variable block cg algorithms for solving large sparse symmetric positive definite linear systems on parallel computers, 1: General iterative scheme. *SIAM J. Matrix. Anal. Appl.*, 16(4):1135–1153, 1995.
12. D. P. O'Leary. The block conjugate gradient algorithm and related methods. *Linear Algebra Applics.*, 29:293–322, 1980.
13. B. N. Parlett, D. R. Taylor, and Z. A. Liu. A look-ahead Lanczos algorithm for unsymmetric matrices. *Math. Comp.*, 44:105–124, 1985.

ITERATIVE Bi-CG TYPE METHODS AND IMPLEMENTATION ASPECTS

H.A. VAN DER VORST and G.L.G. SLEIJPEN
Mathematical Institute, University of Utrecht
Budapestlaan 6, Utrecht, the Netherlands
e-mail: vorst@math.ruu.nl, sleijpen@math.ruu.nl

Abstract

Krylov subspace methods, such as Conjugate Gradients, Bi-CG, and GMRES, have become quite popular for the iterative solution of classes of large sparse linear systems. In a search for further improvement, so-called hybrid methods have emerged. These hybrid methods can be viewed as combinations of the standard Krylov subspace methods, and they attempt to combine attractive properties of these standard methods.
We will start with a brief overview of the standard Krylov subspace methods, in order to establish notation and some basic properties that we need further on. Then we will give more attention to the hybrid methods, including CGS, Generalized CGS, Bi-CGSTAB, Bi-CGSTAB(ℓ), Quasi-Minimal Residual variants, and GMRESR. We will discuss the convergence behavior of these methods, and we will show why it is so important to avoid very large intermediate residual norms. We will also discuss some recent proposals to further improve the numerical stability of these methods.
Iterative methods are often used in combination with so-called preconditioning operators (approximations for the inverses of the operator of the system to be solved) in order to speed up convergence. We will give an overview of the different possibilities for preconditioning.
In this chapter, we will also discuss general implementation aspects of iterative methods, in particular in view of parallel and vector processing.

Notation:
Although most of the ideas carry over to the complex case, we will, for ease of presentation, assume real arithmetic. Matrices are denoted by capitals (A, B, ...), and of order n, unless differently specified. Vectors are represented by lowercast letters (x, y, ...), and Greek symbols are used to represent scalars (α, γ, ...).

1. Krylov subspaces - Projection

1.1. BASIC ITERATION METHODS, KRYLOV SUBSPACES

A very basic idea, that leads to many effective iterative solvers, is to to split the matrix A of a given linear system as the sum of two matrices, one of which a

G. Winter Althaus and E. Spedicato (eds.), Algorithms for Large Scale Linear Algebraic Systems, 217–253.

matrix that would have led to a system that can easily be solved. For the linear system $Ax = b$, the standard splitting $A = I - (I - A)$ leads to the well-known Richardson iteration:

$$x_i = b + (I - A)x_{i-1} = x_{i-1} + r_{i-1}. \tag{1}$$

Multiplication by $-A$ and adding b gives

$$b - Ax_i = b - Ax_{i-1} - Ar_{i-1}$$

or

$$r_i = (I - A)r_{i-1} = (I - A)^i r_0 = P_i(A)r_0. \tag{2}$$

Similarly, the error $x - x_i$ can be expressed as

$$x - x_i = P_i(A)(x - x_0).$$

In these expressions P_i is a (special) polynomial of degree i. The Richardson iteration leads to the polynomial $P_i(t) = (1-t)^i$, and as we will see other methods lead to other polynomials. These polynomials are almost never constructed in explicit form, but it is used in formulas to express the special relation between r_i and r_0.

In the derivation of an iterative scheme we have used a very simple splitting. In many cases better approximations K to the given matrix A are available, which define the splitting $A = K - R$, with $R = K - A$. The corresponding iterative method then reads as

$$Kx_i = b + Rx_{i-1} = Kx_{i-1} + b - Ax_{i-1}.$$

Solving for x_i leads formally to

$$x_i = x_{i-1} + K^{-1}(b - Ax_{i-1}).$$

If we define $B \equiv K^{-1}A$ and $c = K^{-1}b$, then the $K-$splitting is equivalent with the standard splitting for $Bx = c$. Therefore, it is no loss of generality if we further restrict ourselves to the standard splitting. Any other splitting can be incorporated easily in the different schemes by replacing the quantities A and b, by B and c, respectively.

From now on we will assume that $x_0 = 0$. This also does not mean a loss of generality, for the situation $x_0 \neq 0$ can, through a simple linear transformation $z = x - x_0$, be transformed to the system

$$Az = b - Ax_0 = \tilde{b},$$

for which obviously $z_0 = 0$.

For the Richardson iteration it follows that

$$x_{i+1} = r_0 + r_1 + r_2 + \cdots + r_i = \sum_{j=0}^{i} (I - A)^j r_0$$

$$\in \{r_0, Ar_0, \ldots, A^i r_0\} = K_{i+1}(A; r_0).$$

The subspace $K_{i+1}(A; r_0)$ is called the *Krylov subspace* of dimension $i+1$, generated by A *and* r_0. Apparently, the Richardson iteration delivers elements of Krylov subspaces of increasing dimension. Including local iteration parameters in the iteration leads to other elements of the same Krylov subspaces, and hence to other polynomial expressions for the error and the residual.

1.2. ORTHOGONAL BASIS

The standard iteration method 1 generates approximate solutions that belong to Krylov subspaces of increasing dimension. Modern iteration methods construct better approximations in these subspaces with some modest computational overhead as compared with the standard iteration.

In order to identify better solutions in the Krylov subspace we need a suitable basis for this subspace, one that can be extended in a meaningful way for subspaces of increasing dimension. The obvious basis r_0, Ar_0, ..., $A^{i-1}r_0$, for $K^i(A; r_0)$, is not very attractive from a numerical point of view, since the vectors $A^j r_0$ point more and more in the direction of the dominant eigenvector for increasing j (the power method!), and hence the basis vectors become dependent in finite precision arithmetic.

Instead of the standard basis one usually prefers an orthonormal basis, and Arnoldi [1] suggested to compute this basis as follows. Start with $v_1 \equiv r_0/\|r_0\|_2$. Assume that we have already an orthonormal basis v_1, ..., v_j for $K^j(A; r_0)$, then this basis is expanded by computing $t = Av_j$, and by orthonormalizing this vector t with respect to v_1, ..., v_j. In principle the orthonormalization process can be carried out in different ways, but the most commonly used approach is to do this by a modified Gram-Schmidt procedure [24]. This leads to the algorithm in Fig. 1, for the creation of an orthonormal basis for $K^m(A; r_0)$.

It is easily verified that the v_1, ..., v_m form an orthonormal basis for $K^m(A; r_0)$ (that is, if the construction does not terminate at a vector $t = 0$). The orthogonalization leads to relations between the v_j, that can be formulated in a compact algebraic form. Let V_j denote the matrix with columns v_1 up to v_j, then it follows that

$$AV_{m-1} = V_m H_{m,m-1}. \tag{3}$$

The m by $m-1$ matrix $H_{m,m-1}$ is upper Hessenberg, and its elements $h_{i,j}$ are defined by the Arnoldi algorithm.

We see that this orthogonalization becomes increasingly expensive for increasing

$$
\begin{array}{l}
v_1 = r_0/\|r_0\|_2; \\
\textbf{for } j = 1, .., m-1 \\
\quad t = Av_j; \\
\quad \textbf{for } i = 1, ..., j \\
\quad\quad h_{i,j} = v_i^T t; \\
\quad\quad t = t - h_{i,j} v_i; \\
\quad \textbf{end}; \\
\quad h_{j+1,j} = \|t\|_2; \\
\quad v_{j+1} = t/h_{j+1,j}; \\
\textbf{end}
\end{array}
$$

Figure 1: Arnoldi's method with modified Gram–Schmidt orthogonalization

dimension of the subspace, since the computation of each $h_{i,j}$ requires an inner product and a vector update.

Note that if A is symmetric, then so is $H_{m-1,m-1} = V_{m-1}^T A V_{m-1}$, so that in this situation $H_{m-1,m-1}$ is tridiagonal. This means that in the orthogonalization process, each new vector has to be orthogonalized with respect to the previous two vectors only, since all other inner products vanish. The resulting three term recurrence relation for the basis vectors of $K_m(A; r_0)$ is known as the *Lanczos method* and some very elegant methods are derived from it.

2. Krylov subspaces - Solution methods

Since we are looking for a solution x_k in $K^k(A; r_0)$, that vector can be written as a combination of the basis vectors of the Krylov subspace, and hence

$$x_k = V_k y.$$

(Note that y has k components)
There are three projection-type of approaches to find a suitable approximation for the solution x of $Ax = b$:

1. The *Ritz-Galerkin* approach:
   require that $b - Ax_k \perp K^k(A; r_0)$.

2. The *minimum residual* approach:
   require $\|b - Ax_k\|_2$ to be minimal over $K^k(A; r_0)$.

3. The *Petrov-Galerkin* approach:
   require that $b - Ax_k$ is orthogonal to some other suitable k-dimensional subspace.

The Ritz-Galerkin approach leads to such popular and well-known methods as Conjugate Gradients, the Lanczos method, FOM, and GENCG. The minimum residual approach leads to methods like GMRES, MINRES, and ORTHODIR. If we select the k-dimensional subspace in the third approach as $K^k(A^T; s_0)$, then we obtain the Bi-CG, and QMR methods, which both fall in the class of Petrov-Galerkin methods.
More recently, hybrids of the three approaches have been proposed, like CGS, Bi-CGSTAB, BiCGSTAB(ℓ), TFQMR, FGMRES, and GMRESR. We will concentrate on these hybrid methods, but we will also briefly discuss the basic Krylov subspace methods.

2.1. THE RITZ-GALERKIN APPROACH: CG

Let us assume, for simplicity again, that $x_0 = 0$, and hence $r_0 = b$. The Ritz-Galerkin conditions imply that $r_k \perp K^k(A; r_0)$, and this is equivalent with

$$V_k^T(b - Ax_k) = 0.$$

Since $b = r_0 = \|r_0\|_2 v_1$, it follows that $V_k^T b = \|r_0\|_2 e_1$ with e_1 the first canonical unit vector in $\mathbb{R}^k$. With $x_k = V_k y$ we obtain

$$V_k^T A V_k y = \|r_0\|_2 e_1.$$

This system can be interpreted as the system $Ax = b$ projected onto $K^k(A; r_0)$.

Obviously we have to construct the $k \times k$ matrix $V_k^T A V_k$, but this is, as we have seen readily available from the orthogonalization process:

$$V_k^T A V_k = H_{k,k},$$

so that the x_k for which $r_k \perp K^k(A; r_0)$ can be easily computed by first solving $H_{k,k} y = \|r_0\|_2 e_1$, and forming $x_k = V_k y$. This algorithm is known as FOM or GENCG.

When A is symmetric, then $H_{k,k}$ reduces to a tridiagonal matrix $T_{k,k}$, and the resulting method is known as the *Lanczos* method [28]. In clever implementations it is avoided to store all the vectors v_j.
When A is in addition positive definite then we obtain, at least formally, the *Conjugate Gradient* method. In commonly used implementations of this method one forms directly a LU factorization for $T_{k,k}$ and this leads to very elegant short recurrencies for the x_j and the corresponding r_j. The positive definiteness is necessary to guarantee the existence of the LU factorization.

2.2. THE MINIMUM RESIDUAL APPROACH: GMRES

The creation of an orthogonal basis for the Krylov subspace, in Section 1.2, has led to

$$AV_i = V_{i+1}H_{i+1,i}. \tag{4}$$

We will still assume that $x_0 = 0$, and hence $r_0 = b$.
We look for an $x_i \in K^i(A; r_0)$, that is $x_i = V_i y$, for which $\|b - Ax_i\|_2$ is minimal. This norm can be rewritten as

$$\begin{aligned} \|b - Ax_i\|_2 &= \|b - AV_i y\|_2 \\ &= \|b - V_{i+1}H_{i+1,i}y\|_2 \\ &= \|V_{i+1}(\|r_0\|_2 e_1 - H_{i+1,i}y)\|_2. \end{aligned}$$

Now we exploit the fact that V_{i+1} is an orthonormal transformation with respect to the Krylov subspace $K^{i+1}(A; r_0)$:

$$\|b - Ax_i\|_2 = \| \|r_0\|_2 e_1 - H_{i+1,i}y\|_2,$$

and this final norm can simply be minimized by solving the minimum norm least squares problem:

$$H_{i+1,i}y = \|r_0\|_2 e_1.$$

In GMRES [40] this is done with Givens rotations, that annihilate the subdiagonal elements in the upper Hessenberg matrix $H_{i+1,i}$.
Note that when A is symmetric the upper Hessenberg matrix $H_{i+1,i}$ reduces to a tridiagional system. This simplified structure can be exploited in order to avoid storage of all the basis vectors for the Krylov subspace, in a way similar as has been pointed out for CG. The resulting method is known as MINRES [33].

In order to avoid excessive storage requirements and computational costs for the orthogonalization, GMRES is usually restarted after each m iteration steps. This algorithm is referred to as GMRES(m); the not-restarted version is often called 'full' GMRES.

There is an interesting and simple relation between the Ritz-Galerkin approach (FOM and CG) and the minimum residual approach (GMRES and MINRES). In GMRES the projected system matrix $H_{i+1,i}$ is transformed by Givens rotations to an upper triangular matrix (with last row equal to zero). So, in fact, the major difference between FOM and GMRES is that in FOM the last $((i + 1)$-th row is simply discarded, while in GMRES this row is rotated to a zero vector.

In [50] it has been shown how the GMRES-method can be combined (or rather preconditioned) with other iterative schemes. The iteration steps of GMRES (or GCR) are called outer iteration steps, while the iteration steps of the

preconditioning iterative method are referred to as inner iterations. The combined method is called GMRES⋆, where ⋆ stands for any given iterative scheme; in the case of GMRES as the inner iteration method, the combined scheme is called GMRESR[50].
Similar schemes have been proposed recently. In FGMRES[38] the update directions for the approximate solution are preconditioned, whereas in GMRES⋆ the residuals are preconditioned. The latter approach offers more control over the reduction in the residual, in particular breakdown situations can be easily detected and remedied.
In exact arithmetic GMRES⋆ is very close to the Generalized Conjugate Gradient method[3]; GMRES⋆, however, leads to a more efficient computational scheme.

We will discuss practical implementation aspects of these methods in Section 3.1.

2.3. THE PETROV-GALERKIN APPROACH: Bi-CG

For unsymmetric systems we can, in general, not reduce the matrix A to a symmetric system in a lower-dimensional subspace, by orthogonal projections. The reason is that we can not create an orthogonal basis for the Krylov subspace by a 3-term recurrence relation [18]. However, we can try to obtain a suitable non-orthogonal basis with a 3-term recurrence, by requiring that this basis is orthogonal to some other basis.
We start by constructing an arbitrary basis for the Krylov subspace:

$$h_{i+1,i}v_{i+1} = Av_i - \sum_{j=1}^{i} h_{j,i}v_j, \tag{5}$$

which can be rewritten in matrix notation as $AV_i = V_{i+1}H_{i+1,i}$.

Clearly, we can not use V_i for the projection, but suppose we have a W_i for which $W_i^T V_i = D_i$ (an i by i diagonal matrix with diagonal entries d_i), and for which $W_i^T v_{i+1} = 0$. The coefficients $h_{i+1,i}$ can be chosen, we suggest to choose them such that $\|v_{i+1}\|_2 = 1$.
Then

$$W_i^T AV_i = D_i H_{i,i}, \tag{6}$$

and now our goal is to find a W_i for which $H_{i,i}$ is tridiagonal. This means that $V_i^T A^T W_i$ should be tridiagonal too. This last expression has a similar structure as the right-hand side in (6), with only W_i and V_i reversed. This suggests to generate the w_i with A^T.
We start with an arbitrary $w_1 \neq 0$, such that $w_1^T v_1 \neq 0$. Then we generate v_2 with (5), and orthogonalize it with respect to w_1, which means that $h_{1,1} = w_1^T Av_1/(w_1^T v_1)$. Since $w_1^T Av_1 = v_1^T A^T w_1$, this implies that w_2, generated with

$$h_{2,1}w_2 = A^T w_1 - h_{1,1}w_1,$$

is also orthogonal to v_1.
This can be continued, and we see that we can create bi-orthogonal basis sets $\{v_j\}$, and $\{w_j\}$, by making the new v_i orthogonal to w_1 up to w_{i-1}, and then by generating w_i with the same recurrence coefficients, but with A^T instead of A.
Now we have that $W_i^T A V_i = D_i H_{i,i}$, and also that $V_i A^T W_i = D_i H_{i,i}$. This implies that $D_i H_{i,i}$ is symmetric, and hence $H_{i,i}$ is a tridiagonal matrix, which gives us the desired 3-term recurrence relation for the v_j's, and the w_j's.

We may proceed in a similar way as in the symmetric case:

$$AV_i = V_{i+1} T_{i+1,i}, \tag{7}$$

but here we use the matrix $W_i = [w_1, w_2, ..., w_i]$ for the projection of the system

$$W_i^T(b - Ax_i) = 0,$$

or

$$W_i^T A V_i y - W_i^T b = 0.$$

Using (7), we find that y_i satisfies

$$T_{i,i} y = \|r_0\|_2 e_1,$$

and $x_i = V_i y$.
This method is known as the Bi-Lanczos method [28]. Similar as for the symmetric case, there is a variant with short reurrencies. This method is known as Bi-CG.
In [20] a variant of Bi-CG is suggested in which, similar as for GMRES, an overdetermined system with $T_{i+1,i}$ is solved, simply ignoring the fact that the matrix V_{i+1} is not orthogonal. This approach is known as the *Quasi Minimal Residual* method (QMR).
The Bi-CG method has seen some offspring in the form of hybrid methods, methods that can be viewed as combinations of Bi-CG with other methods. These methods include CGS, TFQMR, Bi-CGSTAB, and others. They will be discussed in the next section.

3. Iterative Methods for linear systems

In this section we shall focus on so-called hybrid methods. These are methods that can be viewed as combinations of the standard Krylov subspace methods, such as Bi-CG and GMRES. We shall discuss implementation aspects on vector and parallel computers.

Our focus here is on three groups of methods:

- Variants of GMRES. For general nonsymmetric systems one can construct a full orthogonal basis for the solution space. Unfortunately this leads to a demand in memory space which grows linearly with the number of iteration steps. The problem is circumvented by restarting the iteration procedure after m steps, which leads to GMRES(m), but this may lead to less than optimal convergence behavior. Another interesting approach for obtaining shorter recurrencies is realized in the methods FGMRES and GMRESR.

- CGS. A method that has become quite popular over recent years is the so-called conjugate gradient-squared method. It can be seen as a variant of the Bi-CG method insofar as it also exploits the work carried out with both matrix vector multiplications for improving the convergence behavior. Recent improvements over CGS include TFQMR, and GCGS.

- Bi-CGSTAB and Bi-CGSTAB(ℓ). The two matrix vector operations in Bi-CG can also be used to form, at virtually the same computational costs as for Bi-CG itself, a method that can be viewed as the product of Bi-CG and repeated minimal residual steps (or GMRES(1)-steps). The obvious extension Bi-CGSTAB(ℓ) method, in wich Bi-CG is combined with GMRES(ℓ), offers further possibilities for parallelism, besides better convergence properties.

3.1. VARIANTS OF GMRES

Although GMRES is optimal, in the sense that it leads to a minimal residual solution over the Krylov subspace, it has the disadvantage that this goes with increasing computational costs per iteration step. Of course, with a suitable preconditioner one can hope to reduce the dimension of the required Krylov subspace to acceptable values, but in many cases such preconditioners have not been identified.
The preconditioner K^{-1} is generally viewed as an approximation for the inverse of the matrix A of the system $Ax = b$ to be solved, that means that instead of an optimal update direction $A^{-1}r$, we compute the direction $p = K^{-1}r$. That is, we solve p from $Kp = r$, and K is constructed in such a way that this is easy to do. An attractive idea is to try to get better approximations for $A^{-1}r$, and there are two related approaches to accomplish this. One is to try to improve the preconditioner with updates from the Krylov subspace. This has been suggested first by Eirola and Nevanlinna [17]. Their approach leads to iterative methods that are related to Broyden's method [8], which is a Newton type of method. The Broyden method can be obtained from this update-approach if we do not restrict ourselves to Krylov subspaces. See [52] for a discussion on the relation of these methods.

The updated preconditioners can not be applied right away in the GMRES method, since the preconditioned operator now changes from step to step, and we are not forming a regular Krylov subspace, but rather a subset of a higher dimensional Krylov subspace. However, we can still minimize the residual over this subset.

The idea of variable preconditioning has been exploited in this sense, by different authors. Axelsson and Vassilevski [3] have proposed a Generalized Conjugate Gradient method with variable preconditioning, Saad [38, 39] has proposed a scheme very similar to GMRES, called Flexible GMRES (FGMRES), and Van der Vorst and Vuik have published a GMRESR scheme. FGMRES has received most attention, presumably because it is so easy to implement: only the update directions in GMRES have to be preconditioned, and each update may be preconditioned differently. This means that only one line in the GMRES algorithm has to be adapted. The price to be paid is that the method is not longer robust; it may break down. The GENCG and GMRESR schemes are slightly more expensive in terms of memory requirements and in computational overhead per iteration step. The main difference between the two schemes is that GENCG in [3] works with Gram-Schmidt orthogonalization, whereas GMRESR makes it possible to use Modified Gram-Schmidt. This may give GMRESR an advantage in actual computations. In exact arithmetic GMRESR and GENCG should produce the same results for the same preconditioners. Another advantage of GMRESR over algorithm 1 in [3] is that only one matrix vector product is used per iteration step in GMRESR; GENCG needs two matrix vector products per step.
In GMRESR the residual vectors are preconditioned and if this gives a further reduction then GMRESR does not break down. This gives slightly more control over the method in comparison with FGMRES. In most cases though the results are about the same, and then the efficient scheme for FGMRES has an advantage.

We will briefly discuss the GMRESR method. In [50] it has been shown how the GMRES-method, or more precisely, the GCR-method, can be combined with other iterative schemes. The iteration steps of GMRES (or GCR) are called outer iteration steps, while the iteration steps of the preconditioning iterative method are referred to as inner iterations. The combined method is called GMRES$\star$, where $\star$ stands for any given iterative scheme; in the case of GMRES as the inner iteration method, the combined scheme is called GMRESR[50].
The GMRES$\star$ algorithm is described in Fig. 2.

A sufficient condition to avoid breakdown in this method ($\|c\|_2 = 0$) is that the norm of the residual at the end of an inner iteration is smaller than the norm of the right-hand side residual: $\|Az^{(m)} - r_i\|_2 < \|r_i\|_2$. This can easily be controlled during the inner iteration process. If stagnation occurs, i.e. no progress at all is made in the inner iteration, then it is suggested in [50] to do one (or more) steps of the LSQR method, which guarantees a reduction (but

```
x_0 is an initial guess; r_0 = b - Ax_0;
for i = 0, 1, 2, 3, ...
    Let z^(m) be the approximate solution
    of Az = r_i, obtained after m steps of
    an iterative method.
    c = Az^(m) (often available from the iteration method)
    for k = 0, ..., i-1
        α = (c_k^T c)
        c = c - α c_k
        z^(m) = z^(m) - α u_k
    c_i = c/||c||_2; u_i = z^(m)/||c||_2
    x_{i+1} = x_i + (c_i^T r_i) u_i
    r_{i+1} = r_i - (c_i^T r_i) c_i
    if x_{i+1} is accurate enough then quit
end
```

Figure 2: The GMRES⋆ scheme

this reduction is often only small).

The idea behind these flexible iteration schemes is that we explore parts of high-dimensional Krylov subspaces, hopefully localizing almost the same approximate solution that full GMRES would find over the entire subspace, but now at much lower computational costs. For the inner iteration we may select any appropriate solver, for instance, one cycle of GMRES(m), since then we have also locally an optimal method, or some other iteration scheme, like for instance Bi-CGSTAB.

In [13] it is proposed to keep the Krylov subspace, that is built in the inner iteration, orthogonal with respect to the Krylov basis vectors generated in the outer iteration. Under various circumstances this helps to avoid inspection of already investigated parts of Krylov subspaces in future inner iterations. The procedure works as follows.
In the outer iteration process the vectors c_0, ..., c_{i-1} build an orthogonal basis for the Krylov subspace. Let C_i be the n by i matrix with columns c_0, ..., c_{i-1}. Then the inner iteration process at outer iteration i is carried out with the operator A_i instead of A, and A_i is defined as

$$A_i = (I - C_i C_i^T)A. \tag{8}$$

It is easily verified that $A_i z \perp c_0, ..., c_{i-1}$ for all z, so that the inner iteration process takes place in a subspace orthogonal to these vectors. The additional costs, per iteration of the inner iteration process, are i inner products and i vector updates. In order to save on these costs, one should realize that it is not necessary to orthogonalize with respect to all previous c-vectors, and that "less effective" directions may be dropped, or combined with others. In [32, 4, 13] suggestions are made for such strategies. Of course, these strategies are only attractive in cases where we see too little residual reducing effect in the inner iteration process in comparison with the outer iterations of GMRES$\star$.

Note that if we carry out the preconditioning by doing a few iterations of some other iteration process, then we have inner-outer iteration schemes; these have been discussed earlier in [23, 22].

3.2. CGS AND VARIANTS

For Bi-CG, as for the other methods, we have that $r_j \in K^{j+1}(A; r_0)$, and hence it can be written formally as $r_j = P_j(A)r_0$. Since the basis vectors $\widehat{r}_j$, the scaled vectors w_j, for the dual subspace are generated by the same recurrence relation, we have that $\widehat{r}_j = P_j(A^T)\widehat{r}_0$.
The iteration coefficients for the recurrence relations follow from bi-orthogonality conditions, and may look like:

$$\begin{aligned} \alpha_j &= (Ar_j, \widehat{r}_j) \\ &= (AP_j(A)r_0, P_j(A^T)\widehat{r}_0) \\ &= (AP_j^2(A)r_0, \widehat{r}_0). \end{aligned}$$

Sonneveld [45] observed that we could also construct vectors $\widetilde{r}_j$ with the property that they can be formally written as $\widetilde{r}_j = P_j^2(A)r_0$, and that this can be done with short term recurrences without ever computing explicitly the Bi-CG vectors r_j and $\widehat{r}_j$. The Bi-CG iteration coefficients can then be recovered using the latter form of the innerproduct.
This has two advantages:

1. Since we do not have to form the vectors $\widehat{r}_j$, we can avoid to work with the matrix A^T. This is an advantage for those applications where A^T is not readily available.

2. An even more appealing thought is that, in case of Bi-CG convergence, the formal operator $P_j(A)$ describes how r_0 is reduced to r_j, and naively thinking we might expect that $P_j(A)^2$ would lead to a double reduction. Surprisingly this is indeed often the case. This is certainly a surprise when we come to think of it: the operator $P_j(A)$ depends very much on the starting vector r_0, and delivers a vector r_j that is optimal, in the sense that it is orthogonal to some well-defined subspace. There is no

reason why the vector $\widetilde{r}_j \equiv P_j^2(A)r_0$ should be orthogonal to the subspace $K^{2j}(A^T; \widehat{r}_0)$, or in other words, the optimal polynomial for the starting vector $P_j(A)r_0$ may not be expected to be $P_j(A)$ again.

The resulting method has become well-known under the name CGS, which means Conjugate Gradients Squared. The adjective 'Squared' is obvious from the polynomial expressions, but it would have been more correct to use the name Bi-CGS. CGS can be represented by the algorithm in Fig. 3, in which K defines a suitable preconditioner).

x_0 is an initial guess; $r_0 = b - Ax_0$;
$\tilde{r}_0$ is an arbitrary vector, such that
$(r_0, \tilde{r}_0) \neq 0$,
e.g., $\tilde{r}_0 = r_0$; $\rho_0 = (r_0, \tilde{r}_0)$;
$\beta_0 = \rho_0$; $p_{-1} = q_0 = 0$;
for $i = 0, 1, 2, \ldots$
  $u_i = r_i + \beta_i q_i$;
  $p_i = u_i + \beta_i(q_i + \beta_i p_{i-1})$;
  solve $\widehat{p}$ from $K\widehat{p} = p_i$;
  $\widehat{v} = A\widehat{p}$;
  $\alpha_i = \frac{\rho_i}{(\tilde{r}_0, v)}$;
  $q_{i+1} = u_i - \alpha_i \widehat{v}$;
  solve $\widehat{u}$ from $K\widehat{u} = u_i + q_{i+1}$
  $x_{i+1} = x_i + \alpha_i \widehat{u}$;
  if x_{i+1} is accurate enough then quit;
  $r_{i+1} = r_i - \alpha_i A\widehat{u}$;
  $\rho_{i+1} = (\tilde{r}_0, r_{i+1})$;
  if $\rho_{i+1} = 0$ then method fails to converge !;
  $\beta_{i+1} = \frac{\rho_{i+1}}{\rho_i}$;
end

Figure 3: CGS with preconditioning

In this scheme the $\widetilde{r}_i$ have been represented by r_i.

CGS has become a rather popular method over the years, because of its fast convergence, compared with Bi-CG, and also because of the modest computational costs per iteration step, as compared with GMRES. Comparisons of these three methods have been appeared in many studies (see, e.g., [36, 9, 34, 31]). However, CGS usually shows a very irregular convergence behaviour, which can be understood from the observation that all kinds of irregular convergence effects in Bi-CG are squared. This behaviour can even lead to cancellation and a spoiled solution [49], and one should always check the value of $||b - Ax_j||_2$ for the final approximation x_j. In Section 3.3 we shall discuss the, sometimes disastrous, effects of irregular convergence.

When the matrix A is symmetric positive definite, Bi-CG produces the same x_i and r_i as CG, and hence CGS does not break down then.

In the case of nonsymmetry of A, a thorough convergence analysis has not yet been given. Surprisingly enough the method, applied to a suitably preconditioned linear system, does often not really suffer from nonsymmetry.

Note that the computational costs per iteration are about the same for Bi-CG and CGS, but CGS has the advantage that only the matrix A itself (and its preconditioner) is involved and not its transpose. This avoids the necessity for complicated data-structures so that matrix vector products Ax and A^*y can be evaluated equally fast).
For practical implementations in a parallel, it might be a slight disadvantage that Ap_i and $A(u_i + q_{i+1})$ cannot be computed in parallel, whereas both the matrix-vector products can be done in parallel in the Bi-G method.

Freund [21] suggested a squared variant of QMR, which was called TFQMR. His experiments show that TFQMR is not necessarily faster than CGS, but it has certainly a much smoother convergence behavior.

In a recent paper, Zhou and Walker [54] have shown that the Quasi-Minimum Residual approach can be followed for other methods, such as CGS and Bi-CGSTAB, as well. The main idea is that in these methods the approximate solution is updated as

$$x_{i+1} = x_i + \alpha_i p_i,$$

and the corresponding residual is updated as

$$r_{i+1} = r_i - \alpha_i A p_i.$$

This means that $AP_i = W_i R_{i+1}$, with W_i a lower bidiagonal matrix. The x_i are combinations of the p_i, so that we can try to find the combination $P_i y_i$ for which $||b - AP_i y_i||_2$ is minimal. If we insert the expression for AP_i, and ignore the fact that the r_i are not orthogonal, then we can minimize the norm of the residual in a quasi-minimum least squares sense, similar to QMR.

3.3. EFFECTS OF IRREGULAR CONVERGENCE

As we have mentioned in our discussion on CGS, irregular convergence can have negative effects on the accuracy of the approximate solution. By very irregular convergence we refer to the situation where successive residual vectors in the iterative process differ in orders of magnitude in norm, and some of these residuals may be even much bigger in norm than the starting residual. We will indicate why this is a point of concern, even if eventually the (updated) residual satisfies a given tolerance. For more details we refer to Sleijpen et al [41, 42].

We will say that an algorithm is *accurate* for a certain problem if the *updated residual* r_j and the *true residual* $b - Ax_j$ are of comparable size for the j's of interest.

The best we can hope for is that for each j the error in the residual is only the result of applying A to the update w_{j+1} for x_j in finite precision arithmetic:

$$r_{j+1} = r_j - Aw_{j+1} - \Delta_A w_{j+1} \tag{9}$$

if

$$x_{j+1} = x_j + w_{j+1}, \tag{10}$$

for each j, where Δ_A is an $n \times n$ matrix for which $|\Delta_A| \preceq n_A \overline{\xi} |A|$: n_A is the maximum number of non-zero matrix entries per row of A, $|B| \equiv (|b_{ij}|)$ if $B = (b_{ij})$, $\overline{\xi}$ is the relative machine precision, the inequality $\preceq$ refers to element-wise $\leq$. In the Bi-CG type methods that we consider, we compute explicitly the update Aw_j for the residual r_j from the update w_j for the approximation x_j by matrix multiplication: for this part, (9) describes well the local deviations caused by evaluation errors.

In the "ideal" case (i.e. situation (9) whenever we update the approximation) we have that

$$\begin{aligned} r_k - (b - Ax_k) &= \sum_{j=1}^{k} \Delta_A w_j \\ &= \sum_{j=1}^{k} \Delta_A (e_{j-1} - e_j), \end{aligned} \tag{11}$$

where the perturbation matrix Δ_A may depend on j and e_j is the approximation error in the jth approximation: $e_j \equiv x - x_j$. Hence,

$$\begin{gathered} |\|r_k\| - \|b - Ax_k\|| \leq \\ 2k\, n_A \overline{\xi}\, \||A|\| \sum_{j=0}^{k} \|e_j\| \leq \\ 2\Gamma \overline{\xi} \sum_{j=0}^{k} \|r_j\|, \\ \text{where } \Gamma \equiv n_A \||A|\| \|A^{-1}\|. \end{gathered} \tag{12}$$

Except for the factor Γ, the last upper–bound appears to be rather sharp. We see that approximations with large approximation errors may ultimately lead to an inaccurate result. Such large local approximation errors are typical for CGS, and Van der Vorst[49] describes an example of the resulting numerical inaccuracy is given. If there are a number of approximations with comparable large approximation errors, then their multiplicity may replace the factor k, otherwise it will be only the largest approximation error that makes up virtually the bound for the deviation.

We will first discuss variants of hybrid Bi-CG methods with a smoother and also often faster convergence behavior. The convergence behavior of all these methods can be made very smooth by local minimization techniques, discussed in [54]. These additional smoothing techniques do not esentially change the speed of convergence, however, since they use intermediate results of the unsmoothed procedure, and these may already have been contaminated with errors due to finite precision computations.

The effects of irregular convergence, that is large intermediate residual norms, are felt in all those approaches, although at different degrees. We will see, in Section 4, how these effects can be largely undone by a careful update procedure for the current approximations for the solution as well as the residual. This *reliable updating* can be done by adding only a few additional statements to an existing code.

3.4. GENERALIZED CGS

We have had very good experiences with CGS in the context of solving nonlinear problems with Newton's method. Before we consider the problem how to modify CGS, we will first briefly discuss the effects of using CGS in a Newton solver.

In the Newton method one has to solve a Jacobian system for the correction. This can be done by any method of choice, e.g., CGS or Bi-CGSTAB (see next section). Often fewer Newton steps are required to solve a non-linear problem accurately when using CGS. Although Bi-CGSTAB often solves each of the linear systems (defined by the Jacobi matrices) faster, the computational gain in these inner loops does not always compensate for the loss in the outer loop because of more Newton steps.

This phenomenon can be understood as follows.
For eigenvalues λ that are extremal in the convex hull of the set of all eigenvalues of A (the Jacobian matrix), the values $P_i(\lambda)$ of the Bi-CG polynomials P_i tend to converge more rapidly towards zero than for eigenvalues λ in the interior. Since CGS squares the Bi-CG polynomials, CGS may be expected to reduce extremely well the components of the initial residual r_0 in the direction of the eigenvectors associated with extremal eigenvalues λ: with reduction factor $P_i(\lambda)^2$. Of course, the value $P_i(\lambda)$ can also be large, specifically for interior eigenvalues and in an initial stage of the process. CGS amplifies the associated components, (which also explains the typical irregular convergence behavior of

CGS). The Bi-CGSTAB polynomial Q_i does not have this tendency of favoring the extremal eigenvalues. Therefore, the BiCGstab methods tend to reduce all eigenvector components equally well: on average, the "interior components" of a Bi-CGSTAB residual r_i are smaller than the corresponding components of a CGS residual $\widetilde{r}_i$, while, with respect to the exterior components the situation is the other way around. However, the non-linearity of a non-linear problem seems often to be represented rather well by the space spanned by the "extremal eigenvectors". With respect to this space, and hence with respect to the complete space, Newtons scheme with CGS behaves like an exact Newton scheme.

We would like to preserve the property of quadric reduction in almost converged eigendirections, but at the same time we would like to avoid the squaring effects in the non-converged eigendirections. Several suggestions for this are made by Fokkema et al [19]. They suggest a so-called Generalized CGS scheme, in which on of the two Bi-CG polynomial factors is replaced by a nearby polynomial.

One suggestion is to carry out the recursion for the second Bi-CG polynomila with two different starting vectors. This leads to a Bi-CG polynomial that will have about the same reducing effects in converged eigendirections, but that is different in other directions. Examples are given in [19], that show some improvement for this approach. The computational costs are about the same as for the original CGS scheme.

Another suggestion is to replace one Bi-CG factor $P_i(A)$ by $(I - \omega A)P_{i-1}(A)$, that is the Bi-CG iteration polynomial of the previous iteration step combined with an appropriate linear factor (for instance a steepest descent factor). This approach has turned out a real improvement in some situations, which may be explained by the observation that when $A = A^T$, the zero's of $P_i(t)$ and $P_{i-1}(t)$ intersect. This means that local extrema of $P_i(t)$ are likely to be annihilated by the zero's of $P_{i-1}(t)$.

For more information on the GCGS schemes, as well as for examples, see [19].

3.5. Bi-CGSTAB

Bi-CGSTAB [49] is based on the observation that the Bi-CG vector r_i is orthogonal to the entire subspace $K^i(A^*, w_1)$, so that we can construct r_i by requiring it to be orthogonal to other basis vectors for $K^i(A^T, w_1)$. As a result, we can, instead of squaring the Bi-CG polynomial, construct iteration methods, by which x_i are generated so that $r_i = \tilde{P}_i(A)P_i(A)r_0$ with other i^{th} degree polynomials $\tilde{P}$. An obvious possibility is to take for $\tilde{P}_j$ a polynomial of the form

$$Q_i(x) = (1 - \omega_1 x)(1 - \omega_2 x)...(1 - \omega_i x), \tag{13}$$

and to select suitable constants ω_j. This expression leads to an almost trivial recurrence relation for the Q_i.

In Bi-CGSTAB ω_j in the jth iteration step is chosen as to minimize r_j with respect to ω_j, for residuals that can be written as $r_j = (I - \omega_j A)Q_{j-1}(A)P_j(A)r_0$. The preconditioned Bi-CGSTAB algorithm for solving the linear system $Ax = b$, with preconditioning K reads as in Fig. reffig:bicgstab.

> x_0 is an initial guess; $r_0 = b - Ax_0$;
> $\bar{r}_0$ $(= w_1)$ is an arbitrary vector, such that
> $\quad(\bar{r}_0, r_0) \neq 0$, e.g., $\bar{r}_0 = r_0$;
> $\rho_{-1} = \alpha_{-1} = \omega_{-1} = 1$;
> $v_{-1} = p_{-1} = 0$;
> for $i = 0, 1, 2, \ldots$
> $\quad\rho_i = (\bar{r}_0, r_i)$; $\beta_{i-1} = (\rho_i/\rho_{i-1})(\alpha_{i-1}/\omega_{i-1})$;
> $\quad p_i = r_i + \beta_{i-1}(p_{i-1} - \omega_{i-1}v_{i-1})$;
> $\quad$Solve $\widehat{p}$ from $K\widehat{p} = p_i$;
> $\quad v_i = A\widehat{p}$;
> $\quad\alpha_i = \rho_i/(\bar{r}_0, v_i)$;
> $\quad s = r_i - \alpha_i v_i$;
> $\quad$if $\|s\|$ small enough then
> $\quad\quad x_{i+1} = x_i + \alpha_i\widehat{p}$; quit;
> $\quad$Solve z from $Kz = s$;
> $\quad t = Az$;
> $\quad\omega_i = (t, s)/(t, t)$;
> $\quad x_{i+1} = x_i + \alpha_i\widehat{p} + \omega_i z$;
> $\quad$if x_{i+1} is accurate enough then quit;
> $\quad r_{i+1} = s - \omega_i t$;
> end

Figure 4: Bi-CGSTAB with preconditioner K

The matrix K in this scheme represents the preconditioning matrix and the way of preconditioning [49]. The above scheme carries out the Bi-CGSTAB procedure for the explicitly postconditioned linear system

$$AK^{-1}y = b,$$

but the vectors y_i and the residual have been backtransformed to the vectors x_i and r_i corresponding to the original system $Ax = b$. Compared with CGS, two extra inner products need to be calculated.
In exact arithmetic, the α_j and β_j have the same values as those generated by Bi-CG and CGS.
Bi-CGSTAB can be viewed as the product of Bi-CG and GMRES(1). Of course,

other product methods can be formulated as well. Gutknecht [26] has proposed BiCGSTAB2, which is constructed as the product of Bi-CG and GMRES(2). Sleijpen and Fokkema [43] have proposed variants in which Bi-CG is combined with GMRES(ℓ): Bi-CGSTAB(ℓ). These variants will be discussed in the next section.

3.5.1. *Bi-CGSTAB(2) and variants*

A dubious aspect of Bi-CGSTAB is that the polynomial factor Q_k has only real roots by construction. It is well-known that optimal reduction polynomials for matrices with complex eigenvalues may have complex roots as well. If, for instance, the matrix A is real skew-symmetric, then GCR(1) stagnates forever, whereas a method like GCR(2) (or GMRES(2)), in which we minimize over two combined successive search directions, may lead to convergence, and this is mainly due to the fact that then complex eigenvalue components in the error can be effectively reduced.
This point of view was taken in [26] for the construction of the BiCGSTAB2 method. In the odd-numbered iteration steps the Q-polynomial is expanded by a linear factor, as in Bi-CGSTAB, but in the even-numbered steps this linear factor is discarded, and the Q-polynomial from the previous even-numbered step is expanded by a quadratic $1 - \alpha_k A - \beta_k A^2$. For this construction the information from the odd-numbered step is required. It was anticipated that the introduction of quadratic factors in Q might help to improve convergence for systems with complex eigenvalues, and, indeed, some improvement was observed in practical situations (see also [35]).
It has been shown in [43] that the polynomial Q can also be constructed as the product of ℓ-degree factors, without the construction of the intermediate lower degree factors. The main idea is that ℓ successive Bi-CG steps are carried out, where for the sake of an A^*-free construction the already available part of Q is expanded by simple powers of A. This means that after the Bi-CG part of the algorithm vectors from the Krylov subspace $s, As, A^2s, ..., A^\ell s$, with $s = P_k(A)Q_{k-\ell}(A)r_0$ are available, and it is then relatively easy to minimize the residual over that particular Krylov subspace. There are variants of this approach in which more stable bases for the Krylov subspaces are generated [42], but for low values of ℓ a standard basis satisfies, together with a minimum norm solution obtained through solving the associated normal equations (which requires the solution of an ℓ by ℓ system. In most cases Bi-CGSTAB(2) will already give nice results for problems where Bi-CGSTAB or BiCGSTAB2 may fail. Note, however, that, in exact arithmetic, if no breakdown situation occurs, BiCGSTAB2 would produce exactly the same results as Bi-CGSTAB(2) at the even-numbered steps.

Bi-CGSTAB(2) can be represented by the following algorithm:

x_0 is an initial guess; $r_0 = b - Ax_0$;

$\widehat{r}_0$ is an arbitrary vector, such that $(r, \widehat{r}_0) \neq 0$,
e.g., $\widehat{r}_0 = r$;
$\rho_0 = 1; u = 0; \alpha = 0; \omega_2 = 1$;
for $i = 0, 2, 4, 6, ...$
$\rho_0 = -\omega_2 \rho_0$
even BiCG step: $\rho_1 = (\widehat{r}_0, r_i); \beta = \alpha\rho_1/\rho_0; \rho_0 = \rho_1$
$u = r_i - \beta u$;
$v = Au$
$\gamma = (v, \widehat{r}_0); \alpha = \rho_0/\gamma$;
$r = r_i - \alpha v$;
$s = Ar$
$x = x_i + \alpha u$;
odd BiCG step: $\rho_1 = (\widehat{r}_0, s); \beta = \alpha\rho_1/\rho_0; \rho_0 = \rho_1$
$v = s - \beta v$;
$w = Av$
$\gamma = (w, \widehat{r}_0); \alpha = \rho_0/\gamma$;
$u = r - \beta u$
$r = r - \alpha v$
$s = s - \alpha w$
$t = As$
GCR(2)-part: $\omega_1 = (r, s); \mu = (s, s); \nu = (s, t); \tau = (t, t)$;
$\omega_2 = (r, t); \tau = \tau - \nu^2/\mu; \omega_2 = (\omega_2 - \nu\omega_1/\mu)/\tau$;
$\omega_1 = (\omega_1 - \nu\omega_2)/\mu$
$x_{i+2} = x + \omega_1 r + \omega_2 s + \alpha u$
$r_{i+2} = r - \omega_1 s - \omega_2 t$
if x_{i+2} accurate enough then quit
$u = u - \omega_1 v - \omega_2 w$
end

For more general Bi-CGSTAB(ℓ) schemes see [43, 42].
Another advantage of Bi-CGSTAB(2) over BiCGSTAB2 is in its efficiency. The Bi-CGSTAB(2) algorithm requires 14 vector updates, 9 inner products and 4 matrix vector products per full cycle. This has to be compared with a combined odd-numbered and even-numbered step in BiCGSTAB2, which requires 22 vector updates, 11 innerproducts, and 4 matrix vector products, and with two steps of Bi-CGSTAB which require 4 matrix vector products, 8 innerproducts and 12 vector updates. The numbers for BiCGSTAB2 are based on an implementation described in [35].
Also with respect to memory requirements, Bi-CGSTAB(2) takes an intermediate position: it requires 2 n-vectors more than Bi-CGSTAB and 2 n-vectors less than BiCGSTAB2.

For distributed memory machines the inner products may cause communication overhead problems (see, e.g., [11]). We note that the Bi-CG steps are very similar to conjugate gradient iteration steps, so that we may consider all

kind of tricks that have been suggested to reduce the number of synchronization points caused by the 4 inner products in the Bi-CG parts. For an overview of these approaches see [5]. If on a specific computer it is possible to overlap communication with communication, then the Bi-CG parts can be rescheduled as to create overlap possibillities:
1. The computation of ρ_1 in the even Bi-CG step may be done just before the update of u at the end of the GCR part.
2. The update of x_{i+2} may be delayed until after the computation of γ in the even Bi-CG step.
3. The computation of ρ_1 for the odd Bi-CG step can be done just before the update for x at the end of the even Bi-CG step.
4. The computation of γ in the odd Bi-CG step has already overlap possibillities with the update for u.
For the GCR(2) part we note that the 5 innerproducts can be taken together, in order to reduce start-up times for their global assembling. This gives the method Bi-CGSTAB(2) a (slight) advantage over Bi-CGSTAB. Furthermore, we note that the updates in the GCR(2) may lead to more efficient code than for BiCGSTAB, since some of them can be combined.

3.6. ACCURATE ITERATION COEFFICIENTS

The Bi-CGSTAB methods are designed in order to obtain better convergence, by combining Bi-CG with the smooth convergence of GMRES. It is, of course, important to avoid loss of local bi-orthogonality in the underlying Bi-CG process, in order to maintain the full power of the Bi-CG part. However, the local bi-orthogonality may be disturbed by, for instance, evaluation errors in the Bi-CG coefficients α and β. They are quotients of scalars $\rho \equiv (r_i, \widehat{r}_0)$ and $\gamma \equiv (Ap, \widehat{r}_0)$ (see the algorithms for BiCGSTAB and BiCGSTAB(2)) and they will be inaccurate if ρ or γ is relatively small (see (15)). The question is, when does this occur and how can it be avoided? Here, we will concentrate on ρ only, but similar arguments apply to γ as well.

As we have seen, The Bi-CGSTAB residual r_i can be expressed as $r_i = Q_i(A)P_i(A)r_0$ where Q_i is an appropriate polynomial of degree i with $Q_i(0) = 1$. We define a quantity ρ_i by

$$\rho_i \equiv (Q_i(A)P_i(A)r_0, \widehat{r}_0). \tag{14}$$

The scalar ρ_i can be small if the underlying Bi-Lanczos process has a near break-down, that is when

$$|(Q_i(A)P_i(A)r_0, \widehat{r}_0)| \ll \|Q_i(A)P_i(A)r_0\|_2 \|\widehat{r}_0\|_2$$

independent of the choice for Q_i. Also a poor choice for Q_i may lead to a small ρ_i (which occurs in Bi-CGSTAB if the GCR(1) part (nearly) stagnates).
We will focus on this last possibillity, that is we will not consider the typical breakdown situations inherent to the Bi-CG part; these have to be repaired by

other means. In other words, we will assume that the Bi-Lanczos process itself (and the LU decomposition) does not (nearly) break down.

The relative rounding error ϵ_i in ρ_i can relatively and sharply be bounded by

$$|\epsilon_i| \leq n\overline{\xi}\frac{(|r_i|, |\widehat{r}_0|)}{|(r_i, \widehat{r}_0)|} \leq n\overline{\xi}\frac{\|r_i\| \, \|\widehat{r}_0\|}{|(r_i, \widehat{r}_0)|}. \tag{15}$$

For a small relative error the expression at the right-hand side should be small.

Since the Bi-CG residual $P_i(A)r_0$, here to be denoted by s_i, is orthogonal to $K^i(A^T; \widehat{r}_0)$ it follows that

$$(r_i, \widehat{r}_0) = \theta_i (A^i s_i, \widehat{r}_0)$$

where θ_i is the leading coefficient in $Q_i(A)$:

$$Q_i(A) = \theta_i A^i + \theta_{i-1}^{(i)} A^{i-1} + \ldots.$$

Therefore, since $\|\widehat{r}_0\|/|(A^i s_i, \widehat{r}_0)|$ does not depend on Q_i, minimizing the right-hand side of (15) is equivalent to minimizing

$$\frac{\|Q_i(A)s_i\|}{|\theta_i|} \tag{16}$$

with respect to all polynomials Q_i of exact degree i with $Q_i(0) = 1$. It can be shown that this minimization problem is solved by the FOM polynomial P_i^F, associated with the initial residual s_i: P_i^F is the ith degree polynomial for which $r_i^F = P_i^F(A)s_i$ (cf. Sect. 2.2). This polynomial is characterized by:

$$P_i^F(A)s_i \perp K^i(A; s_i) \text{ and } P_i^F(0) = 1.$$

For optimally accurate coefficients, Q_i should be the FOM polynomial associated with the current Bi-CG residual s_i, which means that it should be a different polynomial in each iteration step. It is clear that this is not very practical.

For efficiency reasons, we have used products of first degree polynomials in Bi-CGSTAB and products of degree ℓ polynomials in Bi-CGSTAB(ℓ). Of course, our arguments can also be applied to such low degree factors. Therefore, suppose that $s = Q_{i-\ell}(A)P_i(A)r_0$ (as in Bi-CGSTAB(ℓ)) has been computed and that the vectors s, As, ..., $A^\ell s$ are available. The suggestion for Bi-CGSTAB(ℓ) to minimize the residual over this particular Krylov subspace is equivalent to selecting a polynomial factor $q_\ell^{(i)}$ ($Q_i = q_\ell^{(i)} Q_{i-\ell}$), of exact degree ℓ with $q_\ell^{(i)}(0) = 1$ such that

$$\|q_\ell^{(i)}(A)s\| \text{ is minimal.} \tag{17}$$

However, for optimal accurate coefficients, one should attempt to minimize

$$\frac{\|q_\ell^{(i)}(A)s\|}{|\theta_i|} \tag{18}$$

where θ_i is the leading coefficient in $q_\ell^{(i)}(A) = \theta_i A^\ell + \ldots$.
The GMRES polynomial q_ℓ^G of degree ℓ solves (17), the FOM polynomial q_ℓ^F solves (18). The FOM polynomial is not optimal for minimizing the norm of the residual, since

$$\|q_\ell^G(A)s\| = |c_\ell| \, \|q_\ell^F(A)s\|$$

with c_ℓ as the cosine in the Givens reduction for the elimination of the element $h_{\ell+1,\ell}$ in GMRES.
On the other hand, the GMRES polynomial is not optimal for realizing small evaluation errors in the iteration coefficients, since [41]:

$$\frac{\|q_\ell^F(A)s\|}{|\theta_\ell^F|} = |c_\ell| \frac{\|q_\ell^G(A)s\|}{|\theta_\ell^G|}.$$

For degree 1 factors, as in Bi-CGSTAB, we have that

$$c_1 = \frac{(s, As)}{\|s\| \, \|As\|}. \tag{19}$$

(in the BiCGSTAB algorithm, t represents As).

Clearly, for extremely small $|c_i|$, say $|c_i| \leq \sqrt{\xi}$ (in the $\ell = 1$ case, this means that s and As are almost orthogonal), GMRES polynomials for the degree ℓ factors may lead to large inaccuracies in the coefficients ρ_i, α and β, while FOM polynomials on the other hand will lead to large residuals. In both situations, this will have noticeable effects on the speed of convergence.
The same phenomena can be observed when in a consecutive number of sweeps $|c_i|$ is small, but not necessarily extremely small (say, it takes k sweeps before $|c_{i-k}c_{i-k+1}\cdots c_i| \leq \sqrt{\xi}$). In other words, the inaccuracies seem to accumulate. This seems to occur quite often in actual computation. E.g., for linear systems from PDEs with large advection terms, Bi-CGSTAB often stagnates, although all c_i may be larger than, say .1, and none of the ω_i are relatively small ($\omega_i = c_i\|s\|/\|As\|$).

Both Bi-CGSTAB and BiCGstab(ℓ) are built on top of the same Bi-CG process, but for almost the same computational costs, one sweep of BiCGstab(ℓ) covers the same Bi-CG traject as ℓ sweeps of Bi-CGSTAB. In one sweep of Bi-CGstab(ℓ), GMRES(ℓ) is applied once, while in ℓ sweeps of BiCGSTAB GMRES(1) is applied ℓ times all together.
For two reasons it is preferable to use GMRES(ℓ) instead of $\ell\times$GMRES(1):

1. GMRES(ℓ) usually gives a much better residual reduction than ℓ times GMRES(1).

2. In ℓ steps of GMRES(1), ℓ small c_1's may contribute to inaccuracies in the coefficients α and β, where GMRES(ℓ) contributes to this only once. Furthermore, it is the reduction obtained by GMRES(ℓ), wit respect to GMRES($\ell - 1$), that counts, and because of the superlinear convergence behavior of GMRES this reduction will in general be smaller than the reductions obtained with GMRES(1) steps.

However, we do not recommend to take ℓ large; $\ell = 2$ or $\ell = 4$ will usually lead already to almost optimal efficiency. The computational costs increase slightly by increasing ℓ (i.e. $2\ell + 10$ vector updates and $\ell + 7$ inner products per 4 matrix multiplications), and more vectors have to be stored ($2\ell + 5$ vectors). Moreover, the method is less accurate for larger ℓ due to the fact that intermediate residuals (as r and $r - \omega_1 s$ in the Bi-CGSTAB(2) algorithm) can be large, with similar negative effects as in Section 3.3.

For Bi-CGSTAB there is a simple strategy that relaxes the danger of error amplification in consecutive sweeps with small $|c_i|$: replace in the Bi-CGSTAB algorithm the line

$$\text{`}\omega = (s,t)/(t,t)\text{'}$$

by the piece of code in Fig. 5.

$$\vdots$$

$$c = (s,t)/(\|s\|\,\|t\|);$$

$$\omega = \text{sign}(c)\max\left(|c|, \frac{1}{2}\sqrt{2}\right)\|s\|/\|t\|;$$

$$\vdots$$

Figure 5: Modification for ω in Bi-CGSTAB

In this way we limit the influence of $|c|$. The constant $\frac{1}{2}\sqrt{2}$ represents a compromise between two orthogonal vectors (the FOM-vector and a vector that gives no reduction: these vectors are orthogonal and an intermediate vector under an angle of 45 degrees has been chosen), it may be replaced by any other fixed non-small constant less than 1. A similar simple strategy can be designed for Bi-CGSTAB(ℓ); see [41] for details.

4. Matching Updates for x and r

In the Bi-CG related methods we see that the approximations for x and the residual vector r are updated by different vectors, and that the value for x does not influence the further iteration process, whereas the value for r does. In exact arithmetic the updated r is equal to the true residual $b - Ax$, but in rounded arithmetic it is unavoidable that differences between r and $b - Ax$ occur. This means that we may be misled for our stopping criteria, which are usually based upon knowledge of the updated r (and that we may have iterated too far in vain).

In this section we will discuss some techniques that have been proposed recently for the improvement of the updating steps [44]. It turns out that this can be settled by relatively easy means.

Although the techniques in the previous section led to smoother and faster convergence and more accurate approximations the approximation may still not as accurate as possible. Here, we strive for optimal accuracy, i.e the updated r_i should be very close to the values of $b - Ax_i$, while leaving the convergence of the updated r intact.

First, we observe that even if x_m is the exact solution then the residual, computed in rounded arithmetic as $b - Ax_m$, may not be expected to be zero: using the notation of Section 3.3,

$$\begin{aligned} \|b - Ax_m\| &\leq \overline{\xi}\,(\|b\| + n_A\, \||A|\|\, \|x_m\|) \\ &\leq 2\Gamma\,\overline{\xi}\,\|b\|. \end{aligned} \tag{20}$$

Therefore, the best we can strive for is an approximation x_m for which the true residual and the updated one differ in order of magnitude by the initial residual times the relative machine precision ($\mathcal{O}(\overline{\xi}\,\|r_0\|)$; recall that we assumed $x_0 = 0$, and hence $r_0 = b$).

Now it becomes also obvious why it is a bad idea to replace the updated residual in each step by the true one. Except from the fact that this would cost an additional matrix vector multiplication in each step, it also introduces errors in the recursions for the residuals. Although these errors may be expected to be small relatively to r_0, they will be large relatively to r_i if $\|r_i\| \ll \|r_0\|$. This perturbs the local bi-orthogonality of the underlying Bi-CG process and it may significantly slow down the speed of convergence. This observation suggests to replace the updated residual by the true one only if the updated residual has the same order of magnitude as the initial residual. However, meanwhile x_i and r_i may have drifted apart, and replacing r_i by $b - Ax_i$ brings in the "error of x_i" in the recursion (bounded as in (12)), and again the speed of convergence may be affected. Although it is a good idea to use true residuals at strategic places, the approximation x_i should first be 'tied' more closely to the updated residual r_i. We can achieve this by updating x_i cumulatively: if $x_i = x_0 + w_1 + \ldots + w_i$

(cf. (9)) then we actually compute x_i in groups as

$$x_i = x_0 + x'_1 + x'_2 + \ldots \tag{21}$$

where, for some decreasing sequence of indices $\pi(1) = 1, \pi(2), \ldots, x'_j$ represents the sum of a group;

$$x'_j = w_{\pi(j)} + w_{\pi(j)+1} + \ldots + w_{\pi(j+1)-1}, \text{ etc..}$$

Simultaneously, we compute r_i as

$$r_i = r_0 - Ax'_1 - Ax'_2 - \ldots . \tag{22}$$

In this way we can control the size of the updates for x_i and r_i, and we avoid large errors (cf. (12)): for a proper choice of the $\pi(j)$, the x'_j will be small even if some of the w_j are large. This groupwise update can be interpreted as a shift of the given linear system.

In desigining a modification strategy for the algorithms, we kept in mind that we only may allow errors which
(a) are small with respect to the initial residual r_0 (otherwise accuracy will be disturbed) and
(b) are small with respect to the present updated residual r_i (otherwise the local bi-orthogonality may be affected).

It is no restriction to take $x_0 = 0$, since this situation can be forced simply by a shift: shift x by x_0, and b by Ax_0. This shift can be made explicit in the hybrid Bi-CG algorithms by making three changes:
(i) adding as a last line to the initialization phase

$$x = x_0; \quad x' = 0; \quad b' = r_0;$$

(ii) adding as a last line in the algorithms (just after 'end')

$$x = x + x';$$

(iii) replacing all x_i (and x) by x' (skipping the index i).
Even in rounded arithmetic, this modification will not change the value of any of the vectors and scalars in the computational scheme, except for the x's. Since $x + x'$ is the approximation that we are interested in, one also may want to change the termination criterion. We propose to replace the line

if x is accurate enough then quit;

by

if $\|r_{i+1}\|$ is small enough then quit;

To allow for a more accurate way of updating of the residual and the approximation, we suggest to add another few lines just before 'end' in the algorithm, as is shown in Fig. 6.

```
⋮
x = x_0;   x' = 0;   b' = r_0;
for i = 0, 1, 2, ...
    ⋮  Replace all x_i and x by x'.
    ⋮
    if r_{i+1} is small enough then quit;
    set 'compute_res' and 'update_app';
    if 'compute_res' is true
        r_{i+1} = b' - Ax';
        if 'update_app' is true
            x = x + x'; x' = 0; b' = r_{i+1};
        endif
    endif
end
x = x + x';
```

Figure 6: Matching update strategy

We propose to replace the updated residual by the true one on strategically chosen steps (we have to explain when the value of the boolean functions *'compute_res'* is true). However, we also propose to shift the problem once in a while (when the boolean function *'update_app'* is true) in order to let the right-hand decrease (cf. (20)), and this is what happens in equations (21) and (22). We use the property that, in exact arithmetic, also these intermediate shifts do not change the iteration parameters and vectors (except for the vectors x). Observe that the updated residual r_{i+1} is replaced by the true residual $b' - Ax'$ of the shifted problem if *'compute_res'* is true.

For this we propose the following strategy.
Update x and b' only if the residual is significantly smaller than the initial residual, while an intermediate residual was larger (cf. (21), (22) and reminder (a)):

$$\begin{aligned}&\textit{'update\_app'} = \text{true}\\ &\text{if} \quad \|r_{i+1}\| \leq \|b\|/100 \ \ \& \ \ \|b\| \leq \mu \qquad (23)\\ &\text{else } \textit{'update\_app'} = \text{false},\end{aligned}$$

where $\mu \equiv \max \|r_i\|$ and the maximum is taken over all residuals since the previous update of x and b' (since the previous *'update_app'* is true).

The bound in (20) suggests that the norm $\|b\|$ of the initial residual should be used as criterion for shifting the problem (*'update_app'* is true if $\|r_{i+1}\| \leq \|b\|$ & $\|r_i\| \geq \|b\|$). However, if the process converges irregularly this would lead to many shifts. The relaxed version in (23) turns out to work equally well at less costs.
Compute a true residual whenever *'compute_res'* is true and if a previous residual is larger than the initial residual and significantly larger than the present updated residual:

$$\begin{aligned} &\text{'compute\_res'} = \text{true} \\ &\text{if} \quad \|r_{i+1}\| \leq M/100 \ \ \& \ \ \|b\| \leq M \\ &\qquad \text{or 'update\_app' is true} \\ &\text{else 'compute\_res'} = \text{false}, \end{aligned} \tag{24}$$

where $M \equiv \max \|r_i\|$ and the maximum is taken over all residuals since the last computation of the true residual.
Replacing the updated residual by the true one perturbs the recursion for the residuals. If the residual decreases too much since the previous replacement, the perturbation may become large relatively to the present residual (reminder (b)). Therefore, *'compute_res'* may be true more often than *'update_app'*.

We suggest to add the above strategy to an existing code. That means that an additional matrix-vector multiplication has to be performed whenever a true residual has to be computed. The conditions (23) and (24) are chosen as to minimize the number of these additional computations. One also may try to skip a matrix-vector multiplication in one of the preceding lines of the algorithm, which requires some additional care for Bi-CGSTAB(ℓ), but which easily can be accomplished for CGS.

CGS can be modified as suggested above, in a way that the new lines do not require additional matrix vector multiplications, and there is no need to restrict the number of computations of true local residuals. For more information on this, see [44].
Especially for the Bi-CGSTAB methods, the simple strategy of Fig. 6 with update criterions (23) and (24) does not lead to much additional work. The additional computation of a true residual takes place after the process encounters residuals that are (much) larger than the initial residual. Since Bi-CGSTAB(ℓ) tends to show much smoother convergence behavior than CGS, for small ℓ, the additional work in these methods is usually much less than for CGS.

5. Preconditioning

A popular class of preconditioners is known as incomplete factorizations: ILU [29]. They can be thought of as approximating the exact LU factorization of a given matrix A (e.g. computed via Gaussian elimination) by disallowing certain fill-ins. As opposed to other PDE-based preconditioners such as multigrid and domain decomposition, this class of preconditioners are primarily algebraic in nature and can in principle be applied to any sparse matrices. In this paper we will discuss some new viewpoints for the construction of effective preconditioners.

In particular, we will discuss parallelization aspects, including re-ordering, series expansion and domain decomposition techniques. Generally, this class of preconditioner does not possess a high degree of parallelism in its original form. Re-ordering and approximations by truncating certain series expansion will increase the parallelism, but usually with a deterioration in convergence rate. Domain decomposition offers a compromise.

The general problem of finding a preconditioner for a linear system $Ax = b$ is to find a matrix M (the *preconditioner*) with the properties that M is a good approximation to A in some sense and that the system $Mx = b$ is much easier to solve than the original system. We will then solve the *preconditioned system* $M^{-1}Ax = M^{-1}b$ (or the symmetric version $(L^{-1}AL^{-T})(L^Tx) = L^{-1}b$ when both A and $M = LL^T$ are symmetric positive definite) by standard iterative methods such as the conjugate gradient method, in which only the actions of A and M^{-1} are needed.
The choice of M varies from purely "black box" algebraic techniques which can be applied to general matrices to "problem dependent" preconditioners which exploits special features of a particular problem class. Obviously, the more special features we know and can exploit in a problem, the better we can construct a preconditioner for it. However, there is still a practical need for preconditioning techniques that can be applied to a general matrix. Incomplete factorization preconditioners are among the candidates for this. They can be thought of as modifications of Gaussian Elimination in which certain fill-ins are disallowed in order to ensure that the action of M^{-1} is inexpensive. These preconditioners can also be viewed as a bridge between direct methods such as Gaussian Elimination and classical relaxation methods such as Jacobi, Gauss-Seidel and SOR. We will discuss some approaches that have been proposed in the literature to efficiently implement the incomplete factorization methods on parallel computers. A shorter survey can also be found in [16].

There are two general approaches for parallelizing numerical methods:

1. extract maximal parallelism from a method which works well on sequential computer, *without* changing its numerical properties,

2. modify or approximate a good sequential method to increase the parallelism available, thus possibly degrading its numerical properties.

There is a fundamental difficulty when applying the above general principles to incomplete factorization methods. The major parallel bottleneck lies in the backsolves involving the triangular LU factors. The same bottleneck arises in computing the LU factors themselves but this occurs only once in the beginning of the iteration. Even though these backsolves possess some degree of parallelism which can be exploited, this is often not sufficient to efficiently exploit many parallel architectures, especially massively parallel ones. On the other hand, modifying or approximating the sequential method in order to increase the amount of parallelism invariably leads to slower convergence rates. This should not be too surprising; it is just an instance of the fundamental trade-off between parallelism (which prefers locality) and fast convergence rate (which prefers global dependence) which governs many genuinely globally coupled systems (e.g. elliptic PDEs). The goal is to make the right trade-off for a given architecture/algorithm configuration.

There are three basic methodologies to extract or increase the parallelism in ILU methods: re-ordering, series expansions (including polynomial preconditioners), and domain decomposition. We shall briefly discuss them next. We shall assume that A comes from finite difference or finite element discretization of a PDE over a regular grid, and our discussions will focus on point ILU as our example. We note that the issue of parallelization are the same for relatex incomplete factorization methods, such as Modified ILU [25], and relaxed incomplete factorizations [2], since the data dependence for these methods is identical.

5.1. RE-ORDERING

The Hyperplane Ordering: Assume that the ILU factors have been computed and consider now the task of computing the product $y = L^{-1}v$. The goal is to find an ordering with which the components of y can be computed with maximal parallel efficiency. This is accomplished by the so-called hyperpane or wavefront ordering. A similar ordering can be used for computing $U^{-1}v$.
It is obvious that if one generalizes this idea to a $d-$dimensional grid with n grid points in each direction, we can extract $O(n^{d-1})$ degree of parallelism. If the number of parallel processors is of the same order, then one can achieve good parallel efficiency. Thus, for a fixed number of processors, higher dimensional problems are easier to parallelise. On the other hand, potential degradation in performance can be caused by memory addressing with unequal stride for $d > 2$ and cache problems with the indirect addressing needed to access data on the hyperplanes when the grid is stored as a 2D array.
The use of hyperplane ordering has been investigated by Radicati and Vitaletti [37] for the IBM 3090. Numerical experiments on the CM2 can be found in Berryman et al [7]. The hyperplane ordering for regular 3D grids has been described in detail for vector computers in van der Vorst [48]. Recently, Barszcz et al [6] gave a data mapping for the hyperplane ordering in 3D useful for distributed memory architectures and compare the performance on an 8-processor Cray Y-MP, a 128 processor Intel iPSC/860 and a 32K processor CM-2.

Multi-color Orderings: Since the degree of parallelism for ILU methods are limited in the natural ordering, a popular alternative is to use orderings that are designed to be more parallel. However, it must be emphasized that most of these orderings are *not* equivalent to the natural ordering, in the sense that the ILU factors computed using these are generally different from those generated using the natural ordering. Thus, the goal is to tradeoff the relatively fast and well-understood convergence rate of the natural ordering for orderings with a high degree of parallelism.
An example is the well-known red-black ordering for 5-point stencils in 2D. Because the red points depend only on the black points but not on each other, they can all be updated in parallel. Thus the degree of parallelism is $n^2/2$, a substantial increase from $O(n)$ for the natural ordering. However, since the data dependence are completely local and there is no global sharing of information, the convergence rate is poor. In fact, Kuo and Chan [27] proved that the condition number of the preconditioned system in the red-black ordering is only about 1/4 that of the *unpreconditioned* system for ILU, MILU and SSOR, with no asymptotic improvement as h tends to zero.
One way to strike a better balance between parallelism and fast convergence is to use more colors [15]. In principle, since the different colors are updated sequentially, using more colors decreases the parallelism but increases the global dependence and hence the convergence. The key is to choose the number of colors to match the architecture. For example, Doi and Hoshi [14] used up to 75 colors for a 76^2 grid on the NEC SX-3/14 and achieved 2 Gflops performance, which is much better than that for the hyperplane ordering.

Multi-wavefront Orderings: A different approach to increase parallelism is to use several hyperplane wavefronts to sweep through the grid, the idea being that all wavefronts can be updated in parallel. For example, van der Vorst [48] considered starting wavefronts from each of the four corners in a 2D rectangular grid, or from the eight corners in a 3D grid. Earlier, Meurant [30] used a similar idea in which the grid is divided into equal parts (e.g. halves or quadrants) and each part is ordered in its own natural ordering.

5.2. SERIES EXPANSIONS

Instead of using an ordering with more parallelism, a quite different approach to increase the parallelism in the naturally ordered ILU method is to replace it by an *approximation* which can be evaluated more efficiently in parallel.
In order to illustrate this, consider the computation of $(I-L)^{-1}v$, which is needed in applying the preconditioner. Here we have assumed without loss of generality that the diagonal entries of the lower triangular factor has been scaled to unity. It can be easily proved that if the spectral radius $\rho(L)$ satisfies $\rho(L) < 1$ and L is n by n strictly lower triangular, then we have the following

two finite expansions:

$$\textbf{Neumann Expansion:} \quad (I - L)^{-1}v = (I + L + L^2 + ... + L^{n-1})v,$$

$$\textbf{Euler Expansion:} \quad (I - L)^{-1}v = (I + L^{2^s})(I + L^{2^{s-1}})...(I + L)v,$$

where $s = \lceil \log_2 n - 1 \rceil$. Note that $L^n = 0$. Each of the terms on the right-hand-side of the above expansions can be evaluated in parallel efficiently because they only involve repeated multiplication of v by sparse matrices. The idea is to then truncate the expansions but keeping enough terms so that the convergence rate is not too adversely affected.
Van der Vorst [47] used this idea when he applied a truncated Neumann expansion to the diagonal blocks (which correspond to grid lines) in the point ILU factorization in order to increase the degree of vectorization. In the same paper, he also used the Euler expansion (of low order).

Finally, a related method is the class of *polynomial preconditioners* in which A^{-1} is approximated by a low degree polynomial in A, chosen in some optimal manner. In [51] it is shown how GMRES can often be effectively preconditioned by a Chebyshev matrix polynomial, for which the coefficients are obtained from eigenvalue approximations from a limited number of GMRES steps. In particular, the harmonic Ritz values have been employed as useful approximations and a surprisingly simple algorithm is presented in [51] for the computation of the Chebyshev parameters of the Chebyshev polynomial over a piecewise linear contour that encloses the eigenvalue approximations. Using this type of relatively expensive polynomial preconditioners often leads to a significant reduction in GMRES steps and hence the required communication-intensive innerproducts have a lesser degrading effect on the parallel performance of the preconditioned GMRES algorithm on distributed memory machines.

5.3. DOMAIN DECOMPOSITION

In this general approach, the physical domain or grid is decomposed into a number of overlapping or non-overlapping subdomains on each of which an independent incomplete factorization can be computed and applied in parallel. The main idea is to obtain more parallelism at the subdomain level rather than at the grid point level. Usually, the interfaces or overlapping region between the subdomains must be treated in a special manner. The advantage of this approach is that it is quite general and can be used with different methods used within different subdomains.

Radicati and Robert [36] used an algebraic version of this approach by computing ILU factors within overlapping block diagonals of a given matrix A. When applying the preconditioner to a vector v, the values on the overlapped region is averaged from the two values computed from the two overlapping ILU

factors.
The approach of Radicati and Robert has been further perfectioned by De Sturler [12], who studies the effects of overlap from the point of view of geometric domain decompositioning. He introduces artificial mixed boundary conditions on the internal boundaries of the subdomains. In [12]:Table 5.8 experimental results are shown for a decomposition in 20×20 slightly overlapping subdomains of a 200×400 mesh for a discretized convection-diffusion equation (5-point stencil). When taking ILU preconditioning for each subdomain, it is shown that the complete linear system can be solved by GMRES on a 400-processor distributed memory Parsytec system with an efficiency in the order of 80% (compared with $\frac{1}{400}$-th of the CPU time of ILU preconditioned GMRES for the unpartitioned system on 1 single processor).

In [46], Tan studies the interface conditions along subdomains and forces continuity for the solution at the interface up to some degree. He proposes to include also mixed derivatives in these relations. The involved parameters can be determined locally by means of normal mode analysis, and they are adapted to the discretized problem. It is shown that the resulting domain decomposition method defines a standard iterative method for some splitting $A = M - N$, and the llocal coupling aimes at minimizing the largest eigenvalues of $I - AM^{-1}$. Of course, this method can be accelerated and impressive results for GMRES acceleration are shown in [46]. Some attention is paid to the case where the solutions for the subdomains are obtained in only modest accuracy per iteration step.

Chan and Goovaert [10] showed that the domain decomposition approach can actually lead to *improved* convergence rates, at least when the number of subdomains is not too large. The reason derives from the well-known divide and conquer effect when applied to methods with superlinear complexity such as ILU: it is more efficient to applied such methods to smaller problems and piece the global solution together.

Recently, Washio and Hayami [53] employed a domain decomposition approach for a rectangular grid by which one step of SSOR is done for the interior part of each subdomain. In order to make this domain-decoupled SSOR more resemble the global SSOR, the SSOR iteration matrix for each subdomain is modified by premultiplying it with a matrix $(I - X_L)^{-1}$ and postmultiplying it by $(I - X_U)^{-1}$. The matrices X_L and X_U depend on the couplings between adjacent subdomains. In order to further improve the parallel performance, the inverses are approximated by low-order truncated Neumann series. A similar approach is suggested in [53] for a block Modified ILU preconditioner. Experimental results have been shown for a 32-processor NEC-Cenju distributed memory computer.

References

[1] W. E. Arnoldi. The principle of minimized iteration in the solution of the matrix eigenproblem. *Quart. Appl. Math.*, 9:17–29, 1951.

[2] O. Axelsson. *Iterative Solution Methods.* Cambridge University Press, Cambridge, 1994.

[3] O. Axelsson and P. S. Vassilevski. A black box generalized conjugate gradient solver with inner iterations and variable-step preconditioning. *SIAM J. Matrix Anal. Appl.*, 12(4):625–644, 1991.

[4] Z. Bai, D. Hu, and L. Reichel. A Newton basis GMRES implementation. Technical Report 91-03, University of Kentucky, 1991.

[5] R. Barrett, M. Berry, T. Chan, J. Demmel, J. Donato, J. Dongarra,V. Eijkhout, R. Pozo, C. Romine, and H. van der Vorst. *Templates for the Solution of Linear Systems: Building Blocks for Iterative Methods.* SIAM, Philadelphia, PA, 1994.

[6] E. Barszcz, R. Fatoohi, V. Venkatakrishnan, and S. Weeratunga. Triangular systems for CFD applications on parallel architectures. Technical report, NAS Applied Research Branch, NASA Ames Research Center, 1994.

[7] H. Berryman, J. Saltz, W. Gropp, and R. Mirchandaney. Krylov methods preconditioned with incompletely factored matrices on the CM-2. Technical Report 89-54, NASA Langley Research Center, ICASE, Hampton, VA, 1989.

[8] G. C. Broyden. A new method of solving nonlinear simultaneous equations. *Comput. J.*,12:94–99, 1969.

[9] G. Brussino and V. Sonnad. A comparison of direct and preconditioned iterative techniques for sparse unsymmetric systems of linear equations. *Int. J. for Num. Methods in Eng.*, 28:801–815, 1989.

[10] T.F. Chan and D.Goovaerts. A note on the efficiency of domain decomposed incomplete factorizations. *SIAM J. Sci. Stat. Comp.*, 11:794–803, 1990.

[11] L. Crone and H. van der Vorst. Communication aspects of the conjugate gradient method on distributed-memory machines. *Supercomputer*, X(6):4–9, 1993.

[12] E. De Sturler. *Iterative methods on distributed memory computers.* PhD thesis, Delft University of Technology, Delft, the Netherlands, 1994.

[13] E. De Sturler and D. R. Fokkema. Nested Krylov methods and preserving the orthogonality. In N. Duane Melson, T.A. Manteuffel, and S.F. McCormick, editors, *Sixth Copper Mountain Conference on Multigrid Methods*, volume Part 1 of *NASA Conference Publication 3324*, pages 111–126. NASA, 1993.

[14] S. Doi and A. Hoshi. Large numbered multicolor MILU preconditioning on SX-3/14. *Int'l J. Computer Math.*, 44:143–152, 1992.

[15] Shun Doi. On parallelism and convergence of incomplete LU factorizations. *Appl. Num. Math.*, 7:417–436, 1991.

[16] J. J. Dongarra, I. S. Duff, D. C. Sorensen, and H. A. van der Vorst. *Solving Linear Systems on Vector and Shared Memory Computers.* SIAM, Philadelphia, PA, 1991.

[17] T. Eirola and O. Nevanlinna. Accelerating with rank-one updates. *Lin. Alg. and its Appl.*, 121:511–520, 1989.

[18] V. Faber and T. A. Manteuffel. Necessary and sufficient conditions for the existence of a conjugate gradient method. *SIAM J. Numer. Analysis*, 21(2):352–362, 1984.

[19] D.R. Fokkema, G.L.G. Sleijpen, and H.A. van der Vorst. Generalized conjugate gradient squared. Technical Report Preprint 851, Mathematical Institute, Utrecht University, 1994.

[20] R. W. Freund and N. M. Nachtigal. QMR: a quasi-minimal residual method for non-Hermitian linear systems. *Num. Math.*, 60:315–339, 1991.

[21] Roland Freund. A transpose-free quasi-minimal residual algorithm for non-Hermitian linear systems. *SIAM J. Sci. Comput.*, 14:470–482, 1993.

[22] E. Giladi, G.H. Golub, and J.B. Keller. Inner and outer iterations for the Chebyshev algorithm. Technical Report SCCM-95-10, Computer Science Department, Stanford University, Stanford, CA, 1995.

[23] G. H. Golub and M. L. Overton. The convergence of inexact Chebyshev and Richardson iterative methods for solving linear systems. *Numerische Mathematik*, 53, 1988.

[24] G. H. Golub and C. F. van Loan. *Matrix Computations.* The Johns Hopkins University Press, Baltimore, 1989.

[25] I. Gustafsson. A class of first order factorization methods. *BIT*, 18:142–156, 1978.

[26] M. H. Gutknecht. Variants of BICGSTAB for matrices with complex spectrum. *SIAM J. Sci. Comput.*, 14:1020–1033, 1993.

[27] J.C.C. Kuo and T.F. Chan. Two-color fourier analysis of iterative algorithms for elliptic problems with red/black ordering. *SIAM J. Sci. Stat. Comp.*, 11:767–793, 1990.

[28] C. Lanczos. Solution of systems of linear equations by minimized iterations. *J. Res. Natl. Bur. Stand*, 49:33–53, 1952.

[29] J. A. Meijerink and H. A. van der Vorst. An iterative solution method for linear systems of which the coefficient matrix is a symmetric M-matrix. *Math.Comp.*, 31:148–162, 1977.

[30] G. Meurant. Domain decomposition methods for partial differential equations on parallel computers. *Int. J. Supercomputing Appls.*, 2:5–12, 1988.

[31] N. M. Nachtigal, S. C. Reddy, and L. N. Trefethen. How fast are nonsymmetric matrix iterations? *SIAM J. Matrix Anal. Appl.*, 13:778–795, 1992.

[32] N. M. Nachtigal, L. Reichel, and L. N. Trefethen. A hybrid GMRES algorithm for nonsymmetric matrix iterations. Technical Report 90-7, MIT, Cambridge, MA, 1990.

[33] C. C. Paige and M. A. Saunders. Solution of sparse indefinite systems of linear equations. *SIAM J. Numer. Anal.*, 12:617–629, 1975.

[34] C. Pommerell and W. Fichtner. PILS: An iterative linear solver package for ill-conditioned systems. In *Supercomputing '91*, pages 588–599, Los Alamitos, CA., 1991. IEEE Computer Society.

[35] Claude Pommerell. *Solution of large unsymmetric systems of linear equations.* PhD thesis, Swiss Federal Institute of Technology, Zürich, 1992.

[36] G. Radicati di Brozolo and Y. Robert. Parallel conjugate gradient-like algorithms for solving sparse non-symmetric systems on a vector multiprocessor. *Parallel Computing*, 11:223–239, 1989.

[37] G. Radicati di Brozolo and M. Vitaletti. Sparse matrix-vector product and storage representations on the IBM 3090 with Vector Facility. Technical Report 513-4098, IBM-ECSEC, Rome, July 1986.

[38] Y. Saad. A flexible inner-outer preconditioned GMRES algorithm. *SIAM J. Sci. Comput.*, 14:461–469, 1993.

[39] Y. Saad. *Iterative methods for sparse linear systems.* PWS Publishing Company, Boston, 1996.

[40] Y. Saad and M. H. Schultz. GMRES: a generalized minimal residual algorithm for solving nonsymmetric linear systems. *SIAM J. Sci. Statist. Comput.*, 7:856–869, 1986.

[41] G. L. G. Sleijpen and H.A. Van der Vorst. Maintaining convergence properties of BICGSTAB methods in finite precision arithmetic. *Numerical Algorithms*, 10:203–223, 1995.

[42] G. L. G. Sleijpen, H.A. Van der Vorst, and D. R. Fokkema. Bi-CGSTAB(ℓ) and other hybrid Bi-CG methods. *Numerical Algorithms*, 7:75–109, 1994.

[43] G. L. G. Sleijpen and D. R. Fokkema. BICGSTAB(ℓ) for linear equations involving unsymmetric matrices with complex spectrum. *ETNA*, 1:11–32, 1993.

[44] G.L.G. Sleijpen and H.A. van der Vorst. Reliable updated residuals in hybrid Bi-CG methods. *Computing*, 56:141–163, 1996.

[45] P. Sonneveld. CGS: a fast Lanczos-type solver for nonsymmetric linear systems. *SIAM J. Sci. Statist. Comput.*, 10:36–52, 1989.

[46] K.H. Tan. *Local coupling in domain decomposition*. PhD thesis, Utrecht University, Utrecht, the Netherlands, 1995.

[47] H. A. van der Vorst. A vectorizable variant of some ICCG methods. *SIAM J. Sci. Stat. Comput.*, 3:86–92, 1982.

[48] H. A. van der Vorst. High performance preconditioning. *SIAM J. Sci. Statist. Comput.*, 10:1174–1185, 1989.

[49] H. A. van der Vorst. Bi-CGSTAB: A fast and smoothly converging variant of Bi-CG for the solution of non-symmetric linear systems. *SIAM J. Sci. Statist. Comput.*, 13:631–644, 1992.

[50] H. A. van der Vorst and C. Vuik. GMRESR: A family of nested GMRES methods. *Num. Lin. Alg. with Appl.*, 1:369–386, 1994.

[51] M.B. van Gijzen. *Iterative solution methods for linear equations in finite element computations*. PhD thesis, Delft University of Technology, Delft, the Netherlands, 1994.

[52] C. Vuik and H.A. van der Vorst. A comparison of some GMRES-like methods. *Lin. Alg. and its Appl.*, 160:131–162, 1992.

[53] T. Washio and K. Hayami. Parallel block preconditioning based on SSOR and MILU. *Numer. Lin. Alg. with Applic.*, 1:533–553, 1994.

[54] Lu Zhou and Homer F. Walker. Residual smoothing techniques for iterative methods. *SIAM J. Sci. Stat. Comp.*, 15:297–312, 1994.

Problems of breakdown and near-breakdown in Lanczos-based algorithms

C. Brezinski* M. Redivo Zaglia † H. Sadok ‡

1 Lanczos method

In 1950, Lanczos [42] proposed a method for transforming a matrix into a similar tridiagonal one. Since, by the theorem of Cayley-Hamilton, the computation of the characteristic polynomial of a matrix and the solution of a system of linear equations are equivalent problems, Lanczos [43], in 1952, used his method for that purpose.

Let $Ax = b$ be a $n \times n$ system of linear equations $Ax = b$. Let x_0 be an arbitrary vector and let $r_0 = b - Ax_0$ be the corresponding residual vector. Let $E_k = \text{span}\left(r_0, Ar_0, \ldots, A^{k-1}r_0\right)$ and $F_k = \text{span}\left(y, A^T y, \ldots, A^{T^{k-1}} y\right)$ where y is an arbitrary nonzero vector.

For solving this system, Lanczos method consists of constructing a sequence of vectors (x_k) defined by

$$(x_k - x_0) \in E_k$$
$$r_k = b - Ax_k \perp F_k.$$

The first condition can be written as

$$x_k - x_0 = -a_1 r_0 - \cdots - a_k A^{k-1} r_0.$$

Thus, multiplying both sides by A, adding and subtracting b leads to

$$r_k = (I + a_1 A + \cdots + a_k A^k) r_0 = P_k(A) r_0$$

*Université des Sciences et Technologies de Lille, Laboratoire d'Analyse Numérique et d'Optimisation, 59655-Villeneuve d'Ascq cedex, France. Email: brezinsk@omega.univ-lille1.fr

†Dipartimento di Elettronica e Informatica, Università degli Studi di Padova, via Gradenigo, 6/a, 35131-Padova, Italy. Email: michela@dei.unipd.it

‡Laboratoire de Mathématiques Appliquées, Université du Littoral, Centre Universitaire de la Mi-Voix, 50 rue F. Buisson, 62100-Calais, France. Email: sadok@lma.univ-littoral.fr

G. Winter Althaus and E. Spedicato (eds.), Algorithms for Large Scale Linear Algebraic Systems, 255-270.

where

$$P_k(\xi) = 1 + a_1\xi + \cdots + a_k\xi^k.$$

The second condition corresponds to

$$(A^{T^i}y, r_k) = (y, A^i r_k) = (y, A^i P_k(A) r_0) = 0 \quad \text{for } i = 0, \ldots, k-1.$$

Let c be the linear functional on the space of polynomials defined by

$$c\left(\xi^i\right) = c_i = \left(y, A^i r_0\right), \qquad i = 0, 1, \ldots.$$

Then, the preceding orthogonality conditions can also be written as

$$c\left(\xi^i P_k(\xi)\right) = 0 \quad \text{for } i = 0, \ldots, k-1.$$

These relations mean that the polynomials P_k form a family of formal orthogonal polynomials (FOP) with respect to the linear functional c [7]. Such polynomials, which are defined apart a multiplying constant, are normalized, in our case, by the condition $\forall k, P_k(0) = 1$. The orthogonality relations of P_k can be written as

$$c_i + a_1 c_{i+1} + \cdots + a_k c_{i+k} = 0 \quad \text{for } i = 0, \ldots, k-1.$$

Thus P_k exists and is unique if and only if this system is nonsingular that is if and only if the Hankel determinant

$$H_k^{(1)} = \begin{vmatrix} c_1 & \cdots & c_k \\ \vdots & & \vdots \\ c_k & \cdots & c_{2k-1} \end{vmatrix} \neq 0. \tag{1}$$

The polynomials which exist are called *regular*.

Thanks to the normalization condition, the vectors x_k such that $r_k = b - Ax_k$ are given by

$$x_k = x_0 - R_{k-1}(A) r_0$$

where $P_k(\xi) = 1 + \xi R_{k-1}(\xi)$.

An important property of Lanczos method is its finite convergence, namely that $\exists k \leq n$ such that $r_k = 0$ and $x_k = x = A^{-1}b$.

Of course, Lanczos method presents some practical interest only if the polynomials P_k can be recursively computed and, thus, the vectors r_k and x_k. The first algorithm for that purpose was the well-known conjugate gradient algorithm due to Hestenes and Stiefel [38] when the matrix A is symmetric and positive definite. The coefficients of the recurrence relationships for the residuals r_k are given by ratios of scalars products. Since A is symmetric positive definite, their numerators are always nonzero. This algorithm was extended to an arbitrary matrix by Fletcher [29]. It is the biconjugate gradient algorithm. In this algorithm, divisions by zero can occur due to a zero scalar product. Such a situation is called a *breakdown* and the algorithm has to be stopped. Other algorithms for the implementation of the Lanczos method can be derived by using other recurrence relationships for computing the orthogonal polynomials P_k [22]. Such recurrences can be found in [7]. All these relations can suffer from breakdowns. It is not our purpose, in this paper, to discuss all the techniques for treating breakdowns in the recursive algorithms for implementing the method of Lanczos but only to review those based on FOP. The case of a *near-breakdown*, due to a division by a scalar product close to zero, will also be discussed. These techniques will be extended to some other methods related to Lanczos'.

2 The computation of orthogonal polynomials

As in the usual case (that is when the linear functional c can be written as an integral with respect to a positive measure), any family of formal orthogonal polynomials satisfies a three-term recurrence relationship of the form

$$P_{k+1}(\xi) \;=\; (A_{k+1}\xi + B_{k+1})\,P_k(\xi) - C_{k+1}P_{k-1}(\xi) \tag{2}$$

for $k = 0, 1, \ldots$ with $P_{-1}(\xi) = 0$ and $P_0(\xi) = 1$. Imposing the orthogonality conditions leads to

$$A_{k+1}\,c\left(\xi^k P_k\right) - C_{k+1}\,c\left(\xi^{k-1}P_{k-1}\right) = 0 \tag{3}$$

$$A_{k+1}\,c\left(\xi^{k+1} P_k\right) + B_{k+1}\,c\left(\xi^k P_k\right) - C_{k+1}\,c\left(\xi^k P_{k-1}\right) = 0. \tag{4}$$

Then, the normalization condition $P_{k+1}(0) = 1$ gives the third relation

$$B_{k+1} - C_{k+1} = 1.$$

The determinant D_k of the preceding 3×3 system giving A_{k+1}, B_{k+1} and C_{k+1} is equal to

$$D_k = -c\left(\xi^k P_k\right)\left[c\left(\xi^k P_k\right) - c\left(\xi^k P_{k-1}\right)\right] - c\left(\xi^{k-1}P_{k-1}\right)\,c\left(\xi^{k+1}P_k\right).$$

When $k = 1$, we have

$$D_1 = -\frac{c_0^2}{c_1^2}\left(c_2^2 - c_1c_3\right).$$

On the other hand, P_2 exists and is unique if and only if

$$\begin{vmatrix} c_0 & c_1 & c_2 \\ c_1 & c_2 & c_3 \\ 1 & 0 & 0 \end{vmatrix} = c_1c_3 - c_2^2 = H_2^{(1)} \neq 0.$$

Thus, we have $D_1 = H_2^{(1)} c_0^2/c_1^2$.

It follows that, if P_2 does not exist, a breakdown will occur in the three–term recurrence relationship. Such a breakdown, arising if and only if a polynomial does not exist, is called a *true breakdown*. On the other hand, even if $H_2^{(1)} \neq 0$, D_1 can be zero (if $c_0 = 0$) and a breakdown can occur in the recurrence relationship. Such a breakdown is not due to the non–existence of the polynomial P_2, but to the relation used for its computation. This kind of breakdown is called a *ghost breakdown* [12].

Instead of the preceding usual three–term recurrence relationship, there exist many other recurrences for computing the polynomials P_k. One of them consists of computing simultaneously the family $\left\{P_k^{(1)}\right\}$ of polynomials orthogonal with respect to the linear functional $c^{(1)}$ defined by

$$c^{(1)}\left(x^i\right) = c\left(x^{i+1}\right) = c_{i+1} \qquad \text{for } i = 0, 1, \ldots.$$

By definition, these polynomials satisfy

$$c^{(1)}\left(x^i P_k^{(1)}\right) = 0 \qquad \text{for } i = 0, \ldots, k-1.$$

They will be normalized so that their leading coefficient be equal to 1. In this case, they are called *monic*. It is easy to check that they exist and are unique if and only if the condition (1) holds. So, P_k and $P_k^{(1)}$ exist under the same condition.

Since the polynomials $\left\{P_k^{(1)}\right\}$ form a family of FOP, they also satisfy the usual three–term recurrence relationship which becomes, since they are monic

$$P_{k+1}^{(1)}(\xi) = (\xi - a_{k+1})P_k^{(1)}(\xi) - b_{k+1}P_{k-1}^{(1)}(\xi) \tag{5}$$

with $P_0^{(1)}(\xi) = 1$ and $P_{-1}^{(1)}(\xi) = 0$. The coefficients a_{k+1} and b_{k+1} are given by

$$\begin{aligned} b_{k+1} &= c^{(1)}\left(\xi^k P_k^{(1)}\right) / c^{(1)}\left(\xi^{k-1} P_{k-1}^{(1)}\right) \\ a_{k+1} &= \left[c^{(1)}\left(\xi^k P_k^{(1)}\right) - b_{k+1} c^{(1)}\left(\xi^k P_{k-1}^{(1)}\right)\right] / c^{(1)}\left(\xi^k P_k^{(1)}\right). \end{aligned}$$

So, a breakdown occurs in this relation if and only if $c^{(1)}(\xi^k P_k^{(1)}) = 0$ (since, $P_k^{(1)}$ exists it means that $c^{(1)}(\xi^{k-1} P_{k-1}^{(1)} \neq 0$ and, thus, there is no division by zero in the expression giving b_{k+1}). But

$$c^{(1)}(\xi^k P_k^{(1)}) = H_{k+1}^{(1)}$$

and we recover the condition for the existence of P_{k+1}. Thus, a breakdown occurs in (5) if and only if the polynomials P_{k+1} and $P_{k+1}^{(1)}$ do not exist. Such a breakdown is called a *true breakdown*.

Let us now see how to compute the polynomial P_{k+1} from P_k and $P_k^{(1)}$. It can be proved that the following relation holds

$$P_{k+1}(\xi) = P_k(\xi) - \lambda_k \, \xi \, P_k^{(1)}(\xi) \tag{6}$$

with $P_0(\xi) = P_0^{(1)}(\xi) = 1$. The coefficient λ_k is given by

$$\lambda_k = c\left(\xi^k P_k\right) / c^{(1)}\left(\xi^k P_k^{(1)}\right).$$

Thus a breakdown occurs in this relation if and only if

$$c^{(1)}\left(\xi^k P_k^{(1)}\right) = 0.$$

Thus, we see that a breakdown occurs if and only if $H_{k+1}^{(1)} = 0$ or, in other terms, if and only if $P_{k+1}^{(1)}$ and P_{k+1} do not exist.

Using alternately the relations (6) and (5) allows to compute simultaneously the two families $\{P_k\}$ and $\left\{P_k^{(1)}\right\}$. Only true breakdowns can occur in these two relations. All the other recurrence relationships can suffer from ghost breakdowns.

Let us now see how these recurrence relationships can be used for implementing the Lanczos method. Setting

$$r_k = P_k(A)r_0 \quad \text{and} \quad z_k = P_k^{(1)}(A)r_0$$

these two relations give

$$\begin{aligned} r_{k+1} &= r_k - \lambda_k A z_k \\ x_{k+1} &= x_k + \lambda_k z_k \\ z_{k+1} &= A z_k - a_{k+1} z_k - b_{k+1} z_{k-1} \end{aligned}$$

This algorithm is known under the names of *Lanczos/Orthodir* [51] and also *BIODIR* [35]. Among all the recursive algorithms for implementing the Lanczos method, it is the only one which can suffer only from true breakdowns. This algorithm cannot be subject to ghost breakdowns and thus it is the only reliable algorithm for the implementation of the Lanczos method. Moreover, as pointed out in [39] (see also [40]), its convergence properties make it a more interesting algorithm than the others. Other old and new algorithms for implementing the method of Lanczos can be derived from the theory of FOP [22].

3 Avoiding breakdowns

Let us now see how to avoid the true breakdowns which can occur in the recurrence relationships given in the preceding Section.

The treatment of a true breakdown consists of the following operations :

1. recognize the occurrence of such a breakdown, that is the non-existence of the next orthogonal polynomial(s),

2. determine the degree of the next existing (that is regular) orthogonal polynomial,

3. jump over the non-existing orthogonal polynomials and have a recurrence relationship which makes only use of the regular ones.

This problem was completely solved by Draux [27] in the case of monic orthogonal polynomials. Since the polynomials $P_k^{(1)}$ are monic and the conditions for the existence of the polynomials P_k and $P_k^{(1)}$ are the same, we shall apply his results for avoiding true breakdowns.

Let us first change slightly our notations to simplest ones. Up to now, the kth polynomial of the family had exactly the degree k and, thus, it was denoted by $P_k^{(1)}$. Now, since some of the polynomials of the family may not exist, we shall only give an index to the existing ones. Thus, the kth regular polynomial of the family will still be denoted by $P_k^{(1)}$ but, now, its degree will be equal to n_k with $n_k \geq k$. The next regular polynomial will be denoted by $P_{k+1}^{(1)}$ and its degree n_{k+1} will be $n_{k+1} = n_k + m_k$. Thus, m_k is the length of the *jump* in the degrees between the regular polynomial $P_k^{(1)}$ and the next one. This change in the notations means that $P_k^{(1)}$ is, in fact, the polynomial previously denoted by $P_{n_k}^{(1)}$. Since the polynomials of the degrees $n_k+1, \ldots, n_k+m_k-1$ do not exist, we are not giving them a name. The same change of notations will be made for the family $\{P_k\}$.

It was proved by Draux [27] that the length m_k of the jump is given by the conditions

$$c^{(1)}\left(\xi^i P_k^{(1)}\right) = 0 \qquad \text{for } i = 0, \ldots, n_k + m_k - 2 \tag{7}$$

$$\neq 0 \qquad \text{for } i = n_k + m_k - 1. \tag{8}$$

Moreover, these polynomials can be recursively computed by the relationship (for a short proof of this relation, see [11])

$$P_{k+1}^{(1)}(\xi) = \left(\alpha_0 + \cdots + \alpha_{m_k-1}\xi^{m_k-1} + \xi^{m_k}\right) P_k^{(1)}(\xi) - C_{k+1} P_{k-1}^{(1)}(\xi) \tag{9}$$

for $k = 0, 1, \ldots$, with $P_{-1}^{(1)}(\xi) = 0, P_0^{(1)}(\xi) = 1, C_1 = 0$ and

$$C_{k+1} = c^{(1)}\left(\xi^{n_k+m_k-1} P_k^{(1)}\right) / c^{(1)}\left(\xi^{n_k-1} P_{k-1}^{(1)}\right)$$

$$\alpha_{m_k-1} c^{(1)}\left(\xi^{n_k+m_k-1} P_k^{(1)}\right) + c^{(1)}\left(\xi^{n_k+m_k} P_k^{(1)}\right) = C_{k+1} c^{(1)}\left(\xi^{n_k} P_{k-1}^{(1)}\right)$$

$$\cdots\cdots\cdots\cdots\cdots\cdots\cdots\cdots\cdots\cdots$$

$$\alpha_0 c^{(1)}\left(\xi^{n_k+m_k-1} P_k^{(1)}\right) + \cdots + \alpha_{m_k-1} c^{(1)}\left(\xi^{n_k+2m_k-2} P_k^{(1)}\right) + c^{(1)}\left(\xi^{n_k+2m_k-1} P_k^{(1)}\right)$$
$$= C_{k+1} c^{(1)}\left(\xi^{n_k+m_k-1} P_{k-1}^{(1)}\right).$$

Since, by definition of m_k, $c^{(1)}\left(\xi^{n_k+m_k-1} P_k^{(1)}\right) \neq 0$, this system is never singular, and no breakdown (true or ghost) can occur in (9).

For implementing the Lanczos method by the algorithm Lanczos/Orthodir, we also need to compute P_{k+1} from P_k and $P_k^{(1)}$. As proved in [17], we have the following relation which generalizes (6)

$$P_{k+1}(\xi) \quad = \quad P_k(\xi) - \xi\left(\beta_0 + \cdots + \beta_{m_k-1}\xi^{m_k-1}\right)P_k^{(1)}(\xi), \qquad (10)$$

where the β_i's are the solution of the system

$$\begin{array}{l} \beta_{m_k-1}c^{(1)}\left(\xi^{n_k+m_k-1}P_k^{(1)}\right) = c\left(\xi^{n_k}P_k\right) \\ \cdots\cdots\cdots\cdots\cdots\cdots\cdots\cdots\cdots\cdots \\ \beta_0 c^{(1)}\left(\xi^{n_k+m_k-1}P_k^{(1)}\right) + \cdots + \beta_{m_k-1}c^{(1)}\left(\xi^{n_k+2m_k-2}P_k^{(1)}\right) = c\left(\xi^{n_k+m_k-1}P_k\right). \end{array}$$

Again, since, by definition of m_k, $c^{(1)}\left(\xi^{n_k+m_k-1}P_k^{(1)}\right) \neq 0$, this system is never singular, no breakdown (true or ghost) can occur in (10).

Thus, using alternately (9) and (10) gives a breakdown-free algorithm for implementing the Lanczos method. This algorithm, given in [17], was called the MRZ where the initials stand for *Method of Recursive Zoom*. It can only suffer from an *incurable hard breakdown* which occurs when $c^{(1)}\left(\xi^{n-1}P_k^{(1)}\right) = 0$ where n is the dimension of the linear system to be solved. Quite similar algorithms, for treating this kind of breakdowns, were also obtained by Gutknecht [33, 34]. Zero divisor-free Hestenes–Stiefel type conjugate direction algorithms can be found in [37]. Another scheme, based on a modified Krylov subspace approach, is presented in [49]. The problem of breakdown can also be treated by introducing new vectors into Krylov subspaces [45]or by an adaptative block Lanczos algorithm [46]. Necessary and sufficient conditions for look-ahead versions of the block conjugate gradient algorithm to be free from serious and incurable breakdowns are given in [24]. Thus, unstable versions of the algorithm can be identified and stable ones proposed.

A more stable version of the MRZ which also only needs the storage of a fixed number of vectors independent of the length of the jumps was recently proposed in [19]. It is based on Horner's rule for computing polynomials and, for that reason, was called the HMRZ. A quite similar technique is also described in [1].

4 Avoiding ghost breakdowns

Breakdowns can be also avoided in the other recurrence relationships which can be used for implementing the method of Lanczos. However, as explained above, these relations can suffer from ghost breakdowns. In this Section, we shall explain how such breakdowns can be cured.

For example, let us assume that, instead of using the three-term recurrence relationship, $P_{k+1}^{(1)}$ is computed from $P_k^{(1)}$ and P_k. As proved in [15], we have the relation

$$P_{k+1}^{(1)}(\xi) \quad = \quad (\delta_0 + \cdots + \delta_{m_k-1}\xi^{m_k-1} + \xi^{m_k})P_k^{(1)}(\xi) - D_{k+1}P_k(\xi) \qquad (11)$$

where m_k is defined as above. This relation can be used together with (10) for computing recursively the families $\{P_k\}$ and $\left\{P_k^{(1)}\right\}$. Imposing the orthogonality conditions, we have

$$D_{k+1}c(\xi^{n_k}P_k) = c^{(1)}\left(\xi^{n_k+m_k-1}P_k^{(1)}\right)$$
$$D_{k+1}c\left(\xi^{n_k+1}P_k\right) - \delta_{m_k-1}c^{(1)}\left(\xi^{n_k+m_k-1}P_k^{(1)}\right) = c^{(1)}\left(\xi^{n_k+m_k}P_k^{(1)}\right)$$
$$\cdots\cdots\cdots\cdots\cdots\cdots\cdots\cdots\cdots\cdots\cdots\cdots\cdots\cdots\cdots\cdots\cdots\cdots\cdots\cdots$$
$$D_{k+1}c\left(\xi^{n_k+m_k}P_k\right) - \delta_0 c^{(1)}\left(\xi^{n_k+m_k-1}P_k^{(1)}\right) - \cdots - \delta_{m_k-1}c^{(1)}\left(\xi^{n_k+2m_k-2}P_k^{(1)}\right)$$
$$= c^{(1)}\left(\xi^{n_k+2m_k-1}P_k^{(1)}\right).$$

Since $c^{(1)}\left(\xi^{n_k+m_k-1}P_k^{(1)}\right) \neq 0$ by definition of m_k, the preceding system is nonsingular if and only if $c(\xi^{n_k}P_k) \neq 0$. If this condition is not satisfied, then a ghost breakdown will occur in the algorithm. The corresponding algorithm for implementing the Lanczos method was called the SMRZ and is discussed at length in [15].

It is possible to avoid such a ghost breakdown by jumping until polynomials P_k and $P_k^{(1)}$ satisfying, in addition, the condition $c(\xi^{n_k}P_k) \neq 0$ have been found. Thus, now, we must be able to jump not only over non-existing orthogonal polynomials but also over the regular ones which cannot be computed by the recurrence relationship under consideration. The technique to be used will be described in the next Section. The same phenomenon arises when trying to compute $P_{k+1}^{(1)}$ from P_{k+1} and $P_k^{(1)}$. The corresponding algorithm for the Lanczos method was called the BMRZ [15].

Let us mention that Gutknecht proposed an unnormalized version of the BIORES algorithm for curing ghost breakdowns in the BIORES by using a three-term recurrence relationship [35] . Another procedure for treating breakdowns in the classical Lanczos algorithm is described in [6]. Avoiding breakdowns in the other algorithms given in [22] was studied in [2].

It is also possible to cure breakdowns by the techniques given in [28]. All these questions are discussed in more details in [18].

5 Near–breakdowns

As explained above, a breakdown occurs in a recurrence relationship when a quantity arising in the denominator of one of its coefficients is equal to zero. If such a quantity is not exactly zero, but close to it, then the corresponding coefficient can become very large and badly computed and roundoff errors can seriously affect the algorithm. This situation is called a *near-breakdown*. In order to avoid such a numerical instability, it is necessary to jump over all the polynomials which could be badly computed and to compute directly the first regular polynomial following them. Such procedures, which consist of jumping over polynomials which do not exist (or give rise to a ghost breakdown) or could be badly computed, were first introduced by Taylor [48] and Parlett, Taylor and Liu [44] under the name of *look-ahead* techniques. They are based on recurrence relationships allowing to jump over existing polynomials. However, not all the recurrence relations for avoiding a breakdown can be generalized into relations for avoiding a near-breakdown. For instance, it was shown in [15, 16] that the MRZ can be generalized into the algorithm GMRZ, the SMRZ can be generalized into the algorithm BSMRZ, but that the BMRZ cannot be generalized.

Let us now discuss in more details the techniques used for avoiding near-breakdowns. For jumping over regular polynomials, it is necessary to use special recurrence relationships. They can be obtained by the procedure explained in [11] and their coefficients are found by imposing the orthogonality conditions to both sides of the relations. For example, (10) becomes in that case

$$P_{k+1}(\xi) = (1 - \xi v_k(\xi)) P_k(\xi) - \xi w_k(\xi) P_k^{(1)}(\xi) \tag{12}$$

where w_k is a polynomial of the degree $m_k - 1$ at most and v_k a polynomial of the degree $m_k - 2$ at most. For computing the coefficients of these polynomials, it is necessary to consider two cases according whether or not $n_k - m_k + 1$ is greater or equal to zero. The corresponding relations can be found in [15].

For computing the two families of polynomials $\{P_k\}$ and $\left\{P_k^{(1)}\right\}$, a second recurrence relationship is needed. The first possible choice is to use the three-term recurrence relationship (9) which now becomes

$$P_{k+1}^{(1)}(\xi) = q_k(\xi) P_k^{(1)}(\xi) + p_k(\xi) P_{k-1}^{(1)}(\xi) \tag{13}$$

where q_k is a monic polynomial of the degree m_k and p_k a polynomial of the degree $m_k - 1$ at most. Their coefficients are given in [15]. The corresponding algorithm for implementing the method of Lanczos uses alternately (12) and (13) and is called the GMRZ. It is a generalization of the MRZ.

The second choice consists of generalizing the relation (11) which becomes

$$P_{k+1}^{(1)}(\xi) = s_k(\xi) P_k^{(1)}(\xi) + t_k(\xi) P_k(\xi) \tag{14}$$

where s_k is a monic polynomial of the degree m_k and t_k a polynomial of the degree $m_k - 1$ at most whose coefficients can be computed as explained in [15]. Making alternately use of the relations (12) and (14) for implementing the Lanczos method leads to an algorithm named the BSMRZ which generalizes the SMRZ.

Let us mention that a look-ahead technique for avoiding breakdowns and near-breakdowns in the three-term recurrence relationship satisfied by the polynomials P_k was also proposed in [32] under the name of *look-ahead Lanczos algorithm.* It reduces to the classical Lanczos algorithm (that is Lanczos/Orthores) when no jump occurs; see also [31, 30].

In all these algorithms, the main point (which is quite difficult) is the definition of the near-breakdown itself. In other words, it is difficult to decide when and how far to jump. Changing the definition can lead to very different numerical results. Of course, since no rigorous analysis of the numerical stability of the recurrence relationships is available, such tests are only based on heuristics. We saw above that, in the case of a true breakdown, the length m_k of the jump is given by the conditions (7) and (8). Of course, in practice, it is impossible to check a strict equality to zero. So, in our first implementation of the algorithms [15, 16], we chose, for treating the near-breakdown, a threshold value ε and defined the value of m_k by the conditions

$$\left|c^{(1)}\left(\xi^i P_k^{(1)}\right)\right| \begin{array}{ll} \leq \varepsilon & \text{for } i = 0, \ldots, n_k + m_k - 2 \\ > \varepsilon & \text{for } i = n_k + m_k - 1. \end{array}$$

This type of near-breakdown is clearly related to a true breakdown and thus it is called a *true near-breakdown*. Obviously, these conditions force themselves from (7) and (8). In some recurrence relationships, a second type of near-breakdown can occur. It can be called a *ghost near-breakdown* since it is related to the ghost breakdown as defined above.

Using the test given above, the beginnings and the lengths of the jumps were found to be quite sensitive to the choice of ε and so were also the numerical results. It means that this test has to be changed for a more appropriate one which has still to be found.

6 Other Lanczos–based algorithms

There exist several algorithms issued from the method of Lanczos for solving systems of linear equations. The *Conjugate Gradient Squared* algorithm (CGS) was obtained by Sonneveld [47]. It consists of considering the residual vectors given by

$$r_k = P_k^2(A) r_0$$

with P_k as defined above. By computing recursively the polynomials P_k^2, and not the polynomials P_k, the use of A^T can be avoided, a drawback of the Lanczos method. This is possible by squaring the recurrence relationships used for implementing the Lanczos method. Thus, true and ghost breakdowns can appear in the recursive algorithms for implementing the CGS for the same reasons as explained above. This is, in particular, the case for the algorithm given by Sonneveld [47] which consists of squaring the recurrence relationships of Lanczos/Orthomin. Since Lanczos/Orthodir can only suffer from true breakdowns, then squaring (9) and (10) leads to a breakdown-free algorithm for the CGS called the MRZS [20]. This algorithm does not make use of A^T.

Avoiding near-breakdowns in the CGS and still not using A^T requires a much more complicated analysis involving orthogonal polynomials. However, the algorithm obtained is quite simple and easy to program. It was obtained by squaring the relationships of the BSMRS and, thus, was called the BSMRZS. However, as explained in the preceding Section, the main point is to use a good heuristics for the jumps. Since our algorithm was obtained by squaring the relations (12) and (14), a ghost breakdown, due to $c(\xi^{n_k} P_k) = 0$, could also occur. So, the ghost near-breakdown which arises when this quantity is close to zero, has also to be avoided. In the program given in [10], it was not tried to cure this type of ghost near-breakdown and the program stopped in that case, which can explain its numerical instability since we could divide by quantities close to zero. Analyzing the intermediate numerical results obtained, we found that, at the preceding step, the jump was not long enough. So, we used another heuristics based on this observation [13]. The numerical results obtained show that our algorithm is quite satisfactory and seems to be stable. This was confirmed by the analysis of its numerical stability by a stochastic arithmetic [26].

Let us mention that another strategy for avoiding true breakdowns in Lanczos/Orthomin was proposed by Bank and Chan [4, 5]. It is similar to the technique proposed in [48, 44] and improved in [41]. It consists of a 2×2 composite step and the corresponding algorithm was called the CSBCG. This technique was extended to the CGS by Chan and Szeto [25] and the algorithm was named CSCGS.

Another Lanczos-based algorithm is the Bi-CGSTAB of Van der Vorst [50]. This algorithm consists of defining the residuals by

$$r_k = V_k(A) P_k(A) r_0$$

where V_k is a polynomial of degree k such that $V_k(0) = 1$ and $V_k(\xi) = (1 - a_k \xi) V_{k-1}(\xi)$ with $V_0(\xi) = 1$. The parameter a_k is chosen to minimize $\|r_k\|$. This algorithm does not make use of A^T. Again, deriving a transpose-free version of the algorithms for curing near-breakdowns in the Bi-CGSTAB needs an heavy analysis. However, the algorithm obtained is quite simple. Again, the main point in such an algorithm concerns the decisions to be taken about the jumps, namely to decide when and how far to jump. A quite satisfactory answer to these problems, based on the analysis of the intermediate quantities involved in the computations, was given in [14].

Recently, a technique related to ours was proposed in [36]. As ours, this algorithm also needs the storage of a number of intermediate vectors which is related to the maximum length allowed for the jumps.

7 Conclusions

The treatment of breakdowns and near-breakdowns in algorithms for implementing the method of Lanczos has been an outstanding problem for 45 years. The first attempts for solving it make use of linear algebra techniques, a quite natural approach for a question related to that domain. However, these attempts were not completely satisfactory. But, as mentioned by Lanczos himself, his method is also related to polynomials satisfying formal orthogonality conditions. Such polynomials form the basis for the construction and the recursive computation of Padé approximants [7, 3, 23] which are rational fractions whose series expansion in ascending powers of the variable matches a given series as far as possible. Indeed, reversing the numbering of the coefficients in the denominators of such approximants leads to formal orthogonal polynomials. Thus, the problem of breakdown in the recursive computation of formal orthogonal polynomials was quite familiar to those working on Padé approximants.

Thus, the analysis and remedies for treating breakdowns and near-breakdowns presented in this paper came out from the theory of formal orthogonal polynomials which also forms the foundations for obtaining all the recursive algorithms for the implementation of the method of Lanczos. This approach is simple and powerful and it was extended to other methods related to Lanczos'. It could also possibly be used in other algorithms based on biorthogonal polynomials [8, 9] and for implementing the extensions of all these methods to nonlinear systems [21].

We do not pretend that the techniques summarized in this paper are able to cure all the possible near-breakdowns, nor that our codes are for all seasons. But, from the numerical examples performed, it seems that they are, at least, able to bring some more numerical stability to the algorithms.

References

[1] H. Ayachour, Thesis, Université des Sciences et Technologies de Lille, in preparation.

[2] C. Baheux, New implementations of Lanczos method, J. Comput. Appl. Math., 57 (1995) 3–15.

[3] G.A. Baker, Jr., P.R. Graves-Morris, *Padé Approximants*, 2nd Edition, Cambridge University Press, Cambridge, 1996.

[4] R.E. Bank, T.F. Chan, A composite step bi-conjugate gradient algorithm for solving nonsymmetric systems, Numerical Algorithms, 7 (1994) 1–16.

[5] R.E. Bank, T.F. Chan, An analysis of the composite step bi-conjugate gradient algorithm for solving nonsymmetric systems, Numer. Math., 66 (1993) 295–319.

[6] D.L. Boley, S. Elhay, G.H. Golub, M.H. Gutknecht, Nonsymmetric Lanczos and finding orthogonal polynomials associated with indefinite weights, Numerical Algorithms, 1 (1991) 21–44.

[7] C. Brezinski, *Padé-type Approximation and General Orthogonal Polynomials*, ISNM vol. 50, Birkhäuser, Basel, 1980.

[8] C. Brezinski, *Biorthogonality and its Applications to Numerical Analysis*, Marcel Dekker, New York, 1992.

[9] C. Brezinski, Biorthogonality and conjugate gradient-type algorithms, in *Contributions in Numerical Mathematics*, R.P. Agarwal ed., WSSIAA vol.2, World Scientific, Singapore, 1993, pp.55–70.

[10] C. Brezinski, M. Redivo-Zaglia, Treatment of near-breakdown in the CGS algorithm, Publication ANO 257, Université des Sciences et Technologies de Lille, November 1991.

[11] C. Brezinski, M. Redivo-Zaglia, A new presentation of orthogonal polynomials with applications to their computation, Numerical Algorithms, 1 (1991), 207–221.

[12] C. Brezinski, M. Redivo-Zaglia, Breakdowns in the computation of orthogonal polynomials, in *Nonlinear Numerical Methods and Rational Approximation, II*, A. Cuyt ed., Kluwer, Dordrecht, 1994, pp. 49–59.

[13] C. Brezinski, M. Redivo-Zaglia, Treatment of near-breakdown in the CGS algorithm, Numerical Algorithms, 7 (1994) 33–73.

[14] C. Brezinski, M. Redivo-Zaglia, Look-ahead in Bi-CGSTAB and other methods for linear systems, BIT, 35 (1995) 169–201.

[15] C. Brezinski, M. Redivo-Zaglia, H. Sadok, Avoiding breakdown and near-breakdown in Lanczos type algorithms, Numerical Algorithms, 1 (1991), 261–284.

[16] C. Brezinski, M. Redivo-Zaglia, H. Sadok, Addendum to "Avoiding breakdown and near-breakdown in Lanczos type algorithms", Numerical Algorithms, 2 (1992) 133–136.

[17] C. Brezinski, M. Redivo-Zaglia, H. Sadok, A breakdown-free Lanczos type algorithm for solving linear systems, Numer. Math., 63 (1992) 29–38.

[18] C. Brezinski, M. Redivo-Zaglia, H. Sadok, Breakdowns in the implementation of the Lánczos method for solving linear systems, *Inter. J. Comp. Math. with Applics.*, to appear.

[19] C. Brezinski, M. Redivo-Zaglia, H. Sadok, New implementations of some look-ahead Lanczos-type algorithms for solving linear systems, submitted.

[20] C. Brezinski, H. Sadok, Avoiding breakdown in the CGS algorithm, Numerical Algorithms, 1 (1991) 199–206.

[21] C. Brezinski, H. Sadok, Some vector sequence transformations with applications to systems of equations, Numerical Algorithms, 3 (1992) 75–80.

[22] C. Brezinski, H. Sadok, Lanczos–type algorithms for solving systems of linear equations, Appl. Numer. Math., 11 (1993) 443–473.

[23] C. Brezinski, J. Van Iseghem, Padé approximations, in *Handbook of Numerical Analysis, vol. III*, P.G. Ciarlet and J.L. Lions eds., North–Holland, Amsterdam, 1994, pp. 47–222.

[24] C.G. Broyden, Look–ahead block–CG algorithms, Optim. Methods and Soft., to appear.

[25] T.F. Chan, T. Szeto, A composite step conjugate gradients squared algorithm for solving nonsymmetric linear systems, Numerical Algorithms, 7 (1994) 17–32.

[26] J.–M. Chesneaux, A.C. Matos, Breakdown and near–breakdown control in the CGS algorithm using stochastic arithmetic, Numerical Algorithms, 11 (1996) 99–116.

[27] A. Draux, *Polynômes Orthogonaux Formels. Applications*, LNM vol. 974, Springer–Verlag, Berlin, 1983.

[28] A. Draux, Formal orthogonal polynomials revisited. Applications, Numerical Algorithms, 11 (1996) 143–158.

[29] R. Fletcher, Conjugate gradient methods for indefinite systems, in *Numerical Analysis*, G.A. Watson ed., LNM vol. 506, Springer–Verlag, Berlin, 1976, pp. 73–89.

[30] R.W. Freund, Solution of shifted linear systems by quasi–minimal residual iterations, in *Numerical Linear Algebra*, L. Reichel, A. Ruttan and R.S. Varga eds., W. de Gruyter, Berlin, 1993, pp.101–121.

[31] R.W. Freund, G.H. Golub, N.M. Nachtigal, Iterative solution of linear systems, Acta Numerica, 1 (1991) 57–100.

[32] R.W. Freund, M.H. Gutknecht, N.M. Nachtigal, An implementation of the look–ahead Lanczos algorithm for non–Hermitian matrices, SIAM J. Sci. Comput., 14 (1993) 137–158.

[33] M.H. Gutknecht, A completed theory of the unsymmetric Lanczos process and related algorithms, Part I, SIAM J. Matrix Anal. Appl., 13 (1992) 594–639.

[34] M.H. Gutknecht, A completed theory of the unsymmetric Lanczos process and related algorithms, Part II, SIAM J. Matrix Anal. Appl., 15 (1994) 15–58.

[35] M.H. Gutknecht, The unsymmetric Lanczos algorithms and their relations to Padé approximation, continued fractions, and the qd algorithm, in *Preliminary Proceedings of the Copper Mountain Conference on Iterative Methods, April 1–5, 1990.*

[36] M.H. Gutknecht, K.J. Ressel, Look-ahead procedures for Lanczos-type product methods based on three-term recurrences, in *Preliminary Proceedings of the Copper Mountain Conference on Iterative Methods, 1996.*

[37] Cs.J. Hegedüs, Generating conjugate directions for arbitrary matrices by matrix equations, Computers Math. Applic., 21 (1991) 71–85; 87–94.

[38] M.R. Hestenes, E. Stiefel, Methods of conjugate gradients for solving linear systems, J. Res. Natl. Bur. Stand., 49 (1952) 409–436.

[39] K.C. Jea, *Generalized Conjugate Gradient Acceleration of Iterative Methods*, Ph.D. Thesis, Dept. of Mathematics, University of Texas at Austin, 1982.

[40] K.C. Jea, D.M. Young, On the simplification of generalized conjugate gradient methods for nonsymmetrizable linear systems, Linear Algebra Appl., 52 (1983) 299–317.

[41] M. Khelifi, Lanczos maximal algorithm for unsymmetric eigenvalue problems, Appl. Numer. Math., 7 (1991) 179–193.

[42] C. Lanczos, An iteration method for the solution of the eigenvalue problem of linear differential and integral operators, J. Res. Natl. Bur. Stand., 45 (1950) 255–282.

[43] C. Lanczos, Solution of systems of linear equations by minimized iterations, J. Res. Natl. Bur. Stand., 49 (1952) 33–53.

[44] B.N. Parlett, D.R. Taylor, Z.A. Liu, A look-ahead Lanczos algorithm for unsymmetric matrices, Math. Comput., 44 (1985) 105–124.

[45] Qiang Ye, A breakdown-free variation of the nonsymmetric Lanczos algorithm, Math. Comput., 62 (1994) 179–207.

[46] Qiang Ye, An adaptative block Lanczos algorithm, Numerical Algorithms, 12 (1996) 97–110.

[47] P. Sonneveld, CGS, a fast Lanczos-type solver for nonsymmetric linear systems, SIAM J. Sci. Stat. Comp., 10 (1989) 36–52.

[48] D.R. Taylor, *Analysis of the Look-Ahead Lanczos Algorithm*, Ph.D. Thesis, Dept. of Mathematics, University of California, Berkeley, Nov. 1982.

[49] C.H. Tong, Qiang Ye, A linear system solver based on a modified Krylov subspace method for breakdown recovery, Numerical Algorithms, 12 (1996) 233–251.

[50] H.A. Van der Vorst, Bi–CGSTAB: a fast and smoothly converging variant of Bi–CG for the solution of nonsymmetric linear systems, SIAM J. Sci. Stat. Comp., 13 (1992) 631–644.

[51] D.M. Young, K.C. Jea, Generalized conjugate–gradient acceleration of nonsymmetrizable iterative methods, Linear Algebra Appl., 34 (1980), 159–194.

Hybrid methods for solving systems of equations

C. Brezinski*

The aim of this paper is to review the most important results on the hybrid procedures [6] for solving systems of linear equations. Starting from two iterative methods for solving a system of linear equations, the hybrid procedure consists of constructing a new sequence of iterates with better convergence properties. For a more complete treatment, see [3]. Hybrid procedures were extended to systems of nonlinear equations in [4] and multiparameter extensions were studied in [5].

1 The hybrid procedure

Let us consider the $p \times p$ system of linear equations $Ax = b$. The hybrid procedure consists of combining the iterates produced by two methods for obtaining a new sequence of iterates with better convergence properties.

Let (x'_n) and (x''_n) be the sequences given by the two iterative methods. We shall construct a new sequence of iterates (x_n) by

$$x_n = \alpha_n x'_n + (1 - \alpha_n) x''_n$$

where the parameter α_n is chosen to minimize the Euclidean norm of the residual vector $r_n = b - Ax_n$. Such a procedure is called a *hybrid procedure* and it was first proposed in [6].

We have

$$r_n = \alpha_n r'_n + (1 - \alpha_n) r''_n = r''_n + \alpha_n (r'_n - r''_n)$$

where $r'_n = b - Ax'_n$ and $r''_n = b - Ax''_n$. Thus

$$(r_n, r_n) = (r''_n, r''_n) + 2\alpha_n (r''_n, r'_n - r''_n) + \alpha_n^2 (r'_n - r''_n, r'_n - r''_n)$$

*Université des Sciences et Technologies de Lille, Laboratoire d'Analyse Numérique et d'Optimisation, 59655–Villeneuve d'Ascq cedex - France. Email: brezinsk@omega.univ-lille1.fr

G. Winter Althaus and E. Spedicato (eds.), Algorithms for Large Scale Linear Algebraic Systems, 271–290.

and the value of α_n minimizing (r_n, r_n) is

$$\alpha_n = -(r_n'', r_n' - r_n'')/(r_n' - r_n'', r_n' - r_n''). \tag{1}$$

Of course, if $r_n' = r_n''$ we shall take $r_n = r_n'$.

Thus, the residuals r_n and the iterates x_n obtained by the hybrid procedure are

$$r_n = \frac{(r_n', r_n' - r_n'')r_n'' - (r_n'', r_n' - r_n'')r_n'}{(r_n' - r_n'', r_n' - r_n'')} \tag{2}$$

$$x_n = \frac{(r_n', r_n' - r_n'')x_n'' - (r_n'', r_n' - r_n'')x_n'}{(r_n' - r_n'', r_n' - r_n'')}. \tag{3}$$

We also have

$$r_n = r_n' - \frac{(r_n', r_n' - r_n'')}{(r_n' - r_n'', r_n' - r_n'')}(r_n' - r_n'') \tag{4}$$

and

$$r_n = r_n'' - \frac{(r_n'', r_n' - r_n'')}{(r_n' - r_n'', r_n' - r_n'')}(r_n' - r_n''). \tag{5}$$

From (4) and (5), it is easy to see that $r_n = 0$ if $r_n' = 0$ or if $r_n'' = 0$. It is also easy to see that

$$(r_n, r_n) = \frac{(r_n', r_n')(r_n'', r_n'') - (r_n', r_n'')^2}{(r_n' - r_n'', r_n' - r_n'')}$$

which is obviously non–negative by Schwarz inequality.

The most important property of the hybrid procedure is, since α_n minimizes $\|r_n\|^2 = (r_n, r_n)$, that

$$\|r_n\| \leq \min\left(\|r_n'\|, \|r_n''\|\right). \tag{6}$$

More precisely we have

$$(r_n, r_n) = (r_n', r_n') - \frac{(r_n', r_n' - r_n'')^2}{(r_n' - r_n'', r_n' - r_n'')} \tag{7}$$

and

$$(r_n, r_n) = (r_n'', r_n'') - \frac{(r_n'', r_n' - r_n'')^2}{(r_n' - r_n'', r_n' - r_n'')}. \tag{8}$$

These two relations show that the hybrid procedure brings no gain if $r_n' - r_n''$ is orthogonal either to r_n' or to r_n'' or, in other words, if (r_n', r_n'') is equal either to (r_n', r_n') or to (r_n'', r_n''). Obviously this situation is avoided if $(r_n', r_n'') \leq 0$, that is if the angle between r_n' and r_n'' is greater or equal to $\pi/2$.

Let us now give some more geometrical considerations about our hybrid procedure. Multiplying scalarly (4) and (5) by r_n' and r_n'' respectively and comparing to (7) and (8) shows that

$$(r_n, r_n) = (r_n, r_n') = (r_n, r_n'')$$

or, in other words

$$(r_n, r_n - r'_n) = (r_n, r_n - r''_n) = (r_n, r'_n - r''_n) = 0.$$

This means that r_n is the height through the origin of the triangle with sides r'_n and r''_n. We also have

$$r_n = (I - \mathrm{P}_n)r'_n = (I - \mathrm{P}_n)r''_n$$

with

$$\mathrm{P}_n = \frac{(r'_n - r''_n)(r'_n - r''_n)^T}{(r'_n - r''_n, r'_n - r''_n)}.$$

We have $\mathrm{P}_n = \mathrm{P}_n^2$ and $\mathrm{P}_n = \mathrm{P}_n^T$ and, thus, P_n and $I - \mathrm{P}_n$ represent orthogonal projections.

From geometrical considerations or from Schwarz inequality, we see that $r_n = 0$ if and only if r'_n and r''_n are collinear. This choice is optimal but unfeasible in practice because we usually have no control on the angle between these two residual vectors.

Obviously, instead of choosing α_n which minimizes (r_n, r_n), it is possible to minimize $(r_n, Z_n r_n)$ where Z_n is a symmetric positive definite matrix [17].

2 Recursive use of the procedure

We shall now discuss some possible strategies for choosing the two iterative methods used in the hybrid procedure. We can, for example,

1. compute x'_n and x''_n by two different methods
2. compute x'_n by some method and take $x''_n = x'_{n-1}$
3. compute x'_n by some method and take $x''_n = x_{n-1}$
4. compute x'_n from x_{n-1} and take $x''_n = x_{n-1}$
5. compute x'_n by some method and x''_n from x_{n-1}
6. compute x'_n by some method and x''_n from x'_n
7. compute x'_n and x''_n by the same method but with $x'_0 \neq x''_0$

In all the cases where x_{n-1} is used for computing x_n we shall take $x_0 = x'_0$ and $r_0 = r'_0$.

Let us now review these various strategies.

Case 1: x'_n and x''_n are computed by two different methods.

This is the most general case which contains all the others. However it is too general and nothing can be said on its algebraic properties without particularizing the methods. Of course, when the methods producing r'_n and r''_n are completely independent one from each other, the cost of one iteration of the hybrid procedure is the sum of the individual costs of each of the underlying methods. However this case is very convenient for a parallel computation of r'_n and r''_n. If both methods are not independent, then the cost of one iteration of the hybrid procedure can be lowered. This is, in particular, the case for the biconjugate gradient algorithm (BCG) [12] and the CGS [21] since, in both methods, the constants appearing in the recurrence relations are the same and thus they have to be computed only once. Then, a coupled implementation of the BCG and the CGS only requires 3 matrix by vector multiplications (instead of 4) and, moreover, A^* is no more needed (as in the BCG alone). Another possible coupled implementation with no extra cost is that of the quasi-minimal residual (QMR) of Freund [13] and Freund and Nachtigal [15] and of the BCG as explained in [16]. An even simpler coupled implementation of these methods was derived by Zhou and Walker [28].

Case 2: x'_n is computed by some method and we take $x''_n = x'_{n-1}$.

This case corresponds to a semi-iterative method where two successive iterates of a method are combined to obtain a better result.

If the vectors x'_n are constructed by an iterative method of the form

$$x'_{n+1} = Bx'_n + b$$

and if the matrix $A = I - B$ is regular, then

$$r'_n = M^n r'_0$$

where $M = ABA^{-1}$. If the eigenvalues of B satisfy

$$\lambda_1 > \lambda_2 \geq \lambda_3 \geq \cdots \geq \lambda_p$$

and if its eigenvectors are linearly independent, then it is well known that r'_n behaves like λ_1^n. It can be proved that r_n obtained by the hybrid procedure will behave like λ_2^n.

If the matrix A is symmetric positive definite and if the basic method used is the conjugate gradient, then $(r'_n, r''_n) = 0$ and we have

$$\|r_n\| \leq \min(\|r'_n\|, \|r'_{n-1}\|).$$

Case 3: x'_n is computed by some method and we take $x''_n = x_{n-1}$.

The idea here is cycling with the iterates obtained by the hybrid procedure and those given by an arbitrary method. Thanks to the minimization property (6) of the hybrid procedure, the norm of the residual r_n decreases at each iteration since we have

$$\|r_n\| \leq \min(\|r'_n\|, \|r_{n-1}\|)$$

and thus the convergence of the process is smoothed.

This method was introduced in [20] (see also [19, pp. 261–262]) for smoothing the convergence of the BCG. It was called the minimal residual smoothing (MRS). A complete theory of the MRS in the case where x'_n is computed by any iterative method (and not only the BCG) was given by Weiss in his thesis [24]. In particular, he proved that the MRS transforms generalized conjugate gradient methods which minimize the pseudo–residual into methods where the true residual is minimum. Thus it is not useful to apply the MRS to methods that minimize the true residual. A survey of these results can be found in [27]. It was recently proved by Zhou and Walker [28] that a smoothing algorithm (but with a different choice for α_n) transforms the BCG into the QMR and the CGS into the TFQMR [14]. They gave relations between the iterates of these two pairs of methods and extended the procedure to a quasi-minimal residual smoothing (QMRS) which can be applied to any iterative method. Other results related to the smoothing technique can be found in [25] and [26].

Case 4: x'_n is computed from x_{n-1} and we take $x''_n = x_{n-1}$.

This case covers the so–called *extrapolation* methods. Let us consider the regular splitting $A = M - N$. Taking $x''_n = x_{n-1}$ and $x'_n = M^{-1}Nx_{n-1} + M^{-1}b$, we obtain from (1)

$$\alpha_n = \frac{(AM^{-1}r_{n-1}, r_{n-1})}{(AM^{-1}r_{n-1}, AM^{-1}r_{n-1})}$$

and thus

$$\begin{aligned} x_n &= (1-\alpha_n)x_{n-1} + \alpha_n(M^{-1}Nx_{n-1} + M^{-1}b) \\ r_n &= (I - \alpha_n AM^{-1})r_{n-1}. \end{aligned}$$

In our case, α_n is chosen (by (1)) in order to minimize (r_n, r_n) and not, as usual, in order to minimize the spectral radius of the iteration matrix $\alpha_n M^{-1}N + (1-\alpha_n)I$. Several choices of the matrix M are of interest

1. for the choice $M = I$, Richardson's method is recovered
2. the choice $M = D$, where D is the diagonal part of A, corresponds to Jacobi method for x''_n
3. the choice $M = D - E$, where $-E$ is the strictly lower part of A, leads to a method looking like SOR
4. $M = (D - \omega E)/\omega$ corresponds to SOR for x''_n
5. $M = (D - \omega E)D^{-1}(D - \omega F)/[\omega(2-\omega)]$, where $-F$ is the strictly upper part of A, corresponds to SSOR for x''_n.

From (8), we have

$$(r_n, r_n) = (r_{n-1}, r_{n-1}) - \frac{(AM^{-1}r_{n-1}, r_{n-1})^2}{(AM^{-1}r_{n-1}, AM^{-1}r_{n-1})}.$$

Thus, from Schwarz inequality, $(r_n, r_n) = 0$ if and only if $AM^{-1} = aI$ where a is a nonzero scalar. This shows that a good choice for M is analogous to a good choice for a preconditioner, a name often given to M.

Case 5: x'_n is computed by some method and x''_n is computed from x_{n-1}.

This is a variant of the case 3, where now x''_n is not taken directly as x_{n-1} but is computed from it by any procedure.

Case 6: x'_n is computed by some method and x''_n is computed from x'_n.

This is a variant of the case 1, where now x''_n is not obtained by an arbitrary method but is computed from x'_n by any procedure. Using Lanczos method and the CGS can be considered as entering also into this case since $r''_n = V_n(A)r'_n$, V_n being an arbitrary polynomial with $V_n(0) = 1$.

In particular, one can think of setting

$$x''_n = -b_n r'_n + x'_n \quad \text{and} \quad r''_n = (1 + b_n A)r'_n$$

and choosing b_n in order to minimize (r''_n, r''_n). This idea is similar to the idea used for constructing the method called Bi-CGSTAB [22] and its variants [7]. These methods can be exactly recovered in our framework.

Case 7: x'_n and x''_n can be computed by the same method but with two different starting points x'_0 and x''_0.

3 Convergence acceleration

We shall now study the acceleration properties of the hybrid procedure. Additional results can be found in [2].

Let θ_n be the angle between r'_n and r''_n. Using the relation

$$(r'_n, r''_n) = \|r'_n\| \cdot \|r''_n\| \cos\theta_n$$

we have

$$\alpha_n = -\frac{\|r'_n\| \cdot \|r''_n\| \cos\theta_n - \|r''_n\|^2}{\|r'_n\|^2 - 2\|r'_n\| \cdot \|r''_n\| \cos\theta_n + \|r''_n\|^2}$$

and

$$\|r_n\|^2 = \frac{\|r'_n\|^2 \|r''_n\|^2 (1 - \cos^2\theta_n)}{\|r'_n\|^2 - 2\|r'_n\| \cdot \|r''_n\| \cos\theta_n + \|r''_n\|^2}.$$

Setting

$$\delta_n = \|r'_n\| \,/\, \|r''_n\|$$

we obtain

$$\alpha_n = -\frac{\delta_n \cos\theta_n - 1}{\delta_n^2 - 2\delta_n \cos\theta_n + 1}$$

and it follows

$$\begin{aligned}
\frac{\|r_n\|^2}{\|r'_n\|^2} &= \frac{1-\cos^2\theta_n}{\delta_n^2 - 2\delta_n\cos\theta_n + 1} &(9)\\
&= 1 - \frac{(\delta_n - \cos\theta_n)^2}{(\delta_n-\cos\theta_n)^2 + \sin^2\theta_n} &(10)\\
&= \frac{\sin^2\theta_n}{(\delta_n-\cos\theta_n)^2+\sin^2\theta_n} &(11)\\
&= \frac{\sin^2\theta_n}{\delta_n^2 - 2\delta_n\cos\theta_n + 1}. &(12)
\end{aligned}$$

From these relations, we immediately obtain the

Theorem 1
Suppose $\exists\theta$ *such that* $\lim_{n\to\infty}\theta_n = \theta$.

1. *If* $\lim_{n\to\infty}\delta_n = 0$ *then* $\lim_{n\to\infty}\alpha_n = 1$.
2. *If* $\lim_{n\to\infty}\delta_n = 1$ *and* $\theta \neq 0, \pi$ *then* $\lim_{n\to\infty}\alpha_n = 1/2$.
3. *If* $\lim_{n\to\infty}\delta_n = \infty$ *then* $\lim_{n\to\infty}\alpha_n = 0$.

This theorem shows that the hybrid procedure asymptotically selects the best method among the two.

Let us now consider the convergence behavior of $\|r_n\| \,/\, \|r'_n\|$. From (12), we immediately have the

Theorem 2
If $\exists\delta,\theta$ *such that* $\lim_{n\to\infty}\delta_n = \delta$, $\lim_{n\to\infty}\theta_n = \theta$ *and* $\delta^2 - 2\delta\cos\theta + 1 \neq 0$, *then*

$$\lim_{n\to\infty}\frac{\|r_n\|^2}{\|r'_n\|^2} = \frac{\sin^2\theta}{\delta^2 - 2\delta\cos\theta + 1} \leq 1.$$

Remark 1
Obviously if $\delta \leq 1$, *we also have*

$$\lim_{n\to\infty}\|r_n\|^2/\|r''_n\|^2 \leq 1.$$

Thus

$$\lim_{n\to\infty} \frac{\|r_n\|}{\min\left(\|r'_n\|\,,\|r''_n\|\right)}$$

exists and is not greater than 1.

Similar results can be obtained by considering the ratio $\|r_n\|^2 \,/\, \|r''_n\|^2$. It must also be noticed that $\|r_n\|^2 \,/\, \|r'_n\|^2$ tends to 1 if and only if $\delta = \cos\theta$. This result comes directly from (10) and we also get the

Theorem 3
A necessary and sufficient condition that $\exists N$ *such that* $\forall n \geq N$

$$0 \leq \frac{\|r_n\|^2}{\|r'_n\|^2} < 1$$

is that $\forall n \geq N,\ (r'_n - r''_n, r'_n) \neq 0$.

Let us now study some cases where (r_n) converges to zero faster than (r'_n) and (r''_n). From (11), we have the

Theorem 4
If $\exists\delta, \exists N$ *such that* $\forall n \geq N,\ 0 \leq \delta_n \leq \delta < 1$, *then a necessary and sufficient condition that*

$$\lim_{n\to\infty} \frac{\|r_n\|}{\|r'_n\|} = 0$$

is that (θ_n) *tends to* 0 *or* π.

Remark 2
Since $\delta_n < 1$ *then,* $\forall n \geq N,\ \|r'_n\| < \|r''_n\|$ *and we have*

$$\lim_{n\to\infty} \frac{\|r_n\|}{\min\left(\|r'_n\|\,,\|r''_n\|\right)} = 0.$$

The assumption $\delta_n \leq \delta < 1$ *does not restrict the generality since, if not satisfied, the ratio* $\|r''_n\| \,/\, \|r'_n\|$ *can be considered instead.*

Let us now study the case where (δ_n) tends to 1. From (12), we first have the

Theorem 5
If $\lim_{n\to\infty} \delta_n = 1$, *then a sufficient condition that* $\lim_{n\to\infty} \|r_n\|/\|r'_n\| = 0$ *is that* (θ_n) *tends to* π.

Remark 3
Since $\lim_{n\to\infty} \delta_n = 1$, *it follows that*

$$\lim_{n\to\infty} \frac{\|r_n\|}{\min\left(\|r'_n\|, \|r''_n\|\right)} = 0.$$

Another result in the case where (δ_n) tends to 1 is given by the

Theorem 6
If $\|r'_n\| / \|r''_n\| = 1 + a_n$ *with* $\lim_{n\to\infty} a_n = 0$, *then a sufficient condition that*

$$\lim_{n\to\infty} \frac{\|r_n\|}{\|r'_n\|} = 0$$

is that $\theta_n = o(a_n)$.

Remark 4
Since $\lim_{n\to\infty} \delta_n = 1$, *the remark 3 still holds.*

It seems quite difficult to obtain more theoretical results on the convergence of the hybrid procedure in the general case. (r'_n) being an arbitrary sequence of residual vectors, the following particular cases were treated in [2]

Case 1: $r''_n = Br_{n-1}$

Case 2: $r''_n = Br'_n$.

Such a situation arises, for example, if we consider a splitting of the matrix A

$$A = M - N$$

and if x''_n is obtained from y (equal to x_{n-1} or x'_n) by

$$x''_n = M^{-1}Ny + M^{-1}b.$$

In this case the associated residual has the form

$$\begin{aligned} r''_n &= b - Ax''_n \\ &= b - A(M^{-1}Ny + M^{-1}b) \\ &= b - (M - N)(M^{-1}Ny + M^{-1}b) \\ &= NM^{-1}(b - Ay). \end{aligned}$$

Thus we have $B = NM^{-1}$ with $y = x_{n-1}$ (case 1) and $y = x'_n$ (case 2). It must be noticed that $B = I - AM^{-1}$. This situation also holds if $B = I - AC$ with C an arbitrary matrix. In this case, we have

$$x_n = \alpha_n x'_n + (1 - \alpha_n)(y + C(b - Ay)).$$

4 Multiparameter hybrid procedures

In some cases, the various components (or blocks of components) of the vectors x'_n and x''_n can behave quite differently. Such a situation can arise, for example, in decomposition methods, multigrid methods, wavelets, multiresolution methods, inertial manifolds, incremental unknowns, and the nonlinear Galerkin method. In these methods, the unknowns are splitted into various subsets each of them being treated in a different way.

So, instead of using the hybrid procedure, it will be better to take a different value of the parameter α_n for each block. This is the reason why we shall now present a multiparameter extension of the hybrid procedure. This extension was introduced in [5].

Let us partition the vectors x'_n and x''_n into m blocks denoted respectively by $(x'_n)^1, \ldots, (x'_n)^m$ and $(x''_n)^1, \ldots, (x''_n)^m$. We set

$$X'_n = \begin{pmatrix} (x'_n)^1 & & \\ 0 & \ddots & 0 \\ & & (x'_n)^m \end{pmatrix} \quad \text{and} \quad X''_n = \begin{pmatrix} (x''_n)^1 & & \\ 0 & \ddots & 0 \\ & & (x''_n)^m \end{pmatrix}.$$

The multiparameter hybrid procedure (in short, the MPHP) consists of constructing the sequence (x_n) by

$$x_n = x''_n + (X'_n - X''_n)\alpha_n$$

where $\alpha_n \in \mathbb{R}^m$. We have

$$r_n = r''_n - A(X'_n - X''_n)\alpha_n.$$

The vector α_n is chosen to minimize $\|r_n\|$. It is given by the least squares solution of $r''_n - A(X'_n - X''_n)\alpha_n = 0$, that is

$$\alpha_n = \left[(X'_n - X''_n)^T A^T A(X'_n - X''_n)\right]^{-1} (X'_n - X''_n)^T A^T r''_n.$$

So, the computation of α_n requires the solution of a $m \times m$ system of linear equations. However, due to the sparsity of X'_n and X''_n, the construction of the matrix of this system is quite cheap. Indeed, let $p_1, \ldots, p_m$ be the dimension of the blocks in the partition of the vectors x'_n and x''_n. Obviously, $p_1 + \cdots + p_m = p$, the dimension of the system. We shall partition the matrix A into m blocks of respective dimensions $p \times p_i$ for $i = 1, \ldots, m$. Let us denote them by $A_1, \ldots, A_m$. It is easy to see that the matrix $A(X'_n - X''_n)$ is the $p \times m$ matrix whose columns are $A_1((x'_n)^1 - (x''_n)^1), \ldots, A_m((x'_n)^m - (x''_n)^m)$.

It must be noticed that the MPHP is not symmetric with respect to x'_n and x''_n.

5 A hybrid minimal residual smoothing

When mixing two methods by the hybrid procedure, the result obtained is an improvement over both residuals. But, if the convergence of one of the two methods is erratic, the hybrid procedure often exhibits a similar behavior. On the other hand, using the minimal residual method leads to a monotone convergence. So, we shall now try to obtain simultaneously an improvement and a monotone convergence. Of course, several strategies are possible. The one we chose is based on the observation that the MRS provides a good result if the residual of the method to be smoothed is almost collinear to that of the previous iterate of the MRS. So, we first took the residuals of the two methods and combine them in order to obtain a new residual as collinear as possible to that of the MRS, and then we applied the MRS. Let us now be more precise.

Let (x'_n) and (x''_n) be the two sequences of iterates and (r'_n) and (r''_n) the corresponding sequences of residuals. The two methods will first be combined by the hybrid procedure

$$\begin{aligned} x_n &= \beta_n x'_n + (1-\beta_n)x''_n \\ r_n &= \beta_n r'_n + (1-\beta_n)r''_n. \end{aligned}$$

Then, the MRS will be used, that is

$$\begin{aligned} y_n &= \alpha_n x_n + (1-\alpha_n)y_{n-1} \\ \rho_n &= \alpha_n r_n + (1-\alpha_n)\rho_{n-1} \end{aligned}$$

with α_n chosen to minimize (ρ_n, ρ_n), that is, as above

$$\alpha_n = -\frac{(\rho_{n-1}, r_n - \rho_{n-1})}{(r_n - \rho_{n-1}, r_n - \rho_{n-1})}.$$

The parameter β_n will be taken so that r_n be as collinear as possible to ρ_{n-1}. Let θ_n be the angle between these two vectors. Since we have

$$\cos^2\theta_n = \frac{(\rho_{n-1}, r_n)^2}{(r_n, r_n)(\rho_{n-1}, \rho_{n-1})}$$

we shall take for β_n the value maximizing the function

$$f(\beta) = \frac{(\rho_{n-1}, r''_n + \beta(r'_n - r''_n))^2}{(r''_n + \beta(r'_n - r''_n), r''_n + \beta(r'_n - r''_n))}.$$

We shall take for y_0 the result of the hybrid procedure applied to x'_0 and x''_0 (ρ_0 will be the corresponding residual) and we shall start applying the MRS to x_n and y_{n-1} only from $n = 1$.

This procedure will be called the *hybrid minimal residual smoothing*, in short the HMRS.

The numerator of f' is a polynomial of degree two with respect to β. Thus, f has two extrema. At one of them, f is zero and so, since f is non-negative for all β, it is a minimum. So, the maximum is attained for the other zero of f' that is

$$\beta_n = \left(1 + \frac{(\rho_{n-1}, r'_n)(r'_n, r''_n) - (\rho_{n-1}, r''_n)(r'_n, r'_n)}{(\rho_{n-1}, r''_n)(r'_n, r''_n) - (\rho_{n-1}, r'_n)(r''_n, r''_n)}\right)^{-1}.$$

Numerical results for illustrating this procedure can be found in [3].

6 Multiple hybrid procedures

In the hybrid procedure, it is also possible to mix more than two iterative methods and to consider the new sequence of residual vectors $\left(\rho_n^{(k)}\right)$ defined from the sequences $\left(r_n^{(1)}\right), \ldots, \left(r_n^{(k)}\right)$ by

$$\rho_n^{(k)} = a_n^{(1)} r_n^{(1)} + \cdots + a_n^{(k)} r_n^{(k)}$$

with $a_n^{(1)} + \cdots + a_n^{(k)} = 1$. The integer k is called the *rank* of the procedure.
If $y_n^{(k)}$ is the vector such that $\rho_n^{(k)} = b - Ay_n^{(k)}$ then

$$y_n^{(k)} = a_n^{(1)} x_n^{(1)} + \cdots + a_n^{(k)} x_n^{(k)}$$

where the $x_n^{(i)}$ are related to the $r_n^{(i)}$ by $r_n^{(i)} = b - Ax_n^{(i)}$. As before, $a_n^{(1)}, \ldots, a_n^{(k)}$ are chosen to minimize $\left\|\rho_n^{(k)}\right\|$.

We have

$$\rho_n^{(k)} = \frac{\begin{vmatrix} r_n^{(1)} & \cdots & r_n^{(k)} \\ \left(r_n^{(1)} - r_n^{(k)}, r_n^{(1)}\right) & \cdots & \left(r_n^{(1)} - r_n^{(k)}, r_n^{(k)}\right) \\ \vdots & & \vdots \\ \left(r_n^{(k-1)} - r_n^{(k)}, r_n^{(1)}\right) & \cdots & \left(r_n^{(k-1)} - r_n^{(k)}, r_n^{(k)}\right) \end{vmatrix}}{\begin{vmatrix} 1 & \cdots & 1 \\ \left(r_n^{(1)} - r_n^{(k)}, r_n^{(1)}\right) & \cdots & \left(r_n^{(1)} - r_n^{(k)}, r_n^{(k)}\right) \\ \vdots & & \vdots \\ \left(r_n^{(k-1)} - r_n^{(k)}, r_n^{(1)}\right) & \cdots & \left(r_n^{(k-1)} - r_n^{(k)}, r_n^{(k)}\right) \end{vmatrix}}$$

$$= \frac{\begin{vmatrix} r_n^{(1)} & \cdots & r_n^{(k)} \\ \left(r_n^{(2)} - r_n^{(1)}, r_n^{(1)}\right) & \cdots & \left(r_n^{(2)} - r_n^{(1)}, r_n^{(k)}\right) \\ \vdots & & \vdots \\ \left(r_n^{(k)} - r_n^{(k-1)}, r_n^{(1)}\right) & \cdots & \left(r_n^{(k)} - r_n^{(k-1)}, r_n^{(k)}\right) \end{vmatrix}}{\begin{vmatrix} 1 & \cdots & 1 \\ \left(r_n^{(2)} - r_n^{(1)}, r_n^{(1)}\right) & \cdots & \left(r_n^{(2)} - r_n^{(1)}, r_n^{(k)}\right) \\ \vdots & & \vdots \\ \left(r_n^{(k)} - r_n^{(k-1)}, r_n^{(1)}\right) & \cdots & \left(r_n^{(k)} - r_n^{(k-1)}, r_n^{(k)}\right) \end{vmatrix}}.$$

Since the denominator of $\rho_n^{(k)}$ is the Gram determinant of the vectors

$$r_n^{(1)} - r_n^{(k)}, \ldots, r_n^{(k-1)} - r_n^{(k)},$$

it is zero if and only if these vectors are dependent. By construction

$$\left\|\rho_n^{(k)}\right\| \le \min\left(\left\|r_n^{(1)}\right\|, \ldots, \left\|r_n^{(k)}\right\|\right).$$

This procedure is called the *multiple hybrid procedure*, in short the MHP. It was proposed in [6] and was studied in [1] where recursive algorithms for its implementation are discussed in the particular case $r_n^{(i)} = r_{n+i-1}$. The cheapest one is based on the H-algorithm studied in [8] and is as follows

$$H_{k+1}^{(i)} = H_k^{(i)} - \frac{\left(r_{n+k} - r_{n+k-1}, H_k^{(i)}\right)}{\left(r_{n+k} - r_{n+k-1}, H_k^{(i+1)} - H_k^{(i)}\right)} \left(H_k^{(i+1)} - H_k^{(i)}\right)$$

with $H_1^{(j)} = r_{n+j}$. For a fixed index n, we obtain $H_k^{(0)} = \rho_n^{(k)}$, the $H_k^{(i)}$'s for $i \neq 0$ being auxiliary vectors. With this particular choice of the vectors $r_n^{(i)}$, the MHP is identical to the *reduced rank extrapolation* (RRE) due to Mešina [18] and Eddy [11].

Let us give another expression for the vectors $\rho_n^{(k)}$ obtained by the MHP in the general case. This expression could serve as an algorithm for their computation. We have the

Theorem 7
Let $v_n^{(1)}, \ldots, v_n^{(k)}$ be k vectors such that

$$\left(v_n^{(i)}, r_n^{(j)}\right) \begin{cases} = 0 & \text{if } i \neq j \\ \neq 0 & \text{if } i = j. \end{cases}$$

Let us set

$$z_n^{(k)} = \sum_{i=1}^{k} \frac{v_n^{(i)}}{\left(v_n^{(i)}, r_n^{(i)}\right)}. \tag{13}$$

Then, the vectors $\rho_n^{(k)}$ *of the* MHP *can be expressed as*

$$\rho_n^{(k)} = \sum_{i=1}^{k} \left(z_n^{(k)}, r_n^{(i)^*}\right) r_n^{(i)^*} \Big/ \sum_{i=1}^{k} \left|\left(z_n^{(k)}, r_n^{(i)^*}\right)\right|^2$$

and we have

$$\left(\rho_n^{(k)}, \rho_n^{(k)}\right) = 1 \Big/ \sum_{i=1}^{k} \left|\left(z_n^{(k)}, r_n^{(i)^*}\right)\right|^2$$

where the $r_n^{(i)^*}$ *are the* $r_n^{(i)}$ *orthonormalized (in any way).*

From the definition (13) of $z_n^{(k)}$, we have

$$z_n^{(k)} = z_n^{(k-1)} + v_n^{(k)} / \left(v_n^{(k)}, r_n^{(k)}\right)$$

with $z_n^{(0)} = 0$. Thus

$$\left(z_n^{(k)}, r_n^{(i)^*}\right) = \left(z_n^{(k-1)}, r_n^{(i)^*}\right) + \left(v_n^{(k)}, r_n^{(i)^*}\right) / \left(v_n^{(k)}, r_n^{(k)}\right).$$

Let us now assume that $\forall i, r_n^{(i)^*}$ is a linear combination of $r_n^{(1)}, \ldots, r_n^{(i)}$ only (such as given, for example, by the Gram–Schmidt process). Thanks to the orthogonality property of $v_n^{(k)}$ with respect to $r_n^{(1)}, \ldots, r_n^{(k-1)}$, we have $\left(v_n^{(k)}, r_n^{(i)^*}\right) = 0$ for $i = 1, \ldots, k-1$ and thus $\left(z_n^{(k)}, r_n^{(i)^*}\right) = \left(z_n^{(k-1)}, r_n^{(i)^*}\right)$, $i = 1, \ldots, k-1$. Thus, from the preceding theorem, we have

$$\begin{aligned}
\frac{\rho_n^{(k)}}{\left\|\rho_n^{(k)}\right\|^2} &= \sum_{i=1}^{k} \left(z_n^{(k)}, r_n^{(i)^*}\right) r_n^{(i)^*} \\
&= \sum_{i=1}^{k-1} \left(z_n^{(k)}, r_n^{(i)^*}\right) r_n^{(i)^*} + \left(z_n^{(k)}, r_n^{(k)^*}\right) r_n^{(k)^*} \\
&= \sum_{i=1}^{k-1} \left(z_n^{(k-1)}, r_n^{(i)^*}\right) r_n^{(i)^*} + \left(z_n^{(k)}, r_n^{(k)^*}\right) r_n^{(k)^*} \\
&= \frac{\rho_n^{(k-1)}}{\left\|\rho_n^{(k-1)}\right\|^2} + \left(z_n^{(k)}, r_n^{(k)^*}\right) r_n^{(k)^*}
\end{aligned}$$

and

$$\begin{aligned}
\sum_{i=1}^{k} \left|\left(z_n^{(k)}, r_n^{(i)^*}\right)\right|^2 &= \sum_{i=1}^{k-1} \left|\left(z_n^{(k)}, r_n^{(i)^*}\right)\right|^2 + \left|\left(z_n^{(k)}, r_n^{(k)^*}\right)\right|^2 \\
&= \sum_{i=1}^{k-1} \left|\left(z_n^{(k-1)}, r_n^{(i)^*}\right)\right|^2 + \left|\left(z_n^{(k)}, r_n^{(k)^*}\right)\right|^2.
\end{aligned}$$

Thus, we finally have

$$\rho_n^{(k)} = \left\|\rho_n^{(k)}\right\|^2 \left[\rho_n^{(k-1)} / \left\|\rho_n^{(k-1)}\right\|^2 + \left(z_n^{(k)}, r_n^{(k)*}\right) r_n^{(k)*}\right]$$

$$\frac{1}{\left\|\rho_n^{(k)}\right\|^2} = \frac{1}{\left\|\rho_n^{(k-1)}\right\|^2} + \left|\left(z_n^{(k)}, r_n^{(k)*}\right)\right|^2.$$

In practice, if the $r_n^{(i)*}$ are obtained from the $r_n^{(i)}$ by the classical Gram–Schmidt process then the vectors $x_n^{(i)*}$ such that $r_n^{(i)*} = b - Ax_n^{(i)*}$ cannot be computed from the $r_n^{(i)*}$ without using A^{-1} and the preceding algorithm has no interest. Below, we shall give a modification of the Gram–Schmidt process which avoids this drawback. However, even if the vectors $x_n^{(i)*}$ are known, the vectors $y_n^{(k)}$ such that $\rho_n^{(k)} = b - Ay_n^{(k)}$ cannot be obtained without using A^{-1} since the sum of the coefficients in the formula of Theorem 7 for $\rho_n^{(k)}$ is not equal to 1. This second drawback can be avoided by assuming that the $r_n^{(i)}$ are mutually orthogonal and we have the

Theorem 8
If $\forall i \neq j, (r_n^{(i)}, r_n^{(j)}) = 0$ *then*

$$\rho_n^{(k)} = \sum_{i=1}^{k} \frac{r_n^{(i)}}{\left(r_n^{(i)}, r_n^{(i)}\right)} \Big/ \sum_{i=1}^{k} \frac{1}{\left(r_n^{(i)}, r_n^{(i)}\right)},$$

$$y_n^{(k)} = \sum_{i=1}^{k} \frac{x_n^{(i)}}{\left(r_n^{(i)}, r_n^{(i)}\right)} \Big/ \sum_{i=1}^{k} \frac{1}{\left(r_n^{(i)}, r_n^{(i)}\right)}$$

and

$$\frac{1}{\left(\rho_n^{(k)}, \rho_n^{(k)}\right)} = \sum_{i=1}^{k} \frac{1}{\left(r_n^{(i)}, r_n^{(i)}\right)}.$$

This result, which is independent of the choice of the vectors $v_n^{(i)}$, is a generalization of a result obtained by Weiss [24, Lemma 4.11, Theorem 4.1] (see also [27, Theorem 3.1] or [26, Lemma 11, Theorem 12]) in the case of the minimal residual smoothing (in short the MRS). This method consists of setting $\rho_n^{(k)} = a\rho_n^{(k-1)} + (1-a)r_n^{(k)}$ and choosing the value of a which minimizes $\left\|\rho_n^{(k)}\right\|^2$. The results of Weiss follow directly from the orthogonality of the vector $r_n^{(k)}$ with the vector $\rho_n^{(k-1)}$ which is a linear combination of $r_n^{(1)}, \ldots, r_n^{(k-1)}$.

Obviously, from the preceding theorem, we have

$$\frac{1}{\left(\rho_n^{(k)}, \rho_n^{(k)}\right)} = \frac{1}{\left(\rho_n^{(k-1)}, \rho_n^{(k-1)}\right)} + \frac{1}{\left(r_n^{(k)}, r_n^{(k)}\right)}.$$

Thus we obtain the following recursive algorithm for implementing the MRS.

For $k = 2, 3, \ldots$, and $n = 0, 1, \ldots$

$$\frac{1}{\left\|\rho_n^{(k)}\right\|^2} = \frac{1}{\left\|\rho_n^{(k-1)}\right\|^2} + \frac{1}{\left\|r_n^{(k)}\right\|^2}$$

$$\rho_n^{(k)} = \left\|\rho_n^{(k)}\right\|^2 \left[\frac{\rho_n^{(k-1)}}{\left\|\rho_n^{(k-1)}\right\|^2} + \frac{r_n^{(k)}}{\left\|r_n^{(k)}\right\|^2}\right]$$

$$y_n^{(k)} = \left\|\rho_n^{(k)}\right\|^2 \left[\frac{y_n^{(k-1)}}{\left\|\rho_n^{(k-1)}\right\|^2} + \frac{x_n^{(k)}}{\left\|r_n^{(k)}\right\|^2}\right]$$

with $\rho_n^{(1)} = r_n^{(1)}$ and $y_n^{(1)} = x_n^{(1)}$.

As proved by Weiss [24], applying the MRS to an orthogonal residual method leads to a minimal residual method. The relation above can be written as

$$\|r_n^{(k)}\| = \frac{\|\rho_n^{(k)}\|}{\sqrt{1 - \|\rho_n^{(k)}\|^2/\|\rho_n^{(k-1)}\|^2}}.$$

Thus, it can be deduced from such a relation that peaks in the sequence $(\|r_n^{(k)}\|)$ correspond to plateaus in the sequence $(\|\rho_n^{(k)}\|)$ [23, 10].

For using the preceding algorithm, we are now faced to the problem of constructing mutually orthogonal vectors $r_n^{(i)*}$ from the vectors $r_n^{(i)}$. As mentioned above, the problem with the usual Gram–Schmidt process is that it is not possible to recover the vectors $x_n^{(i)*}$ such that $r_n^{(i)*} = b - Ax_n^{(i)*}$ from the $r_n^{(i)*}$'s without using A^{-1}. This is the reason why we shall now modify the Gram–Schmidt procedure in order to avoid this problem. We shall look for mutually orthogonal vectors $r_n^{(i)*}$ given by $r_n^{(k)*} = d_k \tilde{r}_n^{(k)}$ with

$$\tilde{r}_n^{(k)} = r_n^{(k)} - \sum_{i=1}^{k-1} b_i^{(k)} r_n^{(i)*}$$

and $\tilde{r}_n^{(1)} = r_n^{(1)}$. Obviously d_k and the $b_i^{(k)}$ also depend on n. We have

$$r_n^{(k)*} = b - Ax_n^{(k)*} = d_k \left(\left(b - Ax_n^{(k)}\right) - \sum_{i=1}^{k-1} b_i^{(k)} \left(b - Ax_n^{(i)*}\right) \right).$$

Thus, it follows

$$x_n^{(k)*} = d_k \left(x_n^{(k)} - \sum_{i=1}^{k-1} b_i^{(k)} x_n^{(i)*} \right)$$

if and only if

$$d_k = \left(1 - \sum_{i=1}^{k-1} b_i^{(k)}\right)^{-1}$$

and $d_1 = 1$. Thus, it is easy to check that these vectors are obtained by the following modification of the Gram–Schmidt process

1. set $r_n^{(1)*} = r_n^{(1)}$,

2. for $k = 2, 3, \ldots$ compute

$$\tilde{r}_n^{(k)} = r_n^{(k)} - \sum_{i=1}^{k-1} \frac{\left(r_n^{(k)}, r_n^{(i)*}\right)}{\left(r_n^{(i)*}, r_n^{(i)*}\right)} r_n^{(i)*}$$

$$r_n^{(k)*} = \tilde{r}_n^{(k)} \Big/ \left(1 - \sum_{i=1}^{k-1} \frac{\left(r_n^{(k)}, r_n^{(i)*}\right)}{\left(r_n^{(i)*}, r_n^{(i)*}\right)}\right).$$

Of course, this algorithm breaks down if $1 - \sum_{i=1}^{k-1} \left(r_n^{(k)}, r_n^{(i)*}\right) / \left(r_n^{(i)*}, r_n^{(i)*}\right) = 0$. In that case, it is possible to skip over the residual $r_n^{(k)}$ which is responsible for the breakdown and to consider the following one. The procedure can be repeated until a residual which does not produce a breakdown has been found.

A smoothing algorithm similar to the preceding one can be applied even if the vectors $r_n^{(k)}$ are not mutually orthogonal and we have for $k = 2, 3, \ldots$, and $n = 0, 1, \ldots$

$$\begin{aligned}
\eta_n^{(k)} &= \left\| r_n^{(k)} \right\| \\
\frac{1}{\tau_n^{(k)2}} &= \frac{1}{\tau_n^{(k-1)2}} + \frac{1}{\eta_n^{(k)2}} \\
\rho_n^{(k)} &= \tau_n^{(k)2} \left[\frac{\rho_n^{(k-1)}}{\tau_n^{(k-1)2}} + \frac{r_n^{(k)}}{\eta_n^{(k)2}} \right] \\
y_n^{(k)} &= \tau_n^{(k)2} \left[\frac{y_n^{(k-1)}}{\tau_n^{(k-1)2}} + \frac{x_n^{(k)}}{\eta_n^{(k)2}} \right]
\end{aligned}$$

with $\rho_n^{(1)} = r_n^{(1)}, y_n^{(1)} = x_n^{(1)}$ and $\tau_n^{(1)} = \left\| r_n^{(1)} \right\|$.

Thus, for this algorithm, the quantities $\tau_n^{(k)}$ are no longer equal to the quantities $\left\| \rho_n^{(k)} \right\|$ (except for $k = 1$ or when the $r_n^{(i)}$'s are mutually orthogonal) and the relation with the MHP no longer holds. When $n = 0$ and the $r_0^{(k)}$'s are the successive residual vectors produced by an arbitrary iterative method, then the general *quasi-minimal residual smoothing* procedure (in short the QMRSm) introduced by Zhou and Walker [28] is recovered (an extra "m" was introduced in the name of this procedure for avoiding confusion with the QMR Squared algorithm). In particular, if the iterates $r_0^{(k)}$ are those given by the biconjugate gradient algorithm (BCG) then Zhou and Walker [28] proved that the $\rho_0^{(k)}$'s given by the QMRSm are the iterates of the QMR of Freund and Nachtigal [13] and the same type of results holds (see eq. (3.1.9) of [28]). This result was implicitly contained in [16] where it is shown that the QMR iterates can be obtained from those of the BCG. When the matrix A is symmetric positive definite and the vectors $r_0^{(k)}$ are those obtained by the conjugate gradient algorithm (they are mutually orthogonal) then the theoretical results given above are recovered, $\tau_n^{(k)} = \left\|\rho_n^{(k)}\right\|$ and the procedure reduces to the MRS. Zhou and Walker [28] also proved that, if the QMRSm is applied to the algorithm CGS of Sonneveld [21] then the TFQMR of Freund [14] is recovered and that the QMRCGSTAB [9] can be similarly obtained from the Bi–CGSTAB of Van der Vorst [22]. A synthetic review of these results and the connections between these algorithms is presented in [23].

We have

$$\eta_n^{(k)} = \|r_n^{(k)}\| = \frac{\tau_n^{(k)}}{\sqrt{1 - \tau_n^{(k)2} / \tau_n^{(k-1)2}}}.$$

So, peaks in the sequence $(\eta_n^{(k)})$ correspond to plateaus in the sequence $(\tau_n^{(k)})$ [23].

In practice, there often exists a discrepancy between the residual vectors $r_n^{(k)}$ and their actual values $b - Ax_n^{(k)}$ which causes the vectors $\rho_n^{(k)}$ computed recursively to differ significantly from the exact residuals $b - Ay_n^{(k)}$. A mathematically equivalent algorithm for avoiding this drawback was formulated by Zhou and Walker [28, Algo. 3.2.2]. This algorithm also applies in our case.

For treating the case of several iterative methods, it is possible, instead of the MHP to use the hybrid procedure of rank 2 *in cascade* as explained in [6]. This procedure (in short the CHP) consists of applying the hybrid procedure of rank 2 to $r_n^{(1)}$ and $r_n^{(2)}$. Denoting by $\rho_n^{(2)}$ the result, the same procedure is then applied to $\rho_n^{(2)}$ and $r_n^{(3)}$. Denoting the result by $\rho_n^{(3)}$, the same procedure is applied again to $\rho_n^{(3)}$ and $r_n^{(4)}$ and so on.

Thus, the CHP is given by, for $k = 2, 3, \ldots$

$$\rho_n^{(k)} = a_n^{(k)} \rho_n^{(k-1)} + \left(1 - a_n^{(k)}\right) r_n^{(k)}$$

with $a_n^{(k)} = -\left(\rho_n^{(k-1)} - r_n^{(k)}, r_n^{(k)}\right) / \left(\rho_n^{(k-1)} - r_n^{(k)}, \rho_n^{(k)} - r_n^{(k)}\right)$ and $\rho_n^{(1)} = r_n^{(1)}$.

The MRS and the CHP are related by the

Theorem 9

If the vectors $r_n^{(i)}$ are mutually orthogonal, then the MRS *and the* CHP *are identical.*

Acknowledgements: I would like to thank Michela Redivo Zaglia for several pertinent comments about this paper.

References

[1] A. Abkowicz, *Etude de la Procédure Hybride Appliquée à la Résolution des Systèmes Linéaires*, Thèse, Université des Sciences et Technologies de Lille, 1995.

[2] A. Abkowicz, C. Brezinski, Acceleration properties of the hybrid procedure for solving linear systems, Applicationes Mathematicae, 23 (1996) 417–432.

[3] C. Brezinski, *Projection Methods for Systems of Equations*, North–Holland, Amsterdam, 1997.

[4] C. Brezinski, J.–P. Chehab, Nonlinear hybrid procedures and fixed point iterations, Appl. Math. Optimization, to appear.

[5] C. Brezinski, J.–P. Chehab, Multiparameter iterative schemes for the solution of linear and nonlinear equations, SIAM J. Sci. Comput., to appear.

[6] C. Brezinski, M. Redivo Zaglia, Hybrid procedures for solving linear systems, Numer. Math., 67 (1994) 1–19.

[7] C. Brezinski, M. Redivo Zaglia, Look–ahead in Bi–CGSTAB and other product methods for linear systems, BIT, 35 (1995) 169–201.

[8] C. Brezinski, H. Sadok, Vector sequence transformations and fixed point methods, in *Numerical Methods in Laminar and Turbulent Flows, vol. I*, C. Taylor et al. eds., Pineridge Press, Swansea, 1987, pp. 3–11.

[9] T.F. Chan, E. Gallopoulos, V. Simoncini, T. Szeto, C.H. Tong, A quasi–minimal residual variant of the Bi–CGSTAB algorithm for nonsymmetric systems, SIAM J. Sci. Comput., 15 (1994) 338–347.

[10] J.K. Cullum, Peaks, plateaus, numerical instabilities in Galerkin minimal residual pair of methods for solving $Ax = b$, Appl. Numer. Math., 19 (1995) 255–278.

[11] R.P. Eddy, Extrapolation to the limit of a vector sequence, in *Information Linkage between Applied Mathematics and Industry*, P.C.C. Wang ed., Academic Press, New York, 1979, pp. 387–396.

[12] R. Fletcher, Conjugate gradient methods for indefinite systems, in *Numerical Analysis, Dundee 1975*, G.A. Watson ed., LNM 506, Springer, Berlin, 1976, pp.73–89.

[13] R.W. Freund, Conjugate gradient–type methods for linear systems with complex symmetric coefficient matrices, SIAM J. Sci. Stat. Comput., 13 (1992) 425–448.

[14] R.W. Freund, A transpose–free quasi–minimal residual algorithm for non–Hermitian linear systems, SIAM J. Sci. Stat. Comput., 14 (1993) 470–482.

[15] R.W. Freund, N.M. Nachtigal, QMR: a quasi–minimal residual method for non–Hermitian linear systems, Numer. Math., 60 (1991) 315–339.

[16] R.W. Freund, T. Szeto, A quasi–minimal residual squared algorithm for non-Hermitian linear systems, Technical Report, RIACS, NASA Ames Research Center, Moffett Field, CA, 1992.

[17] M. Heyouni, H. Sadok, On a variable smoothing procedure for conjugate gradient type methods, submitted.

[18] M. Mes̆ina, Convergence acceleration for the iterative solution of $x = Ax + f$, Comput. Methods Appl. Mech. eng., 10 (1977) 165–173.

[19] W. Schönauer, *Scientific Computing on Vector Computers*, North–Holland, Amsterdam, 1987.

[20] W. Schönauer, H. Müller, E. Schnepf, Numerical tests with biconjugate gradient type methods, Z. Angew. Math. Mech., 65 (1985) T400–T402.

[21] P. Sonneveld, CGS, a fast Lanczos–type solver for nonsymmetric linear systems, SIAM J. Sci. Stat. Comp., 10 (1989) 36–52.

[22] H.A. Van der Vorst, Bi–CGSTAB: a fast and smoothly converging variant of Bi–CG for the solution of nonsymmetric linear systems, SIAM J. Sci. Stat. Comput., 13 (1992) 631–644.

[23] H.F. Walker, Residual smoothing and peak/plateau behavior in Krylov subspace methods, Appl. Numer. Math., 19 (1995) 279–286.

[24] R. Weiss, *Convergence Behavior of Generalized Conjugate Gradient Methods*, PhD thesis, University of Karlsruhe, 1990.

[25] R. Weiss, Error–minimizing Krylov subspace methods, SIAM J. Sci. Stat. Comput., SIAM J. Sci. Comput., 15 (1994) 511–527.

[26] R. Weiss, Properties of generalized conjugate gradient methods, Num. Lin. Algebra with Appl., 1 (1994) 45–63.

[27] R. Weiss, W. Schönauer, Accelerating generalized conjugate gradient methods by smoothing, in *Iterative Methods in Linear Algebra*, R. Beauwens and P. de Groen eds., North–Holland, Amsterdam, 1992, pp.283–292.

[28] L. Zhou, H.F. Walker, Residual smoothing techniques for iterative methods, SIAM J. Sci. Comput., 15 (1994) 297–312.

ABS ALGORITHMS FOR LINEAR EQUATIONS AND APPLICATIONS TO OPTIMIZATION

E. SPEDICATO
University of Bergamo
Z.Q. XIA and L.W. ZHANG
Dalian University of Technology
K. MIRNIA
University of Tabriz

Abstract We first review some of the main results concerning the ABS class for linear systems, introduced in 1984 by Abaffy, Broyden and Spedicato. Then we describe a new algorithm in this class, which solves a general n by n nonsingular linear system in $n^3/3 + O(n^2)$ multiplications without the assumption that the coefficient matrix be regular. The method can be viewed as a variation of the implicit LU algorithm of the ABS class, whose associated factorization contains a factor which is not triangular (but can be reduced to triangular form after suitable row permutations). We describe properties of the method, including in particular an efficient way of updating the Abaffian matrix after column interchanges. The algorithm has a natural application to the simplex algorithm for the LP problem in standard form, where it provides a faster technique for the pivoting operation than the methods based upon the standard LU factorization when the number m of equality constraints is greater than $n > 2$. We present some numerical results showing that the implicit LX algorithm can solve very accurately ill conditioned problems. Then we consider several methods for KT equations, some of them faster or cheaper in storage than classical methods. Finally, we give a unified formulation of feasible direction methods for linearly constrained optimization, including the LP problem, in terms of the parameters of the ABS class.

Key Words Linear Equations, LP Problem, ABS Methods, KT Equations, Feasible Direction Methods, Implicit LX Algorithm

G. Winter Althaus and E. Spedicato (eds.), Algorithms for Large Scale Linear Algebraic Systems, 291–319.

1. The Basic ABS Class

ABS methods have been introduced by Abaffy, Broyden and Spedicato (1984), originally for solving linear equations. They have been later extended to linear least squares, nonlinear equations and optimization problems, see Abaffy and Spedicato (1989) and Spedicato (1993). The ABS literature consists presently of over 350 papers. Here we only consider the so called *basic or unscaled ABS class* for solving the system

$$Ax = b \qquad x \in R^n, \ \ b \in R^m, \ \ m \leq n \tag{1.1}$$

or

$$a_i^T x - b_i = 0, \quad i = 1, \ldots, m \tag{1.2}$$

where

$$A = \begin{bmatrix} a_1^T \\ \cdots \\ a_m^T \end{bmatrix} \tag{1.3}$$

and $rank(A)$ is arbitrary. The recursions of the basic ABS class are the following:

(A) Let $x_1 \in R^n$ be arbitrary, $H_1 \in R^{n,n}$ be nonsingular arbitrary, set $i = 1$

(B) Compute $s_i = H_i a_i$. If $s_i \neq 0$, go to (C). If $s_i = 0$ and $\tau = a_i^T x_i - b_i = 0$, then set $x_{i+1} = x_i$, $H_{i+1} = H_i$ and go to (F), otherwise stop, the system has no solution.

(C) Compute the search vector p_i by

$$p_i = H_i^T z_i \tag{1.4}$$

where $z_i \in R^n$ is arbitrary save for the condition

$$a_i^T H_i^T z_i \neq 0 \tag{1.5}$$

(D) Update the estimate of the solution by

$$x_{i+1} = x_i - \alpha_i p_i \tag{1.6}$$

where the stepsize α_i is given by

$$\alpha_i = (a_i^T p_i - b_i)/a_i^T p_i \tag{1.7}$$

(E) Update the matrix H_i by

$$H_{i+1} = H_i - H_i a_i w_i^T H_i / w_i^T H_i a_i \tag{1.8}$$

where $w_i \in R^n$ is arbitrary save for the condition

$$w_i^T H_i a_i \neq 0 \tag{1.9}$$

(F) Stop if $i = m$, x_{m+1} solves the system, otherwise increment i by one and go to (B).

At the first ABS international conference in Luoyang, 1991, the name *Abaffians* was proposed for the matrices H_i and this name will be used in this paper. Of course recursions of the form (1.8) have appeared before in the literature, the earliest reference being possibly Wedderburn (1934).
Among the properties of the basic ABS class, we quote the following ones, see Abaffy and Spedicato (1989) for proofs.

- Define A_i and W_i by

$$A_i = (a_1, \ldots, a_i)^T \tag{1.10}$$

$$W_i = (w_1, \ldots, w_i)^T \tag{1.11}$$

then

$$H_i A_i = 0 \tag{1.12}$$

$$H_i^T W_i = 0 \tag{1.13}$$

- Let a_i be linearly independent from $a_i, \ldots, a_{i-1}$. Then

$$H_i a_i \neq 0 \tag{1.14}$$

$$H_i^T w_i \neq 0 \tag{1.15}$$

- Define P_i by

$$P_i = (p_1, \ldots, p_i) \tag{1.16}$$

Then, for $rank(A_i) = i$, the following implicit factorization property holds

$$A_i^T P_i = L_i \tag{1.17}$$

where L_i is nonsingular lower triangular.

- Let $rank(A_i) = i$ and define S_i and R_i by

$$S_i = (s_1, \ldots, s_i) \tag{1.18}$$

$$R_i = (r_1, \ldots, r_i) \tag{1.19}$$

where

$$s_i = H_i a_i \tag{1.20}$$

$$r_i = H_i^T w_i \tag{1.21}$$

Then the Abaffian can be written in the form

$$H_{i+1} = H_1 - S_i R_i^T \tag{1.22}$$

and the vectors s_i, r_i can be built via Gram-Schmidt type iterations involving the previous vectors (the search vector p_i can be built in a similar way).

Another useful representation of the Abaffian is the following

$$H_{i+1} = H_1 - H_1 A_i (W_i^T H_1 A_i)^{-1} W_i^T H_1 \tag{1.23}$$

The *scaled ABS class* is obtained by applying (implicitly) the unscaled ABS algorithm to the scaled (or preconditioned) system $V^T Ax = V^T b$, where V is an arbitrary nonsingular matrix of order m. Letting v_i be the i-th column of V, then the recursions of the scaled ABS class are obtained by substituting in the scaled ABS class a_i by $A^T v_i$ and b_i by $b^T v_i$, where $r_i \in R^m$ is the residual vector $Ax_i - b$. The well definiteness conditions are $z_i^T H_i A^T v_i \neq 0$ and $w_i^T H_i A^T v_i \neq 0$. The associated factorization becomes

$$V^T AP = L \tag{1.24}$$

If $m = n$, from (1.24) one obtains

$$A^{-1} = PL^{-1}V \tag{1.25}$$

Since in many cases L is diagonal, and without loss of generality it can always be made to be diagonal by a triangular scaling of V (which does not affect the sequence x_i), (1.25) shows that ABS methods provide an essentially explicit factorization of the inverse, which is a main difference with respect to classical methods. Explicit preconditioners could therefore be obtained by relation (1.25) with approximate evaluation of P and V, see for instance Benzi, Meyer and Tuma (1994).

Among the subclasses of the scaled ABS class we quote the following.

(a) The *conjugate direction subclass.* This class is well defined under the condition (sufficient but not necessary) that A is symmetric and positive definite. It contains the implicit LU algorithm, the Hestenes-Stiefel and the Lanczos algorithms. This class generates all possible algorithms whose search directions are A-conjugate. The vector x_{i+1} minimizes the A-weighted Euclidean norm of the error over $x_1 + Span(p_1, \ldots, p_i)$.

(b) The *orthogonally scaled subclass.* This class is well defined if A has full row rank (m can be greater than n). It contains the ABS formulation of the QR algorithm (the so called *implicit QR algorithm*), of the GMRES and of the conjugate residual algorithms. The scaling vectors are orthogonal and the search vectors are AA^T-conjugate. The vector x_{i+1} minimizes the Euclidean norm of the residual over $x_1 + Span(p_1, \ldots, p_i)$. In general the methods in this class can be applied to overdetermined system to obtain the solution in the least squares sense. A version of the implicit QR algorithm, with reprojection on both the search vector and the scaling vector, tested by Spedicato and Bodon (1992), has outperformed other ABS algorithms for linear least squares methods and methods in the LINPACK and NAG library based upon the classical QR factorization via the Householder matrices.

(c) The *optimally scaled subclass.* This class is obtained by the choice $v_i = A^{-T} p_i$, the inverse of A^T disappearing in the actual formulas, after the transformation $z_i = A^T u_i$, u_i now the parameter defining the search vector. For $u_i = e_i$ the Huang method is obtained, for $u_i = r_i$ a method equivalent to the Craig's algorithm. From (1.24) one obtains $P^T P = D$ or

$$V^T AA^T V = D \tag{1.26}$$

Relation (1.26) was shown by Broyden (1985) to characterize the optimal choice of the scaling parameters in terms of minimizing the effect of a single error in x_i on the final computed solution. Such a property is therefore satisfied by the Huang (and the Craig) algorithm, but not, for instance, by the implicit LU or the implicit QR algorithms. Galantai (1991) has shown that the condition characterizing the optimal choice of the scaling parameters in terms of minimizing the final residual Euclidean norm is $V^T V = D$, a condition satisfied by the implicit QR algorithm, the GMRES method, the implicit LU algorithm and again by the Huang algorithm, which therefore satisfies both conditions. The methods in the optimally stable subclass have the property that x_{i+1} minimizes the Euclidean norm of the error over $x_1 + Span(p_1, \ldots, p_i)$.

In the following, for simplicity of formulation, we will assume that A has full rank, implying that the above stated properties are valid up to the index $i = m$. We will call a matrix $A \in R^{m,n}$, $m \leq n$, *regular* if all its principal submatrices $A^{(i)}$, $i = 1, \ldots, m$, are nonsingular.

2. The Implicit LU Algorithm and the Huang Algorithm

Specific algorithms are obtained by choosing the parameters H_1, z_i and w_i. The *implicit LU algorithm* is given by the choices

$$H_1 = I, \quad z_i = w_i = e_i, \tag{2.1}$$

where e_i is the i-th unit vector in R^n. We quote the following properties of the implicit LU algorithm.

(a) The algorithm is well defined iff A is regular. Otherwise pivoting of the columns has to be performed (or of the equations, if $m = n$).

(b) The Abaffian H_{i+1} has the following structure

$$H_{i+1} = \begin{bmatrix} 0 & 0 \\ \cdots & \cdots \\ 0 & 0 \\ K_i & I_{n-i} \end{bmatrix} \tag{2.2}$$

where $K_i \in R^{n-i,i}$ has the form

$$(K_i)^T = -[A^{(i)}]^{-1} A'^{(n-i)} \tag{2.3}$$

and $A'^{(n-i)}$ is the matrix consisting of the last $n - i$ columns of A.

(c) Only the first i components of p_i can be nonzero and the i-th component is unity. Hence in the implicit factorization (1.17) P_i is unit upper triangular.

(d) Only K_i has to be updated. The algorithm needs $nm^2 - 2m^3/3$ multiplications plus lower order terms, hence, for $m = n$, $n^3/3$ multiplications plus lower order terms.

If $m = n$ from (1.17) we obtain the factorization $A = LU$, with $L = L_n$, $U = P_n^{-1}$ unit upper triangular. This property justifies the name *implicit LU algorithm.* It should be pointed out however, as already stressed by Benzi (1993), that the implicit LU algorithm cannot be reduced to a reordering of the operations performed by the classical LU algorithm, as is the case with Gaussian elimination, but it is an essentially different algorithm (the matrix P, for instance, is an explicit factor of A^{-1}).
The implicit LU algorithm, implemented in the case $m = n$ with pivoting of the equations, has been shown in experiments of Bertocchi and Spedicato (1989) to be numerically stable and in experiments of Bodon (1993) on the vector processor Alliant FX 80 with 8 processor to be about twice faster than the LAPACK implementation of the classical LU algorithm.
Formulations of the implicit LU algorithm for special matrices like banded, block angular or ND matrices, have also been shown to be competitive with the special formulations of the classical LU algorithm, see for instance Bodon (1992) or Yang and Zhu (1994).
A generalization of the implicit LU algorithm where z_i and w_i are similarly defined but H_1 is essentially arbitrary, has been considered by Spedicato and Zhu (1994). The obtained *generalized implicit LU algorithm* is well defined iff AH_1^T is regular. Notice that the implicit LU algorithm with column pivoting can be imbedded in the generalized implicit LU algorithm by taking H_1 as a suitable permutation matrix. For these algorithms the Abaffian has the form

$$H_{i+1} = \begin{bmatrix} 0 \\ K_i \end{bmatrix} \tag{2.4}$$

and the required number of multiplications is no more than $2mn^2 - m^2n$ plus lower order terms, or, for $m = n$, n^3 plus lower order terms. It can be shown that by setting $H_1^T = (z_1, \ldots, z_n)$ one obtains an algorithm equivalent to the algorithm of the basic ABS class with parameters z_i, $w_i = z_i$ and $H_1 = I$.
The *Huang algorithm* is obtained by the parameter choices $H_1 = I$, $z_i = w_i = a_i$. A mathematically equivalent, but numerically more stable, formulation of this algorithm is the so called *modified Huang algorithm* where the search vectors and the Abaffian are given by formulas $p_i = H_i(H_i a_i)$ and $H_{i+1} = H_i - p_i p_i^T / p_i^T p_i$. Some properties of this algorithm are the

following.

- The search vectors are orthogonal and are the same as the vectors obtained by applying the classical Gram-Schmidt orthogonalization procedure to the rows of A. The modified Huang algorithm is related, but is not numerically identical, with the reorthogonalized Gram-Schmidt algorithm of Daniel, Gragg, Kaufmann and Stewart (1977).

- If x_1 is the zero vector, then the vector x_{i+1} is the solution of least Euclidean norm of the first i equations and the solution x^+ of least Euclidean norm of the whole system is approached monotonically and from below by the sequence x_i. Zhang (1995a) has shown how to apply the Huang algorithm, via an active set strategy as in Goldfarb and Idnani (1983), to systems of linear inequalities, where in a finite number of steps it finds the solution of least Euclidean norm or it determines that the system has no solution.

- While the error growth in the Huang algorithm is controlled by the square of the number $\eta_i = \|a_i\|/\|H_i a_i\|$, which is certainly large for some i if A is ill conditioned, the error growth depends only on η_i if p_i or H_i are defined as in the modified Huang algorithm and, at first order, there is no error growth for the modified Huang algorithm.

In this paper we consider the modification of the implicit LU algorithm where H_1 is still the identity matrix but z_i and w_i are not generally multiples of the i-th unit vector. The obtained algorithm is well defined without the assumption that A be regular. The structure of H_{i+1} is more complicated than in (2.2), but can be reduced to that structure after row and column permutations. Similarly P is no more unit upper triangular in general, but can be reduced to such a form after suitable row permutations. The number of multiplications required by the modified algorithm is the same as in the original algorithm, at first order, hence the algorithm, not requiring in general exchanges of rows or of columns, can be considered to be faster than the classical LU algorithm.

3. The Implicit LX Algorithm

The *implicit LX algorithm* is defined by the choices

$$H_1 = I, \; z_i = w_i = e_{k_i}, \tag{3.1}$$

where k_i is an integer, $1 \leq k_i \leq n$, such that

$$e_{k_i}^T H_i a_i \neq 0 \tag{3.2}$$

We have assumed that A has full rank. Hence from (1.14) there is at least one index k_i such that (3.2) is satisfied. For stability reasons it may be recommended to select k_i such that $\eta_i = | e_{k_i}^T H_i a_i |$ is maximized.
The following properties are a straightforward consequence of the general properties of the ABS algorithm. Let N be the set of integers from 1 to n, $N = (1, 2, \ldots, n)$. Let B_i be the set of indexes $k_1, \ldots, k_i$ chosen for the parameters of the implicit LX algorithm up to the step i. Let N_i be the set $N \backslash B_i$. From (1.13) we have

$$H_{i+1}^T e_{k_j} = 0, \quad k_j \in B_i \tag{3.3}$$

which implies:

(a) the index k_i is selected in the set N_{i-1}

(b) the rows of H_{i+1} of index $k \in \mathbf{B_j}$ are null rows.

From the definition of the search vector

$$p_i = H_i^T e_{k_i} \tag{3.4}$$

and the update formula of the Abaffian

$$H_{i+1} = H_i - H_i a_i p_i^T / p_i^T a_i \tag{3.5}$$

the following properties follow easily

(c) the vector p_i has $n - i$ zero components; its k_i-th component is equal to one, i.e.

$$p_i^T e_{k_i} = 1 \tag{3.6}$$

(d) if $x_1 = 0$, then x_{i+1} is a basic type solution of the first i equations, whose nonzero components may lie only in the positions corresponding to the indexes $k \in B_i$

(e) the columns of H_{i+1} of index $k \in N_i$ are the unit vectors e_k, i.e.

$$H_{i+1} e_k = e_k \tag{3.7}$$

while the columns of H_{i+1} of index $k \in B_i$ have zero components in the j-th position, with $j \in B_i$, implying that only $i(n-i)$ elements of such columns have to be computed

(f) at the i-th step $i(n-i)$ multiplications are needed to compute $H_i a_i$ and $i(n-i)$ to update the nontrivial part of H_i. Hence the total number of multiplications is the same as for the implicit LU algorithm, but no pivoting is necessary

(g) by applying a sequence of row permutations that interchange the k-th row with the j_k-th row, $k = 1, \ldots, i$, and the k-th column with the j_k-th column, H_{i+1} is reduced to the structure (2.2), i.e.

$$Q_i H_{i+1} Q_i^T = \begin{bmatrix} 0 & 0 \\ \cdots & \cdots \\ 0 & 0 \\ K_i & I_{n-i} \end{bmatrix} \tag{3.8}$$

where Q_i is the product of the considered elementary permutation matrices

(h) by applying the same sequence of row permutations as in **(g)** the matrix P_i is changed into a matrix whose first i rows define a unit upper triangular matrix, i.e.

$$Q_i P_i = \begin{bmatrix} U_i \\ 0 \end{bmatrix} \tag{3.9}$$

For $m = n$, we have, with $U = U_n$ and $Q = Q_n$

$$U = QP \tag{3.10}$$

Hence the implicit factorization $AP = L$ implies

$$AQ^T U = L \tag{3.11}$$

In other terms the implicit LX algorithm is equivalent to the implicit LU algorithm applied on the matrix AQ^T, i.e. it is equivalent to the implicit LU algorithm with suitable column pivoting

(i) in terms of the representation (1.22) the j_k-th null row of H_{i+1} corresponds to a j_k-th unit vector in $S_i R_i^T$, while the j_k-th column of H_{i+1}, which is a j_k-th unit vector, corresponds to a null column.

We suppose now that the implicit LX algorithm has been applied for m steps. Let $H = H_{m+1}$ and define the matrices $I_B \in R^{n,m}$ and $I_N \in R^{n,n-m}$ by

$$I_B = (e_{k_1}, \ldots, e_{k_m}), \quad k_i \in B_m \tag{3.12}$$

$$I_N = (e_{k_1}, \ldots, e_{k_{n-m}}), \quad k_i \in N_m \tag{3.13}$$

Notice that $I_B^T I_B = I_m$, $I_B^T I_N = 0_{m,n-m}$, $I_N^T I_N = I_{n-m}$, $I_N^T I_B = 0_{n-m,m}$. Here, with reference to the natural application of the implicit LX algorithm to the simplex method, B relates to the basic variables, N to the nonbasic variables.

Matrices I_B and I_N are useful to express in a formal way the structure of the Abaffian generated by the implicit LX algorithm. Let $W = W_m$ and let $Diag(\alpha_i)$ be the diagonal matrix whose i-th diagonal element is α_i. The matrix W associated with the implicit LX algorithm is just the matrix I_B.

Define now $A_B \in R^{m,m}$ and $A_N \in R^{m,n-m}$ by

$$A_B = AI_B \tag{3.14}$$

$$A_N = AI_N \tag{3.15}$$

Notice that A_B consists of the columns of A associated with the basic variables, while A_N consists of the columns of A associated with the nonbasic variables.

Let us now consider the representations of the Abaffian (1.22) and (1.23), for $i = m$. By comparison we have that $S_m R_m^T = A^T(W^T A^T)^{-1} W^T = A^T[I_B^T A^T]^{-1} I_B^T = A^T(A_B^T)^{-1} I_B^T = (I_B A_B^{-1} A)^T$. Hence we can write (1.23) as follows for $i = m$

$$H = I - A^T(A_B^T)^{-1} I_B^T \tag{3.16}$$

Define H_B by

$$H_B = I_B^T H \tag{3.17}$$

Then from (3.16)

$$H_B = 0_{m,n} \tag{3.18}$$

which is equivalent to the fact, observed before, that the m rows of H of index k in B are null rows. Multiplying on the left by I_N^T and on the right by I_N we get

$$I_N^T H I_N = I_{n-m} \tag{3.19}$$

which is equivalent to the fact, observed before, that the $n - m$ columns of H of index k in N are k-th unit vectors. Multiplying on the right by I_B and on the left by I_N^T we get

$$I_N^T H I_B = -\,(A_B^{-1} A_N)^T \tag{3.20}$$

which gives the structure of the nontrivial part of H in terms of the basic and nonbasic columns of A, generalizing relation (2.3).

From the identity $S_m U_m^T = A^T(A_B^T)^{-1} I_B^T$ and by defining

$$S_B = I_B^T S_m \tag{3.21}$$

$$S_N = I_N^T S_m \tag{3.22}$$

$$U_B = I_B^T U_m \tag{3.23}$$

$$U_N = I_N^T U_m \tag{3.24}$$

one gets in a similar way

$$S_B U_B^T = I_m \tag{3.25}$$

$$S_N U_B^T = (A_B^{-1} A_N)^T \tag{3.26}$$

$$S_m U_N^T = 0_{n,n-m} \tag{3.27}$$

From general properties of the ABS algorithm, the matrix $W_i^T H_1 A_i$ is regular, hence $W^T A^T$ is regular. Since $W^T A^T = A_B$ it follows that the matrix A_B associated with the implicit LX algorithm is regular.

4. Implicit column interchanges in the implicit LX algorithm and applications to the simplex method

We have seen in the previous section that determining the elements of B_m is equivalent to determining which columns of A should form the matrix A_B. Changes in A_B are equivalent to changes in B_m. Several algorithms, particularly the simplex method for linear programming, are characterized by exchanges of columns between A_B and A_N. Disregarding stability considerations these exchanges are possible iff the new matrix A_B is nonsingular. In the simplex method, a column of original index N^* from the matrix A_N is interchanged with a column of original index B^* in the matrix A_B. The column in A_N is often taken as the column with minimal relative cost. In terms of the ABS formulation of the simplex method, see Spedicato, Xia and Zhang (1995), this is equivalent to minimize with respect to $i \in N_m$ the scalar η_i defined by

$$\eta_i = c^T H^T e_i \tag{4.1}$$

where $c \in \mathbf{R}^n$ is the vector defining the linear function to be minimized, $f(x) = c^T x$. Notice that $H^T e_i \neq 0$, since only the rows of H of index $i \in B_m$ are zero rows. The column in A_B to be exchanged is usually chosen with the criterion of the maximum displacement along an edge which keeps the basic variables nonnegative. Define the scalar ω_i by

$$\omega_i = x^T e_i / e_i^T H^T e_{N^*} \tag{4.2}$$

where x is the current basic feasible solution. Then the above criterion is equivalent to minimize ω_i with respect the set of indices $i \in B_M$ such that

$$e_i^T H^T e_{N^*} > 0 \tag{4.3}$$

Notice that $H^T e_{N^*} \neq 0$ and that an index i such that (4.3) is satisfied always exists, unless x is a solution of the LP problem. We call B^* the index defined in such a way and $B^* - 1$ and $B^* + 1$ the elements of B that respectively precede and follow B^*. We similarly define $N^* - 1$ and $N^* + 1$. Then the following inequality is satisfied

$$e_{B^*}^T H^T e_{N^*} > 0 \tag{4.4}$$

From the results in the previous section interchanging the B^*-th column with the N^*-th column is equivalent to constructing an Abaffian H' via the sequence of unit vectors $e_{B_1}, \ldots, e_{B^*-1}, e_{N^*}, e_{B^*+1}, \ldots, e_{B_m}$. Therefore, from (1.23), H' has the form

$$H' = I - A^T (W'^T A^T)^{-1} W'^T \tag{4.5}$$

where the i-th column of W', $i = 1, \ldots, m$, has the form

$$W' e_{k_i} = e_{k_i} \tag{4.6}$$

and e_{k_i} is the i-th column of the following matrix $I_{B'} \in R^{n,m}$

$$I_{B'} = (e_{B_1}, \ldots, e_{B^*-1}, e_{N^*}, e_{B^*+1}, \ldots, e_{B_m}) \tag{4.7}$$

Then W' has the form

$$W' = I_{B'} \tag{4.8}$$

From (4.5) and (4.8) the following formula can be obtained, see Spedicato, Xia and Zhang (1995) or Zhang (1995b), where H' is shown to be generated by a rank-one correction to H

$$H' = H - (He_{B^*} - e_{B^*}) e_{N^*}^T H / e_{N^*}^T H e_{B^*} \tag{4.9}$$

Formula (4.12) is of great interest in itself and for its applications to the simplex method. Here we make the following remarks:

- only the indices of the columns to be exchanged appear, not the columns themselves
- stability reasons suggest that pivoting in the simplex method should take into account also the size of $e_{N\bullet}^T H e_{B\bullet}$

Let us now consider the computational cost of update (4.9). Since $He_{B\bullet}$ has at most $n - m$ nonzero components, while $H^T e_{N\bullet}$ has at most m, no more than $m(n-m)$ multiplications are required. The update is most expensive for $m = n/2$ and gets cheaper the smaller m is or the closer it is to n. The number of multiplications for the traditional formulation of the revised simplex method, see Broyden (1975), is $3m^2$. Formula (4.9) is therefore faster if $m > n/4$. In the dual steepest edge method of Forrest and Goldfarb (1992) the overhead for replacing a column is m^2, hence formula (4.9) is faster for $m > n/2$. Therefore formula (4.9) is recommended on overhead considerations for m sufficiently large. Updates of the representation (1.22) have also been obtained, see Spedicato, Xia and Zhang (1995) or Zhang (1996). Their cost is order m^2, as for the classical simplex formulations using the LU factorization. Comparison of the numerical performance of the new formula versus the classical formulas in LP problems has yet to be done.

5. Iterative Refinement in the Implicit LX Method

Iterative refinement is a process which under certain conditions allows to obtain a full digit solution. The process consists in solving by approximate Newton algorithm the equation $r(x) = 0$, where $r(x) = Ax - b$. The Newton iteration is $x' = x - d$, where the search direction d satisfies the linear system $Jd = r(x)$, J being the Jacobian matrix of $r(x)$, which in this case is just $J = A$. The approximate Newton algorithm, which defines the classical iterative refinement, is obtained by solving an approximate Newton equation using the information, affected by roundoff, from the previous solution of the linear system $Ax = b$. Hence, for classical factorization methods, one represents J in terms of the already computed factors of A, so that the solution of the approximate Newton equation takes only order n^2 multiplications. If the error in the factorization is sufficiently small and if the residual $r(x)$ is computed in double precision and then rounded to single precision, then iterative refinement converges to the solution and a full digit approximation of the solution is typically obtained in a few steps. Notice that a small error in the factorization is not expected when the condition number of A is large. For ABS methods one can proceed in a similar way, solving the approximate Newton equation using the search vectors obtained when solving $Ax = b$. Each cycle of the iterative refinement uses order n^2 multiplications, hence the main cost is the storage of the search vectors. Extensive numerical experiments have been performed by Mirnia (1996) on several ABS algorithms, including the implicit LX, the Huang, the modified Huang and the implicit QR algorithm and for both the cases of determined and underdetermined systems. The results show that again iterative refinement converges fast when the condition number of A is not too large, for determined systems, and that convergence under such conditions is also obtained when the system is underdetermined, if the implicit LX algorithm is used. In the following Table we give the results for problem P1, where the matrix is given by $A_{i,j} = |i - j|$ and the right hand side is calculated (exactly) by assigning as a solution the vector with all components equal to one, and for problem P2, where the matrix is the square of the previously defined matrix, with similarly defined right hand side. The performance of the methods is measured by the relative errors ES and ER, defined as follows, with x the computed solution and x^+ the exact solution:

$$ES = \|x - x^+\| / \|x^+\| \tag{5.1}$$

$$ER = \|Ax - b\| / \|b\| \tag{5.2}$$

The first number under the headings EX,ER gives the result before refinement, the second number the result after refinement (FAIL indicates failure of the refinement procedure, usually associated with divergence).

Table 5.1. Iterative refinement with the implicit LX and the modified Huang algorithms

n	Prob.	$EX(LX)$	$ER(LX)$	$EX(HUANG)$
100	$P1$	$3.6E-4/0.$	$3.3E-7/0.$	$6.7E-4/0.$
	$P2$	$1.4E+0/0.$	$1.6E-7/0.$	$3.4E+0/FAIL$
200	$P1$	$2.3E-3/0.$	$2.2E-7/0.$	$3.4E-3/0.$
	$P2$	$6.5E+2/FAIL$	$3.2E-5/FAIL$	$1.7E+8/FAIL$
400	$P1$	$1.5E-2/0.$	$3.8E-7/0.$	$1.8E-2/FAIL$
	$P2$	$2.1E+4/FAIL$	$1.6E-3/FAIL$	$4.6E+8/FAIL$
800	$P1$	$8.0E-2/0.$	$5.4E-7/0.$	
	$P2$	$8.8E+3/FAIL$	$1.1E-3/FAIL$	

The condition number of problem P1 is quite large (for problem P2 it is the square of it). For problem P1, where the implicit LX algorithm without refinement gives an already very good solution, a full digit solution is given by the iterative refinement. For problem P2 iterative refinement converges to a full digit solution for $n = 100$, but there is no convergence for the higher values of n. Notice that, while the error in x is quite large, the error in the residual remains small. Notice also that on these problems the implicit LX algorithm performs better than the modified Huang algorithm.

6. ABS Methods for KT Equations

The KT (Kuhn-Tucker) equations, which should more appropriately be named KKKT (Kantorovich-Karush-Kuhn-Tucker) equations, are the special linear system $Ax = b$, where the matrix A has the form

$$A = \begin{bmatrix} G & C^T \\ C & 0 \end{bmatrix} \tag{6.1}$$

where $G \in R^{n,n}$, $C \in R^{m,n}$. If G is nonsingular, then A is nonsingular iff $CG^{-1}C^T$ is nonsingular. Usually G is nonsingular, symmetric and positive definite. In several cases arising in interior point methods G is diagonal with positive elements. Here we will assume that G is symmetric positive definite, even if some of the methods considered later can be applied under less restrictive assumptions. Notice that in such a case A is symmetric, but indefinite. KT equations are important because they characterize the first order stationarity conditions in linearly constrained optimization.

We will consider some classical and some ABS methods that take into account the structure of A. We will compare the storage and complexity of these methods, showing that for the case when m is close to n, i.e. when there are few degrees of freedom in the context of a constrained optimization problem, the ABS methods are faster, up to a factor more than four. Numerical experiments moreover show that ABS methods are more accurate than both the Aasen method, considered to be the marginally most accurate of the methods for indefinite systems, and the special QR based method usually utilized for this problem.

KT equations can be solved by general methods for indefinite symmetric systems. A method shown in numerical experiments by Barwell and George (1976) to be the marginally most accurate is the method of Aasen (1971), based upon the following factorization of A

$$PAP^T = LTL^T \tag{6.2}$$

where P is a permutation matrix related to row-column interchanges, L is lower triangular and T is symmetric tridiagonal.

Another method commonly used, see Gill, Murray and Wright (1991), is based upon a QR factorization of C and use of the structure of A. In the version that we have tested the QR factorization has been implemented with the use of Householder matrices.

Now we consider a class of methods for solving the KT equations, whose parameters are the parameters H_1, z_i, v_i and w_i used in dealing with the subsystem of the KT equations containing m equations. Let us consider the KT equations in the following form, with obvious partitioning of x and b

$$Gp + C^T z = g \tag{6.3}$$

$$Cp = c \tag{6.4}$$

We first deal with the subsystem (6.4) obtaining the following expression for the general solution

$$p = p_{m+1} + H_{m+1}^T q \tag{6.5}$$

with q arbitrary. The parameter choices made to construct p_{m+1} and H_{m+1} are arbitrary and define a class of algorithms.

Since the KT equations have a unique solution, there must be a choice of q in (6.5) which makes p be the unique first $n-$dimensional subvector of the solution x. Notice that since H_{m+1} is singular, q is not uniquely defined.

By multiplying equation (6.4) on the left by H_{m+1} and noticing that by properties of the Abaffians $H_{m+1}C^T = 0$, we obtain

$$H_{m+1}Gp = H_{m+1}g \tag{6.6}$$

an equation not containing z. Now there are two possibilities to determine p:

Case 1

Consider the system formed by equations (6.4) and (6.6). Such a system is solvable but overdetermined. Since $\text{rank}(H_{m+1}) = n - m$, m equations are eliminated in step (B) of any ABS algorithm applied to this system. If we construct H_{m+1} via a generalized implicit LU algorithm, then H_{m+1} has the form

$$H_{m+1} = \begin{bmatrix} 0 \\ S \end{bmatrix} \tag{6.7}$$

and the first m equations of (6.6) are trivially satisfied, leaving a determined system consisting of (6.4) and the equation

$$SGp = Sg \tag{6.8}$$

Case 2

In equation (6.6) substitute p with the expression of the general solution (6.5) obtaining

$$H_{m+1}GH_{m+1}^T q = H_{m+1}g - H_{m+1}Gp_{m+1} \tag{6.9}$$

The above system can be solved by any ABS method for a particular solution q, m equations being again removed at step (B) of the ABS algorithm as linearly dependent. If again we use the generalized implicit LU algorithm, then we have

$$H_{m+1}GH_{m+1}^T = \begin{bmatrix} 0 & 0 \\ 0 & SGS^T \end{bmatrix} \tag{6.10}$$

and, letting $q = \begin{bmatrix} q_1 \\ q_2 \end{bmatrix}$, equation (6.9) is solved by taking q_1 arbitrary and q_2 satisfying the equation

$$SGS^T q_2 = Sg - SGp_{m+1} \tag{6.11}$$

Once p is determined, there are two approaches to determine z, namely:

Case 3

Solve by any ABS method the overdetermined compatible system

$$C^T z = g - Gp \tag{6.12}$$

by removing at step (B) of the ABS algorithm $n - m$ dependent equations.

Case 4

Let $P = (p_1, \ldots, p_m)$ be the matrix whose columns are the search vectors generated on the system $Cp = c$. Now $CP = L$, with L nonsingular lower diagonal. Multiplying equation (6.12) on the left by P^T we obtain the following triangular system, which defines z uniquely

$$L^T z = P^T g - P^T Gp \tag{6.13}$$

Extensive numerical testing has been performed to evaluate the accuracy of the above considered ABS algorithms for KT equations, and of some other ABS algorithms for this problem not considered here, see Bodon and Spedicato (1995) and Spedicato, Chen and Bodon (1996). Here we give some of the results, which indicate that the tested ABS algorithms are not only cheaper in terms of overhead for the case when m is close to n, but they are usually more accurate, up to three orders in single precision, than both the Aasen and the QR method.

Some of the tested methods are the following:

AASEN - the method of Aasen

QRGAU - the QR method, using Householder rotation matrices with column pivoting to enhance stability. The indefinite system that arises is solved with Gaussian elimination with partial pivoting

IMLU1 - the ABS method where H_{m+1} is built using the implicit LU algorithm and steps (A), (C) are used

IMLU2 - as above, but using steps (B) and (C)

MHUA1 - the ABS method where H_{m+1} is built using the modified Huang algorithm in the standard form and steps (A) and (C). In solving the overdetermined system $H_{m+1}Ax = H_{m+1}c$ at the i-th step that equation is chosen which has the largest residual

MHUA2 - the new ABS method where H_{m+1} is built with the modified Huang algorithm in the form using $2i$ vectors

Storage requirement and the number of multiplications (only higher order terms) required by the above methods are given in the following Tables.

Table 6.1. Storage requirement

Method	general m	$m=0$	$m=n$
AASEN	$(n+m)^2/2$	$n^2/2$	$2n^2$
QRGAU	$nm+n^2/2+(n-m)^2/2$	n^2	$3n^2/2$
IMLU1	$3n^2/2$	$3n^2/2$	$3n^2/2$
IMLU2	$n^2+m^2/2$	n^2	$3n^2/2$
MHUA1	$2n^2+m^2/2+nm$	$2n^2$	$7n^2/2$
MHUA2	$3n^2/2+m^2/2+nm$	$3n^2/2$	$3n^2$

From the above Table we see that for small m the ABS methods require more memory than AASEN. For large m methods IMLU1 and IMLU2 have similar requirement as the QR method.

Table 6.2. Overhead

Method	general m	$m=0$	$m=n$
AASEN	$(n+m)^3/6$	$n^3/6$	$4n^3/3$
QRGAU	$5n^2m/2-m^3/3+(n-m)^3/3$	$n^3/3$	$13n^3/6$
IMLU1	$n^3/3+n^2m-nm^2+m^3/6$	$n^3/3$	$n^3/2$
IMLU2	$n^3/3+n^2m/2-m^3/3$	$n^3/3$	$n^3/2$
MHUA1	$4n^3$	$4n^3$	$4n^3$
MHUA2	$5n^3/2+n^2m-nm/2$	$5n^3/2$	$3n^3$

From the above Table we see that for small m the cheapest method is AASEN, while the QR methods and IMLU1, IMLU2 have the same cost. For large m, IMLU1 and IMLU2 are more than twice cheaper than AASEN and more than four times faster than the QR algorithm.

Testing has been performed on 300 problems, having the following features:

- G is nonsingular, symmetric, but not always positive definite
- C is full rank
- the entries of G, C and of the solution are integers and the right hand side is computed exactly

For a full presentation of the results, see Spedicato, Chen and Bodon (1996). Here we only give some of their conclusions.

- there is little difference in the performance of methods AASEN and QRGAU.
- there is little difference between methods IMLU1 and IMLU2. IMLU1 is marginally more accurate than IMLU2, which however has less storage for small m
- method MHUA2 is marginally more accurate than MHUA1 and moreover has less storage and less overhead
- method MHUA2 is marginally more accurate than both IMLU1 and IMLU2 in ER, but is marginally less accurate in ES
- IMLU1, IMLU2, MHUA1, MHUA2 are all more accurate than AASEN and the QR codes, especially in terms of residual error

Therefore we can conclude that in the considered class of ABS methods for KT equations there are methods that are superior to well known classical methods not only in terms of overhead, for large m, but also in terms of accuracy.

An important case, arising in particular in the interior point methods for linear programming, is when G is diagonal, with positive diagonal elements. In such a case storage and overhead are reduced, as shown in the following Tables, where only the order three terms are considered.

Table 6.3 Storage when G is diagonal

Method	overhead	$m = 0$	$m = n$
AASEN	$nm + m^2/2$	0	$3n^2/2$
IMLU1	$3nm - 2m^2 + (n - 3m)^2/4$	$n^2/4$	n^2
IMLU2	$nm + (n - m)^2$	n^2	n^2
MHUA2	$3n^2/2 + nm + m^2/2$	$3n^2/2$	$3n^2$

Table 6.4 Overhead when G is diagonal

Method	overhead	$m = 0$	$m = n$
AASEN	$nm^2/2 + m^3/2$	0	$2n^3/3$
IMLU1	$n^3/6 + n^2m/2 - nm^2/2 + m^3/3$	$n^3/6$	$n^3/2$
IMLU2	$n^3/3 - n^2m/2 + nm^2 - m^3/3$	$n^3/3$	$n^3/2$
MHUA2	$5n^3/2 + n^2m/2 - nm^2/2$	$5n^3/2$	$5n^3/2$

From the above Tables we observe:

- for small m, AASEN has the least storage, while for large m the least storage pertains to methods IMLU1 and IMLU2
- for small m, AASEN is again the cheapest method, while for large m IMLU1 and IMLU2 are the fastest methods, more than twice faster than the QR based method.

An important problem in interior point methods is to compute the solution for a sequence of problems where only G, which is diagonal, changes. In such a case the ABS methods, which initially work on the matrix C, which is unchanged, are advantaged, particularly for the case of large m, where the dominant cubic term decreases with m and disappears for $m = n$, so that the overhead is dominated by the second order terms. For such a case the overheads are given by the following Table (second order terms are omitted).

Table 6.5 Overhead when only (diagonal) G changes

Method	general m	$m = 0$	$m = n$
AASEN	$nm^2/2 + m^3/2$	0	$2n^3/3$
IMLU1	$n^3/6 + n^2m/2 - 3nm^2/2 + 5m^3/3$	$n^3/6$	0
IMLU2	$n^3/3 - n^2m/2 + m^3/6$	$n^3/3$	0
MHUA2	$5n^3/2 - 5nm^2/2$	$5n^3/2$	0

The fact that the KT equations can be solved for large m with only a quadratic overhead was already observed by Spedicato and Yang (1993), using a modification of the QR factorization, and Li (1995). This result is certainly of great interest for the practical implementation of interior point methods. Preliminary numerical experiments indicate that method MHUA2, closely followed by IMLU1, is the most accurate for this type of problems.

7. ABS Unified Formulation of Feasible Direction Methods for Minimization with Linear Constraints

ABS algorithms can be used to provide a unification of feasible point methods for minimization with linear constraints, including methods for the LP problem. Let us first consider the following problem with only linear equality constraints

$$min f(x), \quad x \in R^n$$

subject to

$$Ax = b, \quad A \in R^{m,n}, \quad m \leq n, \quad rank(A) = m$$

Let x_1 be a feasible starting point, then for an iteration procedure of the form $x_{i+1} = x_i - \alpha_i d_i$, the search direction will generate feasible points iff

$$Ad_i = 0 \tag{7.1}$$

Solving the underdetermined equation (7.1) for d_i by the ABS algorithm, the solution can be written in the following form, taking, without loss of generality, the zero vector as a special solution

$$d_i = H_{m+1}^T q \tag{7.2}$$

where the matrix H_{m+1} depends on the arbitrary choice of the parameters H_1, w_i and v_i used while solving (7.1) and $q \in R^n$ is arbitrary. Notice that the choice of such parameters may also change with i. Hence the general feasible direction iteration has the form

$$x_{i+1} = x_i - \alpha_i H_{m+1}^T q \tag{7.3}$$

The search direction is a descent direction iff $d^T \nabla f(x) = q^T H_{m+1} \nabla f(x) > 0$. Such a condition can always be satisfied by choices of q unless $H_{m+1} \nabla f(x) = 0$, which implies, from the null space structure of H_{m+1}, that $\nabla f(x) = A^T \lambda$ for some λ, hence that x_{i+1} is a K-T point and λ is the vector of the Lagrange multipliers. It is immediate to see that the search directions are descent directions if we select q as

$$q = W H_{m+1} \nabla f(x) \tag{7.4}$$

where W is a symmetric and positive definite matrix. It is conjectured that all feasible descent directions can be put in the form (7.3) with q given by (7.4). Particular well-known algorithms from the literature are obtained by the following choices of q, with $W = I$:

a The *reduced gradient method* of Wolfe
Here H_{m+1} is constructed by the implicit LU (or the implicit LX) algorithm

b The *gradient projection method* of Rosen
Here H_{m+1} is build using the Huang algorithm

c The *method of Goldfarb and Idnani*
Here H_{m+1} is build via the modification of the Huang algorithm where H_1 is a symmetric positive definite matrix approximating the inverse Hessian of $f(x)$.

If there are inequalities two approaches are possible:

A The *active set* approach.

In this approach the set of linear equality constraints is modified at every iteration by adding and/or dropping some of the linear inequality constraints. Adding or deleting a single constraint can be done, for every ABS algorithm, in order two operations, see Zhang (1995b) or Xia, Liu and Zhang (1995).

B The *standard form* approach.

In this approach, by introducing slack variables, the problem with both types of linear constraints is written in the equivalent form

$$min f(x)$$

subject to

$$Ax = b, \quad x \geq 0$$

The following general iteration, started with x_1 a feasible point, will generate a sequence of feasible points

$$x_{i+1} = x_i - \alpha_i \beta_i H_{m+1} \nabla f(x) \tag{7.5}$$

where the line search parameter α_i can be chosen by a minimization along the search direction $H_{m+1}\nabla f(x)$, while the relaxation parameter $\beta_i > 0$ is selected to avoid that the new point has some negative components.

If $f(x)$ is nonlinear, then H_{m+1} can be determined once for all at the first step, since $\nabla f(x)$ will generally change from iteration to iteration, therefore modifying the search direction. If however $f(x) = c^T x$ is linear (we have then the LP problem) to modify the search direction we need to change H_{m+1}. As observed before, the simplex method is obtained constructing H_{m+1} using the implicit LX algorithm, every step of the method corresponding to a change of the parameters e_{k_i}. It can be shown, see Xia (1996), that the famous method of Karmarkar, originally due to Evtushenko (1974), corresponds to using the generalized Huang algorithm, with initial matrix $H_1 = Diag(x_i)$ changing from iteration to iteration. Another method, faster than Karmarkar's since it has superlinear convergence against linear and $\sqrt{n}$ against n complexity, again due to Evtushenko originally, is obtained by the generalized Huang algorithm with initial matrix $H_1 = Diag(x_i^2)$

8. Final Remarks and Conclusions

In this paper we have briefly reviewed the scaled ABS algorithm. We have shown the existence of an algorithm, the implicit LX algorithm, that can solve a general determined linear system in $n^3/3$ plus $O(n^2)$ multiplications without the regularity assumption. Such an algorithm has a natural application to the simplex algorithm for the LP problem, where it provides a faster implicit interchange of basic and nonbasic columns than methods based upon the classical LU factorization, when the number of degrees of freedom is small. We have introduced a class of ABS methods for KT equations, containing methods which are competitive in terms of accuracy with the classical methods of Aasen or of the QR factorization and which are faster when the number of degrees of freedom is small. We have introduced a general class of algorithms for linearly constrained optimization, which unifies and generalizes most of the methods proposed in the literature, including, for the LP problem, the simplex method and the interior point methods.

Other fields where ABS methods can unify and produce new algorithms are iterative methods for linear equations, see Li (1996a,b), and Quasi-Newton methods, see Spedicato and Xia (1992), Spedicato and Zhao (1992) and Deng, Li and Spedicato (1996c).

In the field of ABS methods theoretical advances tend to go faster than the necessary numerical testing. It is hoped that a planned project ABSPACK will be funded in a next future to fully test the ABS methods and to provide software competitive and preferable in some cases to the existing one.

References

[1] Aasen J.O. (1971), On the reduction of a symmetric matrix to tridiagonal form, *BIT* 11,233-244.

[2] Abaffy J. and Spedicato E. (1989), *ABS Projection Algorithms: Mathematical Techniques for Linear and Nonlinear Equations*, Ellis Horwood, Chichester.

[3] Abaffy J., Broyden C.G. and Spedicato E. (1984), A class of direct methods for linear systems, *Numerische Mathematik* 45, 361-376.

[4] Barwell W. and George J.A. (1976), A comparison of algorithms for solving the symmetric indefinite systems, *ACM Trans. Mathem. Software* 2, 242-255.

[5] Benzi M. (1993), A direct row-projection method for sparse linear systems, PhD Thesis, North Carolina State University

[6] Benzi M., Meyer C.D. and Tuma M. (1996), A sparse approximate inverse preconditioner for the conjugate gradient method, *SIAM J. Scientific and Statistical Computation*, to appear.

[7] Bertocchi M. and Spedicato E. (1989), Performance of the implicit Gauss-Choleski algorithm of the ABS class on the IBM 3090 VF, in Proceedings of the 10th Symposium on Algorithms, Strbske Pleso, 30-40.

[8] Bodon E. (1992), Numerical experiments with ABS algorithms on banded systems of linear equations, Report DMSIA 92/18, University of Bergamo.

[9] Bodon E. (1993), Numerical experiments on the ABS algorithms for linear systems of equations, Report DMSIA 93/17, University of Bergamo.

[10] Bodon E. and Spedicato E. (1995), Some numerical experiments on KT type indefinite systems of linear equations, Report DMSIA 95/18, University of Bergamo.

[11] Broyden C.G. (1975), *Basic Matrices*, Mc Millan. Broyden C.G. (1985), On the numerical stability of Huang's and related methods, JOTA 47, 401-412.

[12] C.G. Broyden (1994), Linear equations in optimization, in E. Spedicato (ed.), *Algorithms for Continuous Optimization, the State of the Art*, Kluwer.

[13] Daniel J., Gragg W.B., Kaufman L. and Stewart G.W. (1976), Reorthogonalized and stable algorithms for updating the Gram-Schmidt QR factorization, *Mathematics of Computation* 30, 772-795.

[14] Forrest J.J.H. and Goldfarb D. (1992), Steepest edge simplex algorithms for linear programming, *Mathematical Programming* 57, 341-374.

[15] Evtushenko Y. (1974), Two numerical methods of solving nonlinear programming problems, *Soviet Doklady Akademii Nauk* 251, 420-423.

[16] Galantai A. (1991), Analysis of error propagation in the ABS class, *Annals of the Institute of Statistical Mathematics* 43, 597-603.

[17] Gill P.E., Murray W. and Wright M.H. (1991), *Numerical linear algebra and optimization*, Addison-Wesley.

[18] Goldfarb D. and Idnani A. (1983), A numerically stable dual method for solving strictly convex quadratic programming, *Mathematical Programming* 27, 1-33.

[19] Li S. (1995), Fast computation of Karmarkar's projections, Communication, Second ABS Conference, Beijing.

[20] Li Z. (1996a), Restarted and truncated versions of ABS methods for large linear systems: a basic framework, Report DMSIA 96/2, University of Bergamo.

[21] Li Z. (1996b), Truncated block ABS methods for large linear systems, Report DMSIA 96/26, University of Bergamo.

[22] Deng N., Li Z. and Spedicato E. (1996), On sparse Quasi-Newton quasi- diagonally dominant updates, Report DMSIA 96/1, University of Bergamo.

[23] Mirnia K. (1996), Numerical experiments with iterative refinement of solutions of linear equations by ABS methods, preprint, University of Bergamo.

[24] Spedicato E. (1993), Ten years of ABS methods: a review of theoretical results and computational achievements, *Surveys on Mathematics for Industry* 3, 217-232.

[25] Spedicato E. and Bodon (1992), Numerical behaviour of the implicit QR algorithm in the ABS class for linear least squares, *Ricerca Operativa* 22, 43-55.

[26] Spedicato E. and Bodon E. (1993), Solution of linear least squares via the ABS algorithm, *Mathematical Programming 58*, 111-136.

[27] Spedicato E. and Vespucci M.T. (1992), Variations on the Gram-Schmidt and the Huang algorithms for linear systems: a numerical study, *Applications of Mathematics* 2, 81-100.

[28] Spedicato E. and Xia Z. (1992), Finding general solutions of the Quasi-Newton equation in the ABS approach, *Optimization Methods and Software* 1, 273-281.

[29] Spedicato E. and Zhao J. (1992), Explicit general solution of the Quasi-Newton equation with sparsity and symmetry, *Optimization Methods and Software* 2, 311-319.

[30] Spedicato E. and Yang Z. (1993), A faster method for computing Karmarkar's projection when $m > n/2$, Report DMSIA 93/15.

[31] Spedicato E. and Zhu M. (1994), On the generalized implicit LU algorithm of the ABS class, Report DMSIA 94/3, University of Bergamo.

[32] Spedicato E., Chen Z. and Bodon E. (1996), ABS methods for KT equations, in *Nonlinear Optimization and Applications*, Plenum Press, New York (Di Pillo G. and Giannessi F. editors)

[33] Spedicato E., Xia Z. and Zhang L. (1995), Reformulation of the simplex algorithm via the ABS algorithm, preprint, University of Bergamo.

[34] Xia Z. (1996), ABS generalization and formulation of the interior point method, preprint, University of Bergamo.

[35] Xia Z., Liu Y. and Zhang L. (1992), Application of a representation of ABS updating matrices to linearly constrained optimization, *Northeast Operational Research* 7, 1-9.

[36] Yang Z. (1989), ABS algorithms for solving certain systems of indefinite equations, Report DMSIA 89/6.

[37] Yang Z. and Zhu M. (1994), The practical ABS algorithm for large scale nested dissection linear system, Report DMSIA 94/11, University of Bergamo.

[38] Wedderburn J.H.M. (1934), *Lectures on Matrices*, Colloquium Publications, American Mathematical Society, New York.

[39] Zhang L. (1995a), An algorithm for the least Euclidean norm solution of a linear system of inequalities via the Huang ABS algorithm and the Goldfarb-Idnani strategy, Report DMSIA 95/2, University of Bergamo.

[40] Zhang L. (1995b), Updating of Abaffian matrices under perturbation in W and A, Report DMSIA 95/16, University of Bergamo.

[41] Zhang L. (1996), Application of the implicit LX algorithm to the simplex method, preprint, University of Bergamo.

SOLVING INVERSE THERMAL PROBLEMS USING KRYLOV METHODS

G. MONTERO
Department of Mathematics, Las Palmas de Gran Canaria University.
Edificio de Informática y Matemáticas, Campus Univ. de Tafira,
Las Palmas de Gran Canaria 35017, Spain.

1. Introduction

Phenomena encountered in science and engineering are generally modelled as a set of differential equations with some initial and boundary conditions. In fact, the coefficients of these equations, such as properties of the materials or the media and the distribution of sources, should be given, as well as the initial state and the boundary data. The exact geometry of the domain is also needed. These problems are called direct problems. In these problems, the main objective is to find certain results, depending on the type of problem, from a known situation, e.g. initial and boundary conditions. Direct problems has been studied during many years and, nowadays, there is a wealth of knowledge about the solution uniqueness and stability and a large list of numerical tools.

Inverse problems arises from the ignorance of one of those parameters or conditions. In order to solve this kind of problems, an extra condition must be specified. Thus, an inverse problem can be regarded as finding the cause from a known result.

The ill-posed nature of inverse problems comes from their physical origin. At this point, two questions should be considered, in direct problems the damping effect corresponding with the fluctuations of the solution which are much diminished internally in comparison with the surface solution changes; and, in inverse problems, the lagging effect which means that any noise in the measurements is amplified in the projection to the surface.

The definition of well-posed problem can be found in [6]. If our problem is given by the equation:

$$\mathcal{A}x = b \tag{1}$$

with $x \in X$, and $b \in B$, X and B metric spaces, and $\mathcal{A}$, a differential operator so that $\mathcal{A}X \subset B$, a well-posed problem requires the existence and uniqueness of the solution for any $b \in B$, as well as its stability with respect to perturbations on b,

G. Winter Althaus and E. Spedicato (eds.), Algorithms for Large Scale Linear Algebraic Systems, 321–342.

i.e. the operator $\mathcal{A}^{-1}$must be defined throughout the space B, and be continuous.

Let us consider the approximate values b_δ for b, with a bounded error,

$$\|b - b_\delta\|_B \leq \delta \tag{2}$$

where $\|\cdot\|_B$ is the Euclidean norm defined in B. For well-posed problems, we have the following condition,

$$\|x - x_\delta\|_X \to 0 \qquad \text{if } \delta \to 0 \tag{3}$$

where x is the solution of (1) for an accurate b, and

$$x_\delta = \mathcal{A}^{-1} b_\delta \tag{4}$$

For ill-posed problems, this condition does not hold. Now the inverse operator $\mathcal{A}^{-1}$ is not continuous. Thus, the solution of equation (1) does not depend continuously on the data and x_δ, if it exists, may not be an approximate solution even for small δ.

Ill-posed problems appear in Optimal Control, Linear Algebra, Integral Equations, Functionals Optimization, etc., being very sensitive to errors given in the data measurement.

The applications of Inverse Problems in the natural sciences are numerous. Some of them are the determination of epicardial from surface potential distribution, computer tomography, seismology, electrodynamic prospecting, plasma diagnostic, astrophysics, scattering, determination of field sources and the shape of scattering bodies from indirect information about an electromagnetic field, control of pollutant emission, and diverse thermal problems. Inverse Thermal Problems have also many applications in science and engineering like the determination of the outer surface conditions during the re-entry of space vehicle, the surface conditions at the exhaust of a rocket or jet engine, the motion of a projectile over a gun barrel surface, ..., (see Kurpisz et al. [10]).

A detailed classification of Inverse Problems is given by Kozdoba et al. [9]: Inverse Boundary Problems, Parameter Estimation, Inverse Initial Problems, Inverse Geometry Problems and other Problems. In this chapter, we shall only study the Inverse Boundary Problems arising in Heat Transfer, this is to say, the estimation of the surface boundary conditions (for steady-state problems) from measurements of the solution at one or more points of the body. Specifically, here, some measurements on the boundary will be considered.

2. Boundary Value Thermal Problems

Heat transfer is energy in transit due to differences of temperature. There are three ways for heat transfer, conduction, convection and thermal radiation. The difference between conduction and convection is in the type of particle movement. If the medium is not moving, the heat is transported by conduction. Convective heat transfer appears when there is bulk motion of the medium. In this case, there is a transport of energy due to temperature differences but also due to enthalpy transport, viscous dissipation, compression, etc. Radiative heat transfer consists of energy transport through means of electromagnetic waves, though, it is possible even if there is no intervening medium.

Here we will consider only the pure heat transfer due to conduction and suppose that the temperature field does not vary with time (steady-state problems).

2.1. GOVERNING EQUATIONS

Let us assume a heat transfer process in a domain Ω, of boundary $\Gamma = \Gamma_1 \cup \Gamma_2 \cup \Gamma_3$. For a steady-state problem and isotropic bodies with constant thermal conductivity k the governing equation in Ω can be written as follows,

$$-k\Delta u = 0 \tag{5}$$

where u means the temperature and there are not heat sources.

To obtain a unique solution of the boundary value problems of heat transfer one has to prescribe along the boundary Γ appropriate boundary conditions. Here, only Dirichlet and Neumann boundary conditions are considered. Respectively,

$$u = v(\mathbf{r}) \quad \text{on } \Gamma_1 \tag{6}$$

$$-k\frac{\partial u}{\partial \mathbf{n}} = g(\mathbf{s}) \quad \text{on } \Gamma_2 \cup \Gamma_3 \tag{7}$$

where $\mathbf{r} \in \Gamma_1$ and $\mathbf{s} \in \Gamma_2 \cup \Gamma_3$.

2.2. DIRECT AND INVERSE PROBLEMS

When we are solving such a direct thermal problem, all coefficients in the governing equations and boundary conditions must be specified. These are thermal conductivity, capacity and location of internal heat sources, if they exist and also the geometry of the body. An the main objective is to find the temperature distribution within the domain from the known boundary conditions.

The inverse problem arises when one or more of the previous input data are missing. Then, extra information, usually obtained from measurements at several points of the body, is required to allow us to solve the inverse problem. Let us consider the unknown values to be the Dirichlet boundary condition on Γ_1, this is to say, $v(\mathbf{r}) \in \mathbf{L}^2(\Gamma_1)$, and the extra information the temperature distribution on Γ_3, $u_d \in \mathbf{L}^2(\Gamma_3)$. The inverse thermal problem can be formulated as:

"To find v_{opt} defined by

$$v_{opt} = ArgMin\ j(v) \tag{8}$$

over the set of admissible controls V, a bounded, closed, convex subset of $\mathbf{L}^2(\Gamma_1)$ with

$$j(v) = \frac{1}{2} \int_{\Gamma_3} [u(v) - u_d]^2 \, d\Gamma_3 \tag{9}$$

and where $u(v)$ is the solution of the problem defined by equations (5), (6) and (7).

2.3. SENSITIVITY COEFFICIENTS

Many methods for solving inverse problems are base on the so-called sensitivity coefficients. These are defined as derivatives Z of the measured quantity, in our case the temperature u, with respect to the unknown quantity, say v,

$$\{Z\}_{ij} = \frac{\partial u_i}{\partial v_j} \tag{10}$$

where i and j mean the point where temperature and v are considered, respectively.

Now if the governing equations are differentiated, i.e. for the problem defined by equations (5), (6) and (7), and substituting (10), the governing equations for sensitivity coefficients are obtained,

$$\begin{aligned} -k\Delta Z_j &= 0 && \text{in } \Omega && (11)\\ Z_j &= 1 && \text{on } \Gamma_{1_j} && (12)\\ -k\frac{\partial Z_j}{\partial \mathbf{n}} &= 0 && \text{on } \Gamma_2 \cup \Gamma_3 && (13) \end{aligned}$$

One can observe the similar aspect of these equations with the initial problem. Note that this is a direct problem.

2.4. ADJOINT PROBLEMS

We reconsider the inverse problem introduced in 2.1 and formulated in 2.2. The necessary and sufficient condition for optimality becomes (see i.e. chapter IX in Knowles [8], to find this known result),

$$\int_{\Gamma_3} [u(v_{opt}) - u_d] \, [u(v) - u(v_{opt})] \, d\Gamma_3 \geq 0 \quad \forall v \in V \tag{14}$$

If we introduce the following adjoint problem,

$$\begin{aligned} -k\Delta \tau &= 0 && \text{in } \Omega && (15)\\ \tau &= 0 && \text{on } \Gamma_1 && (16) \end{aligned}$$

$$-k\frac{\partial \tau}{\partial \mathbf{n}} = 0 \qquad \text{on } \Gamma_2 \tag{17}$$

$$-k\frac{\partial \tau}{\partial \mathbf{n}} = u(v) - u_d \qquad \text{on } \Gamma_3 \tag{18}$$

and apply the Green's identity, (14) may be rewritten as,

$$-\int_{\Gamma_3} k\frac{\partial \tau(v_{opt})}{\partial \mathbf{n}}\left[u(v) - u(v_{opt})\right]\, d\Gamma_3 = \int_{\Gamma_1} k\frac{\partial \tau(v_{opt})}{\partial \mathbf{n}}\left[u(v) - u(v_{opt})\right]\, d\Gamma_1 +$$

$$+\int_{\Gamma_2} k\frac{\partial \tau(v_{opt})}{\partial \mathbf{n}}\left[u(v) - u(v_{opt})\right]\, d\Gamma_2 - \int_{\Gamma_1} k\,\tau(v_{opt})\frac{\partial\left[u(v) - u(v_{opt})\right]}{\partial \mathbf{n}}\, d\Gamma_1 -$$

$$-\int_{\Gamma_2} k\,\tau(v_{opt})\frac{\partial\left[u(v) - u(v_{opt})\right]}{\partial \mathbf{n}}\, d\Gamma_2 - \int_{\Gamma_3} k\,\tau(v_{opt})\frac{\partial\left[u(v) - u(v_{opt})\right]}{\partial \mathbf{n}}\, d\Gamma_3 --$$

$$-\int_{\Omega} k\,\Delta\left[u(v) - u(v_{opt})\right]\tau(v_{opt})\, d\Omega + \int_{\Omega} k\Delta\tau(v_{opt})\left[u(v) - u(v_{opt})\right]\, d\Omega =$$

$$= \int_{\Gamma_1} k\frac{\partial \tau(v_{opt})}{\partial \mathbf{n}}\left[u(v) - u(v_{opt})\right]\, d\Gamma_1 = \int_{\Gamma_1} k\frac{\partial \tau(v_{opt})}{\partial \mathbf{n}}\left[v - v_{opt}\right]\, d\Gamma_1 \geq 0$$

$$\forall v \in V \tag{19}$$

3. Combined Function Specification and Regularization Methods

The first method studied here is a combination of the well known Function Specification and Regularization methods (FSM and RM, respectively). This technique takes into account the sensitivity coefficients defined in 2.3, which appear in the algorithm.

3.1. FUNCTION SPECIFICATION METHOD

The Function Specification Method was introduced by Beck [1], who first proposed the sequential approach. It is not the objective of this work to develop new ideas about this method, but to apply it in order to solve our problem defined in section 2.2.

The temperature u_i can be expanded in a Taylor series about arbitrary but known values of v_j,

$$u_i = u_i^* + \sum_{j=1}^{J}(v_j - v_j^*)\left[\frac{\partial u_i}{\partial v_j}\right]_{v_j = v_j^*} \tag{20}$$

which can be written using (10) as follows,

$$u_i = u_i^* + \sum_{j=1}^{J} (v_j - v_j^*) \left[\{Z\}_{ij} \right]_{v_j = v_j^*} \tag{21}$$

Assuming that the number of locations I, where we know the temperature, is equal to the number of unknown J, we obtain a systems of algebraic equations for the objective function (9) given in a discrete form,

$$A v = b \tag{22}$$

where

$$\{A\}_{jk} = \sum_{i=1}^{I} \{Z\}_{ij} \{Z\}_{ik} \qquad j, k = 1, 2, ..., I \tag{23}$$

$$\{b\}_j = \sum_{k=1}^{J} \{A\}_{jk} v_k^* + \sum_{i=1}^{I} \{Z\}_{ij} (u_{d_i} - u_i^*) \qquad j, k = 1, 2, ..., I \tag{24}$$

At this point, the formulation of FSM algorithm is straightforward:
FMS Algorithm.
Choose an initial v^*.
Do While $\sum_{i=1}^{I} [u_i^* - u_{d_i}]^2 \geq \varepsilon$
  Obtain u_i^* from equations (5), (6) and (7), with $v = v^*$.
  For $j = 1, J$ Do
    Obtain $\{Z\}_{ij}$ from equations (11), (12) and (13).
  EndDo
  Solve $A v = b$.
  $v^* = v$.
EndDo

Note that to solve the problem given by equations (5), (6) and (7), or by equations (11), (12) and (13), implies to use some discretization technique such as finite differences, finite elements, finite volume or boundary elements methods.

3.2. REGULARIZATION METHOD

This method for solving ill-posed problems was proposed by Tikhonov in [13], [14] and [15], where he introduced the so-called Regularization Operator R_γ (γ is a regularization parameter), which must carry out the following conditions:
a.- R_γ is determined throughout the space X (see introduction section) for any value of γ.
b.- R_γ is continuous throughout the space X.

c.- $R_\gamma A x \to x$, for $\gamma \to 0$, for any $x \in X$.

The most popular method for constructing regularization operators is the one based on the so-called Smoothing functional,

$$\Lambda_\gamma = \|Ax - b_\delta\|_B^2 + \gamma \|x\|_X^2 \tag{25}$$

This functional takes the following form for our inverse problem (see Beck [2]),

$$\Lambda_\gamma = \sum_{i=1}^{I} [u_i - u_{d_i}]^2 + \gamma \|v\|^2 \tag{26}$$

if we consider a zero order of regularization.

To obtain the value of the regularization parameter we apply the discrepancy method. Thus, $\gamma(\delta)$ is a root of the equation,

$$f(\gamma) = \|Ax_\delta^\gamma - b_\delta\|_B^2 + \delta^2 = 0 \tag{27}$$

where x_δ^γ is the extreme of functional (25) at a fixed $\gamma > 0$. As functional of equation (27) is monotonic and continuous for $\gamma > 0, \gamma$ can be defined uniquely. A simple way to find γ is to start from $\gamma = \gamma^*$, with $f(\gamma^*) > 0$, and iterate

$$\gamma_n = \gamma^* \rho^n \tag{28}$$

where $0 < \rho < 1$ and $n = 1, 2, ...$, as long as $f(\gamma_n) < 0$.

3.3. A COMBINED TECHNIQUE

The Combination of FSM and RM of zero order is often more efficient than either of them. Thus, we present a combined technique substituting the smoothing functional given in (26) for the one used in FSM algorithm. It implies the computing of the regularization parameter.

The FSM+RM algorithm is given below:

FMS+RM Algorithm.

Choose an initial v^*.

Do While $\sum_{i=1}^{I} [u_i^* - u_{d_i}]^2 \geq \varepsilon$

  Obtain u_i^* from equations (5), (6) and (7), with $v = v^*$.

  For $j = 1, J$ Do

    Obtain $\{Z\}_{ij}$ from equations (11), (12) and (13).

  EndDo

  Solve equation (27) using (28) to obtain γ.

  Solve $(A + \gamma I) v = b$.

  $v^* = v$.

EndDo

4. Direct Formulation Method

Far from the FSM and RM algorithms based on the sensitivity coefficients, one can think about formulating the whole problem using a discretization technique for the state, adjoint and optimality equations.

4.1. GOVERNING EQUATIONS, ADJOINT PROBLEM AND OPTIMALITY

This leads to three coupled systems of equations, respectively:

$$Au = Bv \tag{29}$$

$$A^t\tau = C^t(Cu - u_d) \tag{30}$$

$$D\tau = 0 \tag{31}$$

where B is the prolongation rectangular matrix from the space of v the one of u, C the restriction rectangular matrix of u over the points of known data u_d, and D the restriction rectangular matrix of the adjoint variable τ over the same part of the boundary. The associated linear system to (29), (30) and (31) is,

$$\begin{bmatrix} A & 0 & -B \\ -C^tC & A^t & 0 \\ 0 & D & 0 \end{bmatrix} \begin{bmatrix} u \\ \tau \\ v \end{bmatrix} = \begin{bmatrix} 0 \\ -C^tu_d \\ 0 \end{bmatrix} \tag{32}$$

with the particular characteristic of having a nonsymmetric sparse matrix. Thus, adequate methods should be used for solving this linear system.

4.2. BLOCK KRYLOV ALGORITHMS

Nowadays, iterative Krylov methods seem to be the best option to obtain the solution of this type of systems. Here, Bi-CGSTAB (van der Vorst [18], Montero et al. [11]), QMRCGSTAB (Chan et al. [3]) and VGMRES (Saad et al.[16], Saad [17], and Galán et al. [4]) methods will be considered. As the matrix and vectors can be split by blocks, these algorithms should take into account the splitting when vectorial computing are done. This allows us to use a packed storage scheme for A and A^t (see i.e. Galán et al. [5]) and for preconditioning matrices M and M^t.

4.2.1. *Block Bi-CGSTAB Algorithm*

The Preconditioned Bi-CGSTAB Algorithm associated to equation (32) can be written as follows,

Block Preconditioned Bi-CGSTAB Algorithm
Choose u_0, τ_0 and v_0.

$$R_0 = \begin{bmatrix} r_0^1 \\ r_0^2 \\ r_0^3 \end{bmatrix} = \begin{bmatrix} -(Au_0 - Bv_0) \\ -C^t u_d + C^t C u_0 - A^t \tau_0 \\ -D\tau_0 \end{bmatrix};$$

$$\hat{R}_0 = \begin{bmatrix} \hat{r}_0^1 \\ \hat{r}_0^2 \\ \hat{r}_0^3 \end{bmatrix} = \begin{bmatrix} r_0^1 \\ r_0^2 \\ r_0^3 \end{bmatrix}; \; P_0 = \begin{bmatrix} p_0^1 \\ p_0^2 \\ p_0^3 \end{bmatrix} = W_0 = \begin{bmatrix} w_0^1 \\ w_0^2 \\ w_0^3 \end{bmatrix} = 0;$$

$$\rho_0 = \alpha_0 = \varpi_0 = 1;$$

Do While $\|R_{i-1}\| / \|R_0\| \geq \varepsilon \quad i = 1, 2, \ldots$

$$Z_i = \begin{bmatrix} z_i^1 \\ z_i^2 \\ z_i^3 \end{bmatrix} = \begin{bmatrix} M^{-1} r_i^1 \\ M^{-t} r_i^2 \\ r_i^3 \end{bmatrix};$$

$$\rho_i = \hat{R}_0^t Z_i;$$

$$\beta_i = (\rho_i / \rho_{i-1})(\alpha_{i-1} / \varpi_{i-1});$$

$$P_i = \begin{bmatrix} z_i^1 + \beta_i (p_{i-1}^1 - \varpi_{i-1} w_i^1) \\ z_i^2 + \beta_i (p_{i-1}^2 - \varpi_{i-1} w_i^2) \\ z_i^3 + \beta_i (p_{i-1}^3 - \varpi_{i-1} w_i^3) \end{bmatrix};$$

$$Y_i = \begin{bmatrix} y_i^1 \\ y_i^2 \\ y_i^3 \end{bmatrix} = \begin{bmatrix} A p_i^1 - B p_i^3 \\ -C^t C p_i^1 + A^t p_i^2 \\ D p_i^2 \end{bmatrix};$$

$$W_i = \begin{bmatrix} M^{-1} y_i^1 \\ M^{-t} y_i^2 \\ y_i^3 \end{bmatrix};$$

$$\alpha_i = \rho_i / (\hat{R}_0^t W_i);$$

$$S = \begin{bmatrix} s^1 \\ s^2 \\ s^3 \end{bmatrix} = \begin{bmatrix} r_i^1 - \alpha_i y_i^1 \\ r_i^2 - \alpha_i y_i^2 \\ r_i^3 - \alpha_i y_i^3 \end{bmatrix};$$

$$Q = \begin{bmatrix} q^1 \\ q^2 \\ q^3 \end{bmatrix} = \begin{bmatrix} z_i^1 - \alpha_i w_i^1 \\ z_i^2 - \alpha_i w_i^2 \\ z_i^3 - \alpha_i w_i^3 \end{bmatrix};$$

$$T = \begin{bmatrix} t^1 \\ t^2 \\ t^3 \end{bmatrix} = \begin{bmatrix} A\,q^1 - B\,q^3 \\ -C^tC\,q^1 + A^t q^2 \\ D\,q^2 \end{bmatrix};$$

$$\varpi_i = (T^t\,S)/(T^t\,T);$$

$$X_i = \begin{bmatrix} u_i \\ \tau_i \\ v_i \end{bmatrix} = \begin{bmatrix} u_{i-1} + \alpha_i\,p_i^1 + \varpi_i\,q^1 \\ \tau_{i-1} + \alpha_i\,p_i^2 + \varpi_i\,q^2 \\ v_{i-1} + \alpha_i\,p_i^3 + \varpi_i\,q^3 \end{bmatrix};$$

$$R_i = \begin{bmatrix} s^1 - \varpi_i\,t^1 \\ s^2 - \varpi_i\,t^2 \\ s^3 - \varpi_i\,t^3 \end{bmatrix};$$

End Do

4.2.2. *Block QMRCGSTAB Algorithm*
In the same way, the Preconditioned QMRCGSTAB Algorithm associated to equation (32) is,
Block Preconditioned QMRCGSTAB Algorithm
Choose u_0, τ_0 and v_0.

$$R_0 = \begin{bmatrix} r_0^1 \\ r_0^2 \\ r_0^3 \end{bmatrix} = \begin{bmatrix} -(Au_0 - Bv_0) \\ -C^t u_d + C^tCu_0 - A^t\tau_0 \\ -D\tau_0 \end{bmatrix};$$

$$\hat{R}_0 = \begin{bmatrix} \hat{r}_0^1 \\ \hat{r}_0^2 \\ \hat{r}_0^3 \end{bmatrix} = \begin{bmatrix} r_0^1 \\ r_0^2 \\ r_0^3 \end{bmatrix}; \quad P_0 = \begin{bmatrix} p_0^1 \\ p_0^2 \\ p_0^3 \end{bmatrix} = W_0 = \begin{bmatrix} w_0^1 \\ w_0^2 \\ w_0^3 \end{bmatrix} = D_0 = \begin{bmatrix} d_0^1 \\ d_0^2 \\ d_0^3 \end{bmatrix} = 0;$$

$$\rho_0 = \alpha_0 = \varpi_0 = 1; \;\; \theta_0 = \eta_0 = 0; \;\; \gamma = \|R_0\|;$$

Do While $\|R_{i-1}\| \,/\, \|R_0\| \geq \varepsilon \quad i = 1, 2, ...$

$$Z_i = \begin{bmatrix} z_i^1 \\ z_i^2 \\ z_i^3 \end{bmatrix} = \begin{bmatrix} M^{-1} r_i^1 \\ M^{-t} r_i^2 \\ r_i^3 \end{bmatrix};$$

$$\rho_i = \hat{R}_0^t\,Z_i;$$

$$\beta_i = (\rho_i/\rho_{i-1})\,(\alpha_{i-1}/\varpi_{i-1});$$

$$P_i = \begin{bmatrix} z_i^1 + \beta_i(p_{i-1}^1 - \varpi_{i-1} w_i^1) \\ z_i^2 + \beta_i(p_{i-1}^2 - \varpi_{i-1} w_i^2) \\ z_i^3 + \beta_i(p_{i-1}^3 - \varpi_{i-1} w_i^3) \end{bmatrix};$$

$$Y_i = \begin{bmatrix} y_i^1 \\ y_i^2 \\ y_i^3 \end{bmatrix} = \begin{bmatrix} A\,p_i^1 - B\,p_i^3 \\ -C^tC\,p_i^1 + A^t p_i^2 \\ D\,p_i^2 \end{bmatrix};$$

$$W_i = \begin{bmatrix} M^{-1} y_i^1 \\ M^{-t} y_i^2 \\ y_i^3 \end{bmatrix};$$

$$\alpha_i = \rho_i / (\hat{R}_0^t\, W_i);$$

$$S = \begin{bmatrix} s^1 \\ s^2 \\ s^3 \end{bmatrix} = \begin{bmatrix} r_i^1 - \alpha_i\, y_i^1 \\ r_i^2 - \alpha_i\, y_i^2 \\ r_i^3 - \alpha_i\, y_i^3 \end{bmatrix};$$

$$\bar{\theta}_i = \|S\| / \gamma; \; c = \frac{1}{\sqrt{1 + \bar{\theta}_i^2}}; \; \bar{\gamma} = \gamma\, \bar{\theta}_i\, c; \; \bar{\eta}_i = c^2 \alpha_i;$$

$$\bar{D}_i = \begin{bmatrix} \bar{d}_i^1 \\ \bar{d}_i^2 \\ \bar{d}_i^3 \end{bmatrix} = \begin{bmatrix} p_i^1 + \frac{\theta_{i-1}^2\, \eta_{i-1}}{\alpha_i} d_{i-1}^1 \\ p_i^2 + \frac{\theta_{i-1}^2\, \eta_{i-1}}{\alpha_i} d_{i-1}^2 \\ p_i^3 + \frac{\theta_{i-1}^2\, \eta_{i-1}}{\alpha_i} d_{i-1}^3 \end{bmatrix};$$

$$\bar{X}_i = \begin{bmatrix} \bar{u}_i \\ \bar{\tau}_i \\ \bar{v}_i \end{bmatrix} = \begin{bmatrix} u_{i-1} + \bar{\eta}_i\, \bar{d}_i^1 \\ \tau_{i-1} + \bar{\eta}_i\, \bar{d}_i^2 \\ v_{i-1} + \bar{\eta}_i\, \bar{d}_i^3 \end{bmatrix};$$

$$Q = \begin{bmatrix} q^1 \\ q^2 \\ q^3 \end{bmatrix} = \begin{bmatrix} z_i^1 - \alpha_i\, w_i^1 \\ z_i^2 - \alpha_i\, w_i^2 \\ z_i^3 - \alpha_i\, w_i^3 \end{bmatrix};$$

$$T = \begin{bmatrix} t^1 \\ t^2 \\ t^3 \end{bmatrix} = \begin{bmatrix} A\,q^1 - B\,q^3 \\ -C^tC\,q^1 + A^t q^2 \\ D\,q^2 \end{bmatrix};$$

$$\varpi_i = (T^t\, S)/(T^t\, T);$$

$$X_i = \begin{bmatrix} u_i \\ \tau_i \\ v_i \end{bmatrix} = \begin{bmatrix} u_{i-1} + \alpha_i\, p_i^1 + \varpi_i\, q^1 \\ \tau_{i-1} + \alpha_i\, p_i^2 + \varpi_i\, q^2 \\ v_{i-1} + \alpha_i\, p_i^3 + \varpi_i\, q^3 \end{bmatrix};$$

$$R_i = \begin{bmatrix} s^1 - \varpi_i t^1 \\ s^2 - \varpi_i t^2 \\ s^3 - \varpi_i t^3 \end{bmatrix};$$

$$\theta_i = \|R_i\| / \gamma; \; c = \frac{1}{\sqrt{1+\theta_i^2}}; \; \gamma = \bar{\gamma}\, \theta_i\, c; \; \eta_i = c^2 \varpi_i;$$

$$D_i = \begin{bmatrix} d_i^1 \\ d_i^2 \\ d_i^3 \end{bmatrix} = \begin{bmatrix} q^1 + \frac{\bar{\theta}_i^2 \bar{\eta}_i}{\varpi_i} \bar{d}_i^1 \\ q^2 + \frac{\bar{\theta}_i^2 \bar{\eta}_i}{\varpi_i} \bar{d}_i^2 \\ q^3 + \frac{\bar{\theta}_i^2 \bar{\eta}_i}{\varpi_i} \bar{d}_i^3 \end{bmatrix};$$

$$X_i = \begin{bmatrix} u_i \\ \tau_i \\ v_i \end{bmatrix} = \begin{bmatrix} \bar{u}_i + \eta_i d_i^1 \\ \bar{\tau}_i + \eta_i d_i^2 \\ \bar{v}_i + \eta_i d_i^3 \end{bmatrix};$$

End Do

4.2.3. *Block VGMRES Algorithm*

Finally, we close this set of methods with a version of GMRES algorithm,

Block Preconditioned VGMRES Algorithm

Choose u_0, τ_0 and v_0; k_{init}, k_{top}, δ; $k = k_{init}$;

$$R_0 = \begin{bmatrix} r_0^1 \\ r_0^2 \\ r_0^3 \end{bmatrix} = \begin{bmatrix} -(Au_0 - Bv_0) \\ -C^t u_d + C^t C u_0 - A^t \tau_0 \\ -D\tau_0 \end{bmatrix};$$

Do While $\|R_{i-1}\| / \|R_0\| \geq \varepsilon \quad i = 1, 2, \ldots$

$$\beta_{i-1} = \|R_{i-1}\|;$$

$$S_1 = \begin{bmatrix} s_1^1 \\ s_1^2 \\ s_1^3 \end{bmatrix} = \begin{bmatrix} r_{i-1}^1 / \beta_{i-1} \\ r_{i-1}^2 / \beta_{i-1} \\ r_{i-1}^3 / \beta_{i-1} \end{bmatrix};$$

If $\|R_{i-1}\| / \|R_0\| \geq \delta$ and $k < k_{top}$ Do $k = k + 1$;
Do $j = 1, \ldots, k$

$$Z_i = \begin{bmatrix} z_i^1 \\ z_i^2 \\ z_i^3 \end{bmatrix} = \begin{bmatrix} M^{-1} s_i^1 \\ M^{-t} s_i^2 \\ s_i^3 \end{bmatrix};$$

$$W = \begin{bmatrix} w^1 \\ w^2 \\ w^3 \end{bmatrix} = \begin{bmatrix} A\, z_j^1 - B\, z_j^3 \\ -C^t C\, z_j^1 + A^t z_j^2 \\ D\, z_j^2 \end{bmatrix};$$

Do $n = 1, ..., j$

$$\{H\}_{nj} = W^t S_n;$$

$$W = \begin{bmatrix} w^1 \\ w^2 \\ w^3 \end{bmatrix} = \begin{bmatrix} w^1 - \{H\}_{nj}\, s_n^1 \\ w^2 - \{H\}_{nj}\, s_n^2 \\ w^3 - \{H\}_{nj}\, s_n^3 \end{bmatrix};$$

EndDo

$$\{H\}_{j+1j} = \|W\|$$

$$S_{j+1} = \begin{bmatrix} s_{j+1}^1 \\ s_{j+1}^2 \\ s_{j+1}^3 \end{bmatrix} = \begin{bmatrix} w^1 / \{H\}_{j+1j} \\ w^2 / \{H\}_{j+1j} \\ w^3 / \{H\}_{j+1j} \end{bmatrix};$$

End Do

Solve $U_k^t \bar{p} = d_k$ and $U_k^t\, p = \bar{p}$ where $\begin{cases} \{d_k\}_m = \{H\}_{1m} \\ \{U_k\}_{lm} = \{H\}_{l+1m} \end{cases} \quad l, m = 1, ..., k;$

$$\lambda_i = \frac{\beta_{i-1}}{1 + d_k^t\, p};$$

$$y_k = \lambda_i\, p;$$

$$X_i = \begin{bmatrix} u_i \\ \tau_i \\ v_i \end{bmatrix} = \begin{bmatrix} u_{i-1} + Z_k^1\, y_k \\ \tau_{i-1} + Z_k^2\, y_k \\ v_{i-1} + Z_k^3\, y_k \end{bmatrix}; \text{ where } \begin{bmatrix} Z_k^1 \\ Z_k^2 \\ Z_k^3 \end{bmatrix} = \begin{bmatrix} z_1^1 & z_2^1 & \dots & z_k^1 \\ z_1^2 & z_2^2 & \dots & z_k^2 \\ z_1^3 & z_2^3 & \dots & z_k^3 \end{bmatrix}$$

$$R_i = \begin{bmatrix} r_i^1 \\ r_i^2 \\ r_i^3 \end{bmatrix} = \begin{bmatrix} -(Au_i - Bv_i) \\ -C^t u_d + C^t C u_i - A^t \tau_i \\ -D\tau_i \end{bmatrix};$$

EndDo

5. Iterative Method using an Uncoupling Technique

Let consider again the coupled problem defined by equations (29), (30) and (31). Introducing them in the objective function, it leads to,

$$J(v) = \frac{1}{2}(CA^{-1}B\, v - u_d)^t (CA^{-1}B\, v - u_d) \tag{33}$$

The residual in Krylov methods for this functional is given by the projection of τ over Γ_3, say the gradient of $J(v)$ with inverse sense. Thus, the main idea in this technique is to make minimal residual (equation (31)) over Γ_3 using such iterative methods:

$$r = -D\tau \tag{34}$$

The computing of τ involves, first, the resolution of (29) for a given v and then the resolution of (30) using the computed u.

5.1. KRYLOV METHODS

In our problem, one can think of Conjugate Gradient method [7] (CG) due the symmetric character of A. This is the first advantage of this technique in front of the Direct one studied in section 4.

Note that the classical matrix by vector product which appears in all these algorithms, may be identified in the same way the residual was. If

$$r_i = u_d - CA^{-1}B\,v_i = -D\tau_i \tag{35}$$

implies to solve (29) and (30), a matrix by vector product will involve the resolution of two similar equations. From now on let represent the solutions of (29) and (30), respectively,

$$u = LAP(A,B,v) \tag{36}$$
$$\tau = RES(A,C,u,u_d) \tag{37}$$

The product $y = CA^{-1}B\,p$, p any vector, may be represented as a set of three steps:

$$\begin{array}{llcl} I. & x & = & LAP(A,B,p). \\ II. & \sigma & = & RES(A,C,x,0). \\ III. & y & = & D\sigma. \end{array}$$

Taking all these comments into account, the Krylov algorithms can be written as below.

5.1.1. *CG Algorithm*
Choose v_0;
Solve $u_0 = LAP(A,B,v_0)$; Solve $\tau_0 = RES(A,C,u_0,u_d)$; $r_0 = -D\tau_0$;
$\rho_0 = 1$; $p_0 = 0$;
Do While $\|r_{i-1}\| \,/\, \|r_0\| \geq \varepsilon$ $(i = 1,2,...)$,
  $\rho_i = r_{i-1}^t r_{i-1}$; $\beta_i = \rho_i/\rho_{i-1}$; $p_i = r_{i-1} + \beta_i\,p_{i-1}$;
  Solve $x_i = LAP(A,B,p_i)$; Solve $\sigma_i = RES(A,C,x_i,0)$; $y_i = D\sigma_i$;

$\alpha_i = \rho_i/(p_i^t y_i)$; $v_i = v_{i-1} + \alpha_i p_i$; $r_i = r_{i-1} - \alpha_i y_i$;
EndDo

5.1.2. *Bi-CGSTAB Algorithm*
Choose v_0;
Solve $u_0 = LAP(A, B, v_0)$; Solve $\tau_0 = RES(A, C, u_0, u_d)$; $r_0 = -D\tau_0$;
Choose $\hat{r}_0$, such that $\hat{r}_0^t r_0 \neq 0$, i.e. $\hat{r}_0^t = r_0$;
$\rho_0 = \alpha_0 = \varpi_0 = 1$; $p_0 = y_0 = 0$;
Do While $\|r_{i-1}\| / \|r_0\| \geq \varepsilon$ $(i = 1, 2, ...)$,
$\rho_i = \hat{r}_0^t r_{i-1}$; $\beta_i = (\rho_i/\rho_{i-1})(\alpha_{i-1}/\varpi_{i-1})$; $p_i = r_{i-1} + \beta_i(p_{i-1} - \varpi_{i-1} y_{i-1})$;
Solve $x_i = LAP(A, B, p_i)$; Solve $\sigma_i = RES(A, C, x_i, 0)$; $y_i = D\sigma_i$;
$\alpha_i = \rho_i/(\hat{r}_0^t y_i)$; $s_i = r_{i-1} - \alpha_i y_i$;
Solve $z_i = LAP(A, B, s_i)$; Solve $\mu_i = RES(A, C, z_i, 0)$; $t_i = D\mu_i$;
$\varpi_i = (t_i^t s_i) / (t_i^t t_i)$; $v_i = v_{i-1} + \alpha_i p_i + \varpi_i s_i$; $r_i = s_i - \varpi_i t_i$;
EndDo

5.1.3. *QMRCGSTAB Algorithm*
Choose v_0;
Solve $u_0 = LAP(A, B, v_0)$; Solve $\tau_0 = RES(A, C, u_0, u_d)$; $r_0 = -D\tau_0$;
Choose $\hat{r}_0$, such that $\hat{r}_0^t r_0 \neq 0$, i.e. $\hat{r}_0^t = r_0$;
$\rho_0 = \alpha_0 = \varpi_0 = 1$; $\theta_0 = \eta_0 = 0$; $\gamma = \|r_0\|$; $p_0 = y_0 = d_0 = 0$;
Do While $\|r_{i-1}\| / \|r_0\| \geq \varepsilon$ $(i = 1, 2, ...)$,
$\rho_i = \hat{r}_0^t r_{i-1}$; $\beta_i = (\rho_i/\rho_{i-1})(\alpha_{i-1}/\varpi_{i-1})$; $p_i = r_{i-1} + \beta_i(p_{i-1} - \varpi_{i-1} y_{i-1})$;
Solve $x_i = LAP(A, B, p_i)$; Solve $\sigma_i = RES(A, C, x_i, 0)$; $y_i = D\sigma_i$;
$\alpha_i = \rho_i/(\hat{r}_0^t y_i)$; $s_i = r_{i-1} - \alpha_i y_i$;
$\bar{\theta}_i = \|s_i\| / \gamma$; $\bar{c} = 1/\sqrt{1 + \bar{\theta}_i^2}$; $\bar{\gamma} = \gamma \bar{\theta}_i \bar{c}$; $\bar{\eta}_i = \bar{c}^2 \alpha_i$;
$\bar{d}_i = p_i + \frac{\theta_{i-1}^2 \eta_{i-1}}{\alpha_i} d_{i-1}$; $\bar{v}_i = v_{i-1} + \bar{\eta}_i d_i$;
Solve $z_i = LAP(A, B, s_i)$; Solve $\mu_i = RES(A, C, z_i, 0)$; $t_i = D\mu_i$;
$\varpi_i = (t_i^t s_i) / (t_i^t t_i)$; $r_i = s_i - \varpi_i t_i$;
$\theta_i = \|r_i\| / \bar{\gamma}$; $c = 1/\sqrt{1 + \theta_i^2}$; $\gamma = \bar{\gamma} \theta_i c$; $\eta_i = c^2 \varpi_i$;
$d_i = s_i + \frac{\bar{\theta}_{i-1}^2 \bar{\eta}_{i-1}}{\varpi_i} \bar{d}_i$; $v_i = \bar{v}_i + \eta_i d_i$;
EndDo

5.1.4. *VGMRES Algorithm*
Choose v_0; k_{init}, k_{top}, δ; $k = k_{init}$;
Solve $u_0 = LAP(A, B, v_0)$; Solve $\tau_0 = RES(A, C, u_0, u_d)$; $r_0 = -D\tau_0$;
Do While $\|r_{i-1}\| / \|r_0\| \geq \varepsilon$ $(i = 1, 2, ...)$,
$\beta_{i-1} = \|r_{i-1}\|$; $s_1 = r_{i-1}/\beta_{i-1}$;
If $\|r_{i-1}\| / \|r_0\| \geq \delta$ and $k < k_{top}$ Do $k = k + 1$;
Do $j = 1, ..., k$
Solve $x_i = LAP(A, B, s_j)$; Solve $\sigma_i = RES(A, C, x_i, 0)$; $w_i = D\sigma_i$;

Do $n = 1,..,j$

$\{H\}_{nj} = w^t s_n; \ w = w - \{H\}_{nj} \ s_n;$

EndDo

$\{H\}_{j+1j} = \|w\|; \ s_{j+1} = w/\{H\}_{j+1j};$

End Do

Solve $U_k^t \bar{p} = d_k$ and $U_k^t p = \bar{p}$ where $\left\{ \begin{array}{l} \{d_k\}_m = \{H\}_{1m} \\ \{U_k\}_{lm} = \{H\}_{l+1m} \end{array} \right. \quad l, m = 1, ..., k;$

$\lambda_i = \frac{\beta_{i-1}}{1+d_k^t p}; \ y_k = \lambda_i p; \ v_i = v_{i-1} + S_k y_k;$ where $S_k = [s_1 s_2 \cdots s_k]$

Solve $u_0 = LAP(A, B, v_0)$; Solve $\tau_0 = RES(A, C, u_0, u_d)$; $r_0 = -D\tau_0$;

EndDo

6. Test Cases

To study the convergence behavior of this set of methods and establish differences between the three strategies, three examples with different levels of discretization and known data will be solved. All cases are referred to a squared domain Ω of side unity, and boundaries Γ_1, Γ_2, Γ_3 and Γ_4, for the following particular case:

$$\Delta u = 0 \tag{38}$$

$$u = v(x) \quad \text{on } \Gamma_1 \tag{39}$$

$$\frac{\partial u}{\partial \mathbf{n}} = 0 \quad \text{on } \Gamma_2 \cup \Gamma_3 \cup \Gamma_4 \tag{40}$$

The known measurement u_d are given on Γ_3, and they are taken from the resolution of the direct problem for certain functions v.

Finite differences have been used for the discretization, and the inner problems formulated in (36) and (37) have been solve using Bi-CGSTAB.

6.1. EXAMPLE A

Let consider the inverse problem when u_d corresponds to the solution of (38), (39) and (40) for $v = 1$. We use a grid of 11×11 points.

Figure 1 shows the evolution of the residual norm and the CPU time for this problem. Iterative formulation (with any Krylov method) obtained the solution in one or two steps. FSM combined with RM has also a good behavior but it takes much more time. Direct formulation does not improve the error from certain values. In figure 2 the solution obtained for each method in drawn. Note that the higher oscillations are associated to direct formulation.

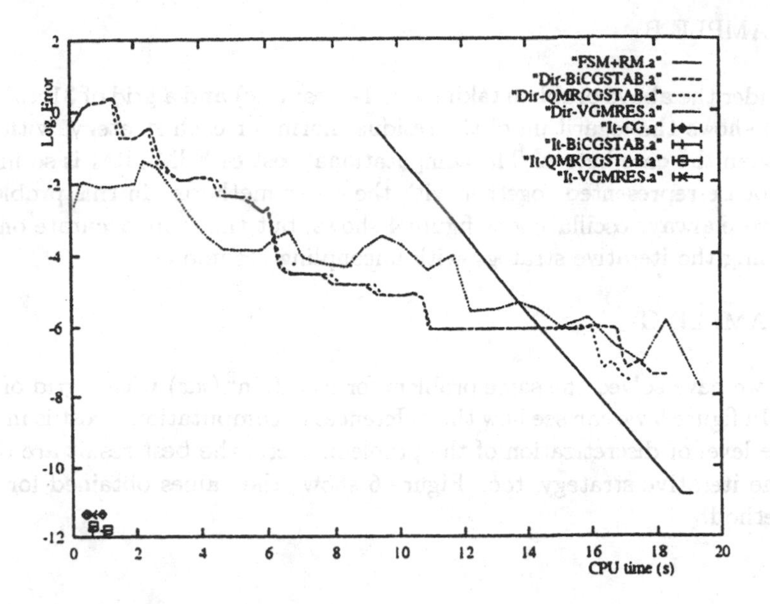

Figure 1. Example A: Log. Residual Norm versus computational cost.

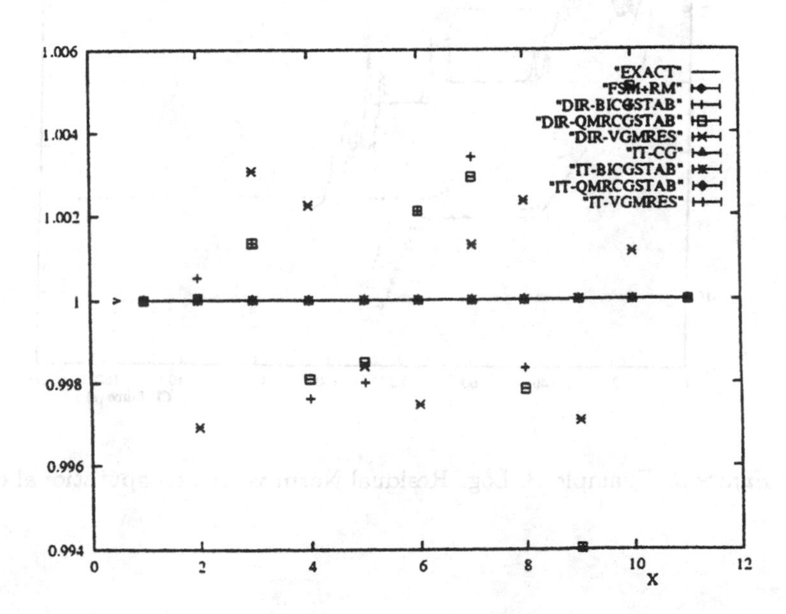

Figure 2. Example A: Results obtained for each method.

6.2. EXAMPLE B

We consider the above problem taking $v = 1+\cos^9(\pi x)$ and a grid of 31×31 points. Figure 3 shows the logarithm of the residual norm for each strategy, with similar results than the case before. The computational cost of FSM+RM is so high that it can not be represented together with the other methods. In this problem, the results were always oscillating as figure 4 shows, but the more accurate ones were found using the iterative strategy with uncoupling technique.

6.3. EXAMPLE C

Finally, we have solved the same problem for $v = 2\sin^2(\pi x)$ with a grid of 51×51 points. In figure 5 we can see how the differences of computational cost is increasing with the level of discretization of the problem. Here, the best result are obtained using the iterative strategy, too. Figure 6 shows the values obtained for v using each method.

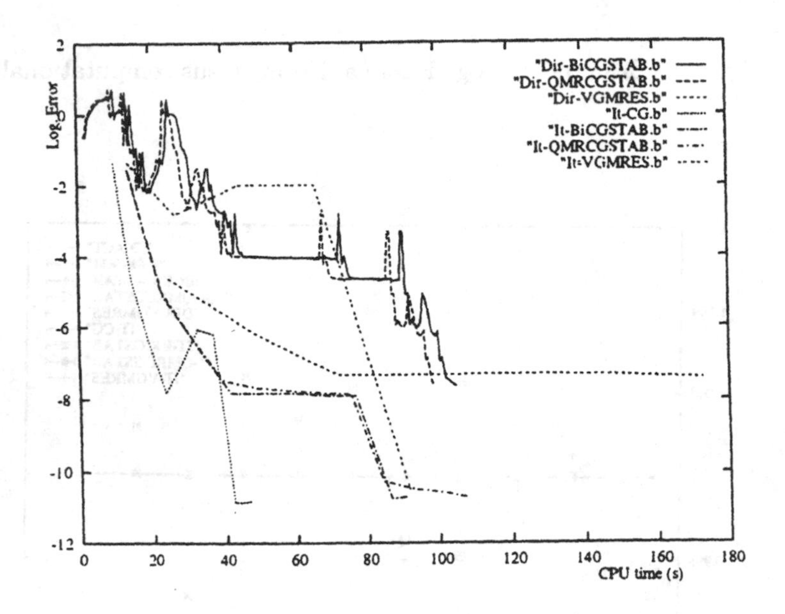

Figure 3. Example B: Log. Residual Norm versus computational cost.

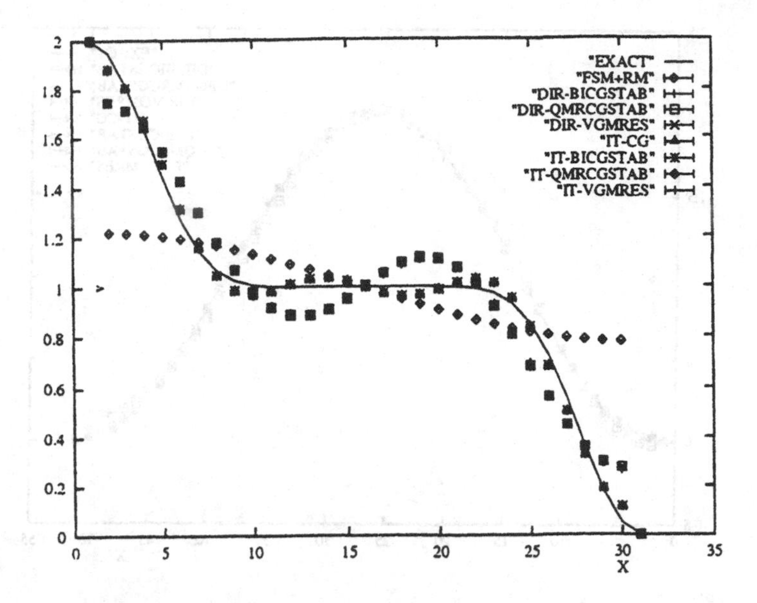

Figure 4. Example B: Results obtained for each method.

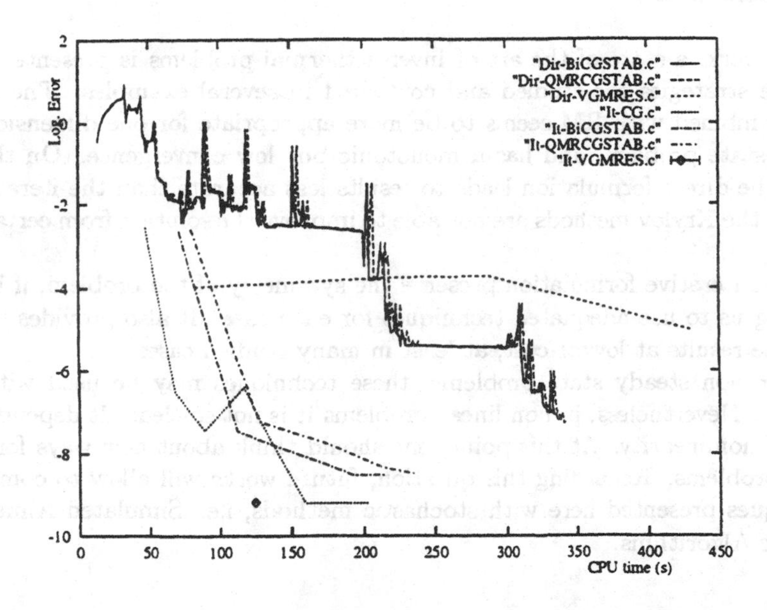

Figure 5. Example C: Log. Residual Norm versus computational cost.

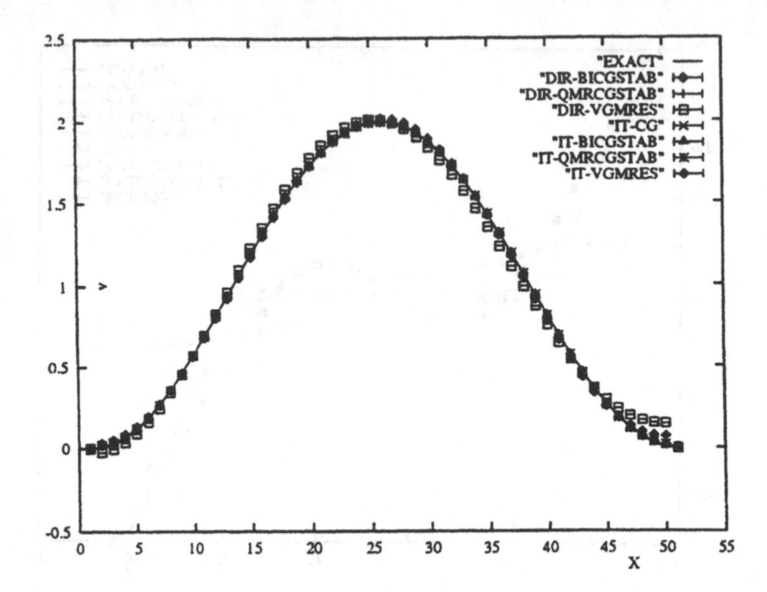

Figure 6. Example C: Results obtained for each method.

7. Conclusions

In this work, a state of the art of inverse thermal problems is presented. Three possible strategies are studied and compared in several examples. The classical FSM combined with RM seems to be more appropriate for one dimensional non steady state problem, and has a monotonic but low convergence. On the other hand, the direct formulation leads to results less accurate than the iterative one. In fact, the Krylov methods are not able to improve the solution from certain steps on.

The iterative formulation preserve the symmetry of the problem, if it exists, allowing us to use adequated techniques for each case. It also provides the most accurate results at lowest cost, at least in many studied cases.

For non steady state problems, these techniques may be used with a few changes. Nevertheless, in non linear problems it is not so clear. It depends on the type of nonlinearity. At this point, one should think about new ways for solving these problems. Regarding this question, future works will allow to combine the techniques presented here with stochastic methods, i.e. Simulated Annealing or Genetic Algorithms.

8. Acknowledgments

I am grateful to G. Winter who encouraged me to participate in this NATO ASI SERIES. I also thank J. Periaux and A. Suárez for providing me with very useful suggestions.

9. References

[1] Beck, J.V. (1962) Calculation of Surface Heat Flux from an Internal Temperature History, ASME Paper 62-HT-46.

[2] Beck, J.V., Backwell, B. and St.Clair, C.R. (1985) *Inverse Heat Conduction: Ill-Posed Problems,* Wiley Intersc., New York.

[3] Chan, T.F., Gallopoulos, E., Simoncini, V., Szeto, T. and Tong, C.H. (1994) A Quasi-Minimal Residual Variant of the Bi-CGSTAB Algorithm for Nonsymmetric Systems, *SIAM J. Sci. Comput.* **15**, 2, 338-247.

[4] Galán, M., Montero, G. and Winter, G. (1994) Variable GMRES: an optimizing self-configuring implementation of GMRES(k) with dynamic memory allocation, Technical Report of CEANI.

[5] Galán, M., Montero, G. and Winter, G. (1994) A Direct Solver for the Least Square Problem Arising From GMRES(k), *Com. Num. Meth. Eng.* **10**, 743-749.

[6] Hadamard, J. (1923) *Lectures on the Cauchy Problem in Linear Partial Differential Equations*, Yale Univ. Press, New Haven.

[7] Hestenes, M.R. and Stiefel, E (1952), Methods of Conjugate Gradients for Solving Linear Systems, *J. Res. Nat. Bur. Standards*, **49**, 409-436.

[8] Knowles, G. (1981) *An Introduction to Applied Optimal Control*, Academic Press, New York.

[9] Kozdoba, L.A. and Krukovsky, P.G. (1982) *Methods for solving Inverse Heat Transfer Problems*, Naukova Dumka, Kiev.

[10] Kurpisz, K. and Nowak, A.J. (1995) *Inverse Thermal Problems*, Computational Mechanics Publications, Southampton.

[11] Montero, G. and Suárez, A. (1995) Left-Right preconditioning versions of BCG-like methods, *Neural, Parallel & Scientific Computations* **3**, 4, 487-501.

[12] Montero, G. and Suárez, A. (1995) Métodos de Krylov en la Resolución de Problemas de Control en la Frontera con Operador de Laplace, IVX CEDYA, IV Congreso de Matemática Aplicada, Vic.

[13] Tikhonov, A.N. (1943) On Stability of Inverse Problems, *Lectures of Academy of Sciences USSR* **39**, 195-198.

[14] Tikhonov, A.N. (1963) On the Regularization of Ill-posed Problems, *Lectures of Academy of Sciences USSR* **153**, 49-52.

[15] Tikhonov, A.N. and Arsenin, V.Y. (1977) *Solutions of Ill-Posed Problems*, Winston & Sons, Washington D.C.

[16] Saad, Y. and Schultz, M. (1986) GMRES: A Generalized Minimal Residual Algorithm for Solving Nonsymmetric Linear Systems, *SIAM J. Sci. Comput.* **7**, 856-869.

[17] Saad, Y. (1993) A Flexible Inner-Outer Preconditioned GMRES Algorithm, *SIAM J. Sci. Comput.* **14**, 2, 461-469.

[18] van der Vorst, H.A. (1992) Bi-CGSTAB: A Fast and Smoothly Converging Variant of Bi-CG for the Solution of Nonsymmetric Linear Systems, *SIAM J. Sci. Comput.* **13**, 2, 631-644.

AN INTRODUCTION ON GLOBAL OPTIMIZATION BY GENETIC ALGORITHMS

G. Winter*, J. Periaux†, M. Galán*, B. Mantel†, I. Sánchez*

Abstract

The purpose of this paper is to influence the researchers working in the field of algorithms for large scale systems about the efficiency of evolution algorithms as optimization techniques to explore robust large search spaces and find near-global optima. These evolutionary algorithms (EAs) can be an alternative to numerical methods in difficult optimization problems like complex systems where the phenomena are difficult to model due to uncertainy, noise or even too little knowledge of the real problem. In such cases EAs are robust procedures to overcome these difficulties. In general these algorithms are not just an alternative to traditional methods, nowadays they are used hybridised form to complement and extend numerical methods. The convergence of heuristic optimization techniques is not affected by the continuity or differentiabililty of the functions to be optimized in the applications. These algorithms only require evaluation of the function in search space points. Applications of these evolutionary algorithms have been more convincing than their theory, which is still weak, though under progress. This paper is divided in two parts. A large number of references are included to enhance the presentation of the material. Finally, we describe the main aspects for solving an optimization problem of interest in Aerospace Industry by Genetic Algorithms. The problem considered is the op timum design of an airfoil shape, which is an inverse problem that consists of finding the shape for a given pressure distribution on the airfoil.

1 Exploring large spaces and optimization by genetic algorithms in complex systems

Traditional optimization tools have been used for years and they have proven to be invaluable in many design problems. However their scope is limited both by their restrictions on the design variables and their numerical overhead. These limitations are increasingly reducing their usability to real life problems. In order to overcome this situation new tools have been developed

* *Universidad de Las Palmas de Gran Canaria*

† *Dassault Aviation*

G. Winter Althaus and E. Spedicato (eds.), Algorithms for Large Scale Linear Algebraic Systems, 343–367.

with the help of modern distributed computer technology. These tools are based on evolution strategies and artificial intelligence. They can be used not only as a replacement of traditional continuous optimization but they can extend and improve them.

Many problems related to aerospace, automotive and energy disciplines are quite complex. This complexity arises due to: (1) size of the problem domain; (2) non-linear interactions between various elements; (3) domain constraints; (4) performance measures with dynamics and many independents and co-dependent elements; (5) inc omplete, uncertain and imprecise information. System of nature routinely encounters and solves such problems. Good examples of robust complex adapted systems include genetic evolution of species and human immune system response to foreign bodies. Many suc h problems are solved successfully using Genetic Algorithms and Evolution Strategies demonstrating how they can significantly help the above industries to get a quick feel of the available technology and its potential application. Aeronautical, Automotive and Energy Industries have to compete in an economically globalized world and share a big amount of richness and employment, on the other hand in a leading position they are as high-technology industry affecting many other technologi cally important companion industries (electronics, materials, automation, etc....). There are many conflicting design variables of different kinds, both quantitative and qualitative which should be put together to get equilibrium among several considerations: economical, security, fiability, environmental design, all of them adapted to a rapidly changing market heavily influenced by the preferences of final users and customers. Global optimization techniques like Genetic Algorithms and Evolution Strategies applied to the Aeronautical, Automotive, Energy Industrial processes are able n ot only to improve the final product making it more competitive in the global world market but also to drastically reduce the total cost; maintaining the technological gap of Europe which should be the base of employment saving economical decisions. Like living organisms subject to Darwin's rules of evolution, these new computer technologies consist of simple lines of code representing a population of candidate individuals which through chance matings, crossing over of digital DNA and mutation evolve sharing information and producing novel combinations which eventually after several generations will reach global optimality.

Genetic Algorithms (GAs) were initially developed in the United States. The initial context of GA by John Holland(75) [34] was the design of robust adaptative systems trying to imitate the mechanism of natural evolution. Genetic Algorithms (GAs) were used as an optimization algorithm by Kenneth De Jong. Evolution Strategies (ESs), developed by Ingo Rechenberg and Hans Paul Schwefel in Berlin, Evolutio nary Algorithms are still forming a young and rich field of scientific research and application oriented experiments with surprises and disadvantages where foundations, useful methods, coding, fitness and their landscapes and experiences by applications and experiments must and will be developed intensively in future years. GAs are iterative adaptative search algorithms (somewhat heuristic), that differ from more standard search algorithms (e.g., gradient methods, controlled random search, hill climbing, simulated annealing, etc) in that the search is conducted using the information of a population of structures named schemata instead of a direction or a single point. GAs exploit similarities among better solutions searching among a population in the way of blind search, not using deterministic rules.

GAs are quite robust in producing near-optimal sequences of trials for many applications also including problems with high levels of uncertainty and problems which cannot easily be reduced to closed form. A good textbook on genetic algorithms can be found in Goldberg(89) [23] and others of interest can be found in Davis L.(Ed.)(91) [9] and Michalewicz Z.(94) [41], too. A book where a compendium of state of the art and some applications is "Genetic Algorithms in Engineering and Computer Science"(G.Winter, J.Périaux, M.Galán, P.Cuesta, John Wiley, 1995) [63].

Given some finite discrete domain D and a function $\tilde{f} : D \longrightarrow R$, the search problem or function optimization problem consists of finding the best or near best point in D under $\tilde{f}$. R is ordered and there is a natural induced total ordering on D even if D is multi-dimensional. As a search algorithm, GAs examine some discrete subsets of a search space S^l and return some individual or string s_i(which correspond to a point of the domain function D by means of a mapping called decoding function) whose objective value $f(s_i)$ is an estimator of $Max_{x \in D}\tilde{f}(x)$, being

$$Max_{x \in D}\tilde{f}(x) = Max_{x \in S^l} f(s)$$

where f is the information of the individual's environment.

In case of minimization problems for $\tilde{f}$ we can consider $g = -\tilde{f}$ and

$$Min\tilde{f}(x) = Max\left(-\tilde{f}(x)\right) = Maxg(x).$$

More specifically the process in GAs is sequential; successive generations are produced with new solutions replacing some of the older ones, and each generation must be produced before it can be used as the basis for the following generation. In each generation the individuals are evaluated using some measure of fitness, and according to this fitness value, the fitter individuals are assigned a higher probability of being selected. A number of genetic operators are applied to the parents to generate new individuals called the offsprings by combining the features of both parents . The offsprings are next evaluated and a new generation is formed by selecting the individuals with higher fitness value, rejecting others and maintaining the population size constant. During iteration t, a GA maintains a population of potential solutions $P(t) = \{s_1^t, s_2^t, ..., s_i^t, ...s_n^t\}$ of individuals or chromosomes s_i^t, that are represented as strings $s_i^t\ (a_1, a_2, ..., a_l)$, of length l, the allele of the *i-th* gene (or *i-th* "locus" position) in a string being denoted by $a_i = b, b \in S$, where S is an alphabet set of symbols, being usually $S = \{0, 1\}$ for binary strings.

Let S^l be the set of the strings of length l. A decoding (encoding) is a function $d : S^l \longrightarrow R$, such that for each string $s \in S^l$corresponds to a value $x \in D$, S^l being the search space, $s_i\,(a_1, ..., a_l)$ in S^l.

The fitness function f is a function $f : S^l \longrightarrow R$ which can be seen as the composition of $\tilde{f}$ and d, thus can be defined as $f\,(s) = \tilde{f}\,(d\,(s))$; to the extent of GAs, f is "blackbox" function. The only procedure on f that Genetic Algorithms (GAs) require is its evaluation at a point of the search space (GAs are blind search algorithms). Thus being the rank r of a string s the number of strings whose objective value is smaller under the induced total ordering.

So, being $r\,(s) = Card\left\{s^* \in S^l : f\,(s) \geq f\,(s^*)\right\}$, we can consider the fitness function f as a set of triples $\left\{(s, x, r)\,/s \in S^l, x \in D, r \in R\right\}$, with $d\,(s) = x$, $\tilde{f}\,(x) = r$, thus $f\,(s) = r$, then any permutation in any component of the triplet will produce a new domain function and correspondingly a new encoding (Rawling 91 [48]).

The objective function or fitness function $f\,(s)$ plays the role of the environment; each individuals s is evaluated according to its fitness. In this way a new population (iteration *t+1*) is formed by selection of the better individuals of the former population, as they will form a new solution by means of applying a selection procedure and crossover and mutation operators. It should be noted that diversity of individuals is required to find good solutions with GA.

The mapping from binary string is for $s_i \in S^l$ and $x_i \in D$, being $A \leq x_i \leq B$, is given by the expressions:

$$\left(\sum_{n=1}^{l} a_n 2^{n-1}\right)_{10} = M$$

$$x_i = A + M\frac{B - A}{2^l - 1}$$

and required precision, i.e. " m " decimal places for the variables values x_i when f is a function of k variables $f\,(x_1, x_2, ..., x_k)$ and being $A_k \leq x \leq B_k$ given by the expression:

$$2^{l-1} \leq (B_i - A_i)\,10^m \leq 2^l$$

where in this case a vector string

$$(a_{11}, a_{12}, ..., a_{1l_1}, a_{21}, a_{22}, ..., a_{2l_2}, ..., a_{k1}, a_{k2}, ..., a_{kl_k})$$

is composed by k substrings of length l_j, $j = 1, 2, ..., k$ and l is given by

$$l = \sum_{j=1}^{k} l_j$$

Bit string encoding have several advantages over other encodings, as they are simple to mate and manipulate. On the other side performance theorems have been proved for bit string chromosomes (theorems for other encoding techniques have been proved from the former ones). There are well known advantages in the use of low cardinality alphabets (binary representation=2).

However, real (floating point) codings are claimed to work well and faster in a number of practical problems (Michalewicz 94 [41]), even if the use of real-coded or floating-point genes has arisen some controversy from a theoretical point of view.

Caruana and Schaffer (88) [8] have empirically found that Gray coding is "often" better than "naïve" binary coding for GAs.

The encoding is very important for the success and performance of a GA. One simple, yet important, observation in practical implementation of GA, to facilitate a fast decoding from the binary string representation, is that the choice of the length of the strings should be a multiple of the size of a "natural" variable in the specific hardware/O.S.

Simple GAs operate on a finite population size n and fixed length binary strings l for each chromosome using selection, crossover and mutation operators. The initial population should include individuals with very high diversity whose "allele" are randomly generated. A selection operator identifies the fittest individuals of the current population to be the chosen parents of the next generation. The fitness function provides the environmental feedback for a selection procedure.

Let $P(0)$ be a randomly generated initial population and $P(t)$ the population at time t, $P(0), P(t) \in \left(S^l\right)^n$. Let $p_s : S^l \longrightarrow [0,1]$ be a function such that for all $P(t)$:

$$\sum_{i=1}^{n} p_s\left(s_i^t\right) = 1,$$

thus determining the selection probabilities of the individuals in a population.

A selection operator is an operator that produces an intermediate population $P'(t)$ from the population $P(t)$. A GA without selection is called a *genetic drift* .

Genetic operators (including selection operator) generate new individuals with potentially higher fitness values. Some different selection schemes are:

-proportional selection schemes (Holland 75 [34]) which are selection schemes that choose individuals for birth according to their fitness function values:

$$p_s\left(s_i^t\right) = \frac{f\left(s_i^t\right)}{\sum_{j=1}^{n} f\left(s_j^t\right)}$$

where $p_s\left(s_i^t\right)$ is the probability of an individual in the *i-th* generation to be sampled. It leads to the expectation of the individual s_i^t to occur $\eta_i^t = np_s\left(s_i^t\right)$ times in generation *t+1*; η_i^t is called generally the expected value of s_i^t.

After few generations the population average fitness may be close to the population best fitness leading to premature convergence to a mediocre generation. This particular scheme can lead to a "superindividual" (individual with very high fitness value compared to the rest of the population) which, in a few generations, will take over the population, leading, sometimes, to premature misconvergence. In order to solve this problem, there are possible alternatives like to use scaled fitness values: among them are the sampling of the probability distribution via stochastic remainder selection (Booker 83 [5] and Brindle 81 [7]), or stochastic universal selection (Baker 87 [4], Grefenstette and Baker 89 [30]). Another posibility is to use scaled fitness values by means of an additional mapping $f^l : R \times \left(S^l\right)^n \longrightarrow R$. There are several proposed scaling types, see i.e.: Goldberg 89 [23] and Bäck and Hoffmeister 91 [1].

Another alternative can be the use of the following selection scheme:

-linear ranking selection (Baker 85 [3]); it is based on the notion of rank. In this case, after a reordering (according to the fitness value) of the population, and being s_1 the fittest individual, the selection probabilities are constant values:

$$p_s\left(s_i^t\right) = \frac{1}{\lambda}\left(\eta_{max} - \left(\eta_{max} - \eta_{min}\right)\frac{i-1}{\lambda - 1}\right)$$

where λ is the number of individuals representing one generation (in standard GA, $\lambda = n$) and $\eta_{min} = 2 - \eta_{max}$, $(1 \leq \eta_{max} \leq 2)$.

The scheme assigns the number of copies that each individual should receive according to a non-increasing function and then performs proportional selection based on the assignment. Baker suggests a maximum expected value of $\eta_{max} = 1.1$.

Experimental results give indications relative to the advantage of ranking selection in case of multimodal functions (Baker 85 [3], Whitley 89 [58]).

-(μ, λ)-uniform ranking selection (Schwefel 81 [49]):

$$p_s\left(s_i^t\right) = \begin{cases} \frac{1}{\mu} & 1 \leq i \leq \mu \\ 0 & \mu \leq i \leq \lambda \end{cases}$$

These last schemes are generalized by Bäck and Hoffmeister (91) [1]:

(μ, λ)-Proportional Selection given by the following expression,

$$p_s\left(s_i^t\right) = \begin{cases} \frac{f\left(s_i^t\right)}{\sum_{j=1}^{\mu} f\left(s_j^t\right)} & 1 \leq i \leq \mu \\ 0 & \mu \leq i \leq \lambda \end{cases}$$

and (μ, λ)-linear ranking given by,

$$p_s\left(s_i^t\right) = \begin{cases} \frac{1}{\mu}\left(\alpha_{max} - 2\left(\alpha_{max} - 1\right)\frac{i-1}{\mu-1}\right) & 1 \leq i \leq \mu \\ 0 & \mu \leq i \leq \lambda \end{cases}$$

these being schemes where only μ individuals are allowed to be selected, thus μ guides the selective pressure.

The fraction $\frac{\mu}{\lambda} = \frac{1}{5}$ seems to give the approximate rate of convergence (speed) for unimodal problems. A much higher value must be considered to explore the search space for multimodal problems. Always

$$\sum_{i=1}^{\lambda} p_s\left(s_i^t\right) = 1$$

must be satisfied.

Here evolution strategies (ESs) can be introduced:

$$P\left(t\right) = \left\{S_1^t, S_2^t, ..., S_k^t\right\} \in I^k$$

where I^k denotes a population of $k \in \{\mu, \lambda\}$ individuals at generation t, the genetic operators being modeled by the mappings $I^k \rightarrow I^\mu$ for solution, $k \in \{\lambda, \mu + \lambda\}$, $I^\mu owI$ for recombination and $I \rightarrow I$ mapping for mutation with $I^\mu \rightarrow I^\mu$ for a single iteration from a population $P\left(t\right)$ towards the next parent $P\left(t+1\right)$. The space $I = R^n \times S$ on the type of evoluti on strategy using the set of srategy parameters to be considered (see Bäck and Schwefel, chapter 6 in GAs meeting and Computer Science, and Schwefel, 1995). When the objective is to balance towards more exploitation and less exploration appears tournament selection (Wetzel). One source of convergence difficulty in GAs results from the stochastic errors in sampling and selection when working with small population. Then it is necessary to take special strategies like sizing populations: methods using sharing functions provide a payoff incentive to maintain selective pressure among competing individuals (Goldberg 89) [23]. In natural systems different species have found different niches. Goldberg and Richardson , 87 [24] proposed a way of imposing niche and specification on strings based on some measure of their distance from each other. Progress about convergence of GAs can be found in Goldberg & Debs, 91 [22] and more recently in Mühlenbein-Schlierkamp Voosen, 93 [43] and Thieren & Goldberg 94 [54]. Based on order statistic, one model to predict the selection pressure is found in the interesting paper by Miller & Goldberg(95) [42].

parent 1	1 0 0 1 0 1 1
parent 2	0 1 0 0 1 1 0
random positions	↑
offspring 1	1 0 0 0 1 1 0
offspring 2	0 1 0 1 0 1 1

Table 1: 1-point crossover

Elitist strategies are based on copying the best individual of each generation into the next one, where some/all parents are allowed to reproduce, thus reducing the possibility that the best individual of the population may fail to produce an offspring in the next generation. Further performance improvements can be obtained using this approach and replacing the worst individuals of the population with newly generated individuals (De Jong 75 [11], Whitley 89 [58]).

The crossover operator randomly chooses a pair of individuals among those previously selected to breed and exchanges or concatenates arbitrary substrings between them. Thus crossover recombines the genetic material from two parents into their children and is controlled by the crossover rate which is defined as the ratio of the number of offsprings produced in each generation to the population size. The crossover rate determines the ratio of the number of searches in subdomains of high average fitness to the number of searches in other subdomains. A high crossover rate allows more exploration in the search space, but if this rate is too high then we use of a lot of runtime exploring unpromising subdomains of the search space. There are different types of crossover operators:

One-point crossover is proposed by Holland. It consists of a random selection of the place to cut the parent chromosomes in two substrings. The children will have their chromosomes composed by two substrings generated by combining the part of one parent to the left of the cut position with the part of the other parent to the right of the cut position (see Table 1).

Two-point crossover operator selects two positions to cut and combines the part between them (see Table 2). It seems to loose less schemata than the traditional one-point crossover operator, see i.e. Schaffer, Caruana, Eshelman and Das 89 [49] and Eshelman, L.J., Caruana, R.A. and Schaffer, J.D. 89 [16].

Uniform crossover operator is a generalization of one-point, two point and multipoint crossover, see i.e. Sysweda (89) [53]. With uniform crossover the allele offspring is obtained randomly from one of the parents (see Table 3).

Other: there are other crossover procedures like segmented, shuffle, etc. Also some recombination operators of interest are proposed by Eshelman and Schaffer (92) [14] and Mühlenbein and Schlierkamp-Voosen (93) [43].

parent 1	1 0 0 1 0 1 1
parent 2	0 1 0 0 1 1 0
random positions	↑ ↑
offspring 1	1 0 0 0 1 1 1
offspring 2	0 1 0 1 0 1 0

Table 2: 2-point crossover

parent 1	1	0	0	1	0	1	1
parent 2	0	1	0	0	1	1	0
random positions	0	1	1	0	1	0	1
offspring 1	1	1	0	1	1	1	0
offspring 2	0	0	0	0	0	1	1

Table 3: uniform crossover

Mutation is an operator used after applying crossover in various individual operator to increase diversity, that is, to take new points in the search space to evaluate. A chromosome's locus is randomly chosen for mutation. In a binary-code the corresponding bits are "flipped" from 0 to 1 or from 1 to 0. A good property of Gray coding is that changing from 0 to a 1 produces a perturbation in either direction (Wright 91 [65]). The mutation rate is defined as the percentage of the total number of genes in the population which are mutated in each generation. The probability of mutation (p_m) must be low and must be appropriated according to the encoding used. Mutation is a secondary operator in GAs, but it seems clear that mutation depends on the length of the string l, thus it seems sensible to consider the probability of mutation depending of l. In the context of binary encoding, Bäck proposes a formula for the approximate evaluation of mutation rate. He also proposes a kind of self-adaptation under certain circumstances (Bäck, T. 92 [2]). The mutation acts as a safeguard against a premature loss of genetic code at a particular position. If the mutation rate is too high, there will have much random perturbation, the offspring will lose their resemblance to the parents and the procedure will lose the ability to learn from the history of the search.

To continue it is very important understanding the behaviour of GAs and to know the following well-known definitions: A schema H is a hyperplane in the metric space $\left(S^l, d\right)$, d being the Hamming distance of one string to another, $x\,(x_1, x_2, ..., x_l)$, $y\,(y_1, y_2, ..., y_l) \in S^l$

$$d : S^l \times S^l \longrightarrow \{0, 1, ..., l\}$$

where,

$$d\,(x, y) = \|i \in \{1, ..., l\} : x_i \neq y_i\| \,.$$

This metric space is represented as strings by $(S \bigcup \{*\})^l$, $* \notin S$ being a wildcard symbol which stands for any symbol in S (Holland 75 [34]), thus unspecified positions, i.e. positions that "don't matter" or "don't care"are filled with *'s. A schema describes a subset of strings with same values at certain string positions, thus three strings (011), (111), (010), may be represented by the schema (*1*). On the other hand the schema (01*0*1) describes the following subset of strings: (010011), (010001), (010011), (011001), (011011). Thus a schema is a set of genes that make up a partial solution.

In a population of size n, among 2^l and $n2^l$ different schemata may be represented.

An interest situation that occurs many times is when a GA can give a population dominated by individuals very close to the optimum in the phenotype space (D), but far in terms of Hamming distance applied to the genotype space $\left(S^l\right)$. In this case one good choice are Gray codes.

Order of a schema, denoted by $O(H)$ is in binary coding the number of 0 and 1 fixed positions in the schema. The notion of $O(H)$is important in calculating survival probabilities of the schema for mutations.

Length of the schema $L(H)$ is the distance between the first and the last fixed string positions.

For example,

$$H_1 = (* * *0110 * *01 * 0 * **)$$

$$H_2 = (* * 10010 * 0 * 0 * 101)$$

$O(H_1) = 7$ $O(H_2) = 10$ $L(H_1) = 13 - 4 = 9$ $L(H_2) = 16 - 3 = 13$.

The length of the schemata defines the compactness of information contained in a schema. Low order and short length schema are known as *building blocks* .

Under proportional selection, simple crossover and mutation, the expected number of copies of a schema H contained in $(t+1)th$ iteration is bounded by the following inequality (Goldberg 89 [23])

$$m(H, t+1) \geq m(H,t)\frac{f(H)}{\overline{f}}\left[1 - p_c\frac{L(H)}{l-1} - p_m O(H)\right]$$

$f(H)$ being the average of the fitness values of all strings which are represented by the schema H,

$$f(h) = \frac{\sum_{s_i^t \in H} f(s_i)}{m(H,t)}$$

and being $\overline{f}$ the average fitness of the entire population,

$$\overline{f} = \frac{\sum_{j=1}^{n} f(s_j^t)}{n}$$

p_c, p_m are crossover and mutation probabilities.

GAs are different from other search methods in the property known as *implicit parallelism* : many schemata may be sampled according to their fitness by the combined actions of survival-of-the-fittest selection and recombination. This property may be stated as large number of the schema in a population being sampled in future trials exponentially overtime and the grow rate is their fitness ratio. Thus Holland's Schema Theorem says that *"above average, short, low-order schemata are given exponentially increasing numbers of trials in successive generations"*.

The estimation for a GA with an optimal population size n about the number of short schemas sampled is on the order of n^3 (Holland 75 [34], Goldberg 89 [23]) and the number of low order schemata sampled is also on the order of n^3, this being an important property of GA with respect to other search algorithms. The cost required to converge by GA when the function is unimodal is given with high probability as $O(nlogn)$ function evaluations (Goldberg *et al.* 91,92 [27], [28]).

The basis for the Schema Theorem is that if the average fitness of the population doesn't change much over the lifetime of an individual, then the expected number of offsprings generated by individual i is proportional to the ratio between the distance of the fitness value to the average fitness and the range of the fitness values. GAs are so successful in complex optimization problems in terms of schemata and the effect of genetic operators.

The behaviour of genetic algorithms is very difficult to model mathematically, but some results exist with phenotypic, genotypic and statistical approaches. A source of loss of diversity results from poor performance of recombination operators in terms of sampling new structures: crossover and mutation operators modify this selective pressure providing diversity and play a critical role in the exploration/exploitation balance. Crossover recombines features of the best chromosomes by combining building blocks of the good ones. GA search process does not form individual chromosomes one by one, rather it builds high utility schema with many fixed bits from high utility schema with few fixed bits (Schema Theorem).

GAs are not guaranteed to find an optimal solution their effectiveness heavily depends on many factors, among them (and largely) of the population size (n). Goldberg 92 theoretically studies the optimal population size. Smith 93 [51] proposes an algorithm of interest, too. Mühlenbein and Schlierkamp-Voosen (1993) [43] have obtained for proportional selection the following relation among the amount of selection S , the standard deviation σ_p of the fitness of the population and the average fitness of the population:

$$\frac{S(t)}{\sigma_p} = \frac{\sigma_p}{\overline{f}(t)}$$

where S is measured by the selection differential $S(t) = \overline{f}_s(t) - \overline{f}(t)$, $\overline{f}_s(t)$ being the average fitness of the select parents. The result is valid for very large popu lations because statistical values are considered. A recombination operator that creates a binomial distribution and for this reason facilitates to obtain some theoretical results is the gene pool recombination (GPR): two parent alleles of an offspring ar e randomly chosen with replacement from the gene pool given by the parent population selected before. Thus the offspring allele is computed using any of the standard recombination schemes such that the allele of the offspring is obtained randomly from one of the parents. This recombination operator facilites to describe mathematically the linkage between the genes at different loci produced by recombination (see Mühlenbein [43], GAs in Engineering and Computer Science, Eurogen Book, 1995). An important practical question is the optimal size of a population according to the amount of real time available. Larger populations are processed in very much less time using Parallel GAs. A simple use of parallelism is the simultaneous production of candidates for the next generation. In order to probabilistically select parents for the crossover operation in case of partition the population of solutions and assignment of one subset of the population to the local memory of each processor is necessary, somehow, to use global information about the fitness. About parallel GAs most studies are still empirical investigations on the choices of topology and migration rates considering the population sizing target as a secondary problem. Recent development by E. Cantu-Paz and D. Goldberg provid es theoretical understanding of parallel GAs focusing on the adequate sizing of demes and attacking the sizing problem using bounding cases for the deme interconnection topologies and migration rates.

Messy genetic algorithms (mGAs) have been proposed by Goldberg, Korb and Deb. 89, 91 [29] [28], opening new possibilities for processing schemata: the chromosomes may be of variable length, redundant genes, free genes to complete, etc. For examples a string in mGAs could be:

$$s_i = \{(3,1),(4,0),(4,1),(3,1)\}$$

where the first number gives a position and the other one the allele. mGAs use "splice" and "cut" operators replacing simple crossover. Both operators are used with defined probabilities and there are two phases in the process, a primordial phase where the population is initialized to contain all possible building blocks of a specified length and a juxtapositional phase where the population of good building blocks is enriched in several generations by means of reproduction like in standard GAs. Recent developments about messy genetic algorithms with test case and results are presented by H. Kargupta (Computational Sciences Methods Group, Los Alamos National Lab) in a NATO Workshop organised by CEANI last July 1996 in Las Palmas. This class of algo rithms provides appropriate relations among members of the search domain in optimization. Another interesting contribution of Kargupta is a systematic approach named SEARCH (Search Envisioned As

Relations & Class Hierarchizing) to introduce domain knowled ge in optimization algorithms by means of investigating the conditions essential for transcending the limits of random enumerative search using a framework developed in terms of relations, classes and partial ordering in blackbox optimization problems.

Also, it is interesting to consider the hybrid genetic algorithms, which combine local search with a genetic algorithm, usually using local search to improve the initial population processed by a genetic algorithm to locate saddle points in t he search space. It is often effective in practice to test the neighbors of a string in random order to make the next movement with a probabilistic decision (Whitley, 95 [61]).

Some general considerations commonly accepted by Genetic Community:

- Generally GAs find nearly global optimal in a variety of problems types, also complex spaces, for example, the search spaces multimodal.
- GAs can be inefficient when the cost is highly dominated by the evaluation of a large number of individual fitness functions. This inefficiency can be softened to a great extent using sensible parallel implementations.
- GAs using binary representation and single-point crossover and binary mutation are robust algorithms, but they are almost never the best algorithms use for any problem.
- Since GAs are stochastic, their performance is a more useful way to view the behaviour of a genetic algorithm than a representation of the behaviour of a genetic algorithm in a single run.
- A GA that satisfies the Schema Theorem is not necessarily a good optimizer. The effectiveness of GAs depends as we have seen from many factors and mainly depends of the appropriated balance between exploitation/exploration or selective pressure/diversity and optimal population size.
- For a fixed GA with the same sets or multisets of values D and R for every domain function with the encoding makes easier to solve there is another domain function that it makes more difficult to solve.
- The GA-behaviour found in many experiments of Genetic Community confirms the well-known contradiction between exploration and exploitation in global optimization. For unimodal functions exploitative search with high convergence rate is desired while for multimodal functions explorative search with a high convergence confidence towards a global optimum point is the objective.

- Never forget that GAs mimic the behaviour of Nature, so they try to get not one "supergood" individual but a population of individuals well adapted to the environment (fitness function).

- Some interesting strategies using heuristic optimization techniques: using GA the majority of the computational cost is spent when after a determined number of populations are produced small improvements are obtained very slowly. This convergence behaviour has motivated some interesting strategies like using GA linked to other methods in a strategic and appropriate way: by GA we find an acceptable approximation to the global minima, it is using initially GA to obtain improves rapidly (few generations) and when it becomes very difficult to get further improvement we continue solving the problem using a traditional or numerical faster method. There is also the possibility to use in the last steps other heuristic methods like Simulated Annealing many times faster than GA in last steps.

- One important difference between simulated annealing algorithms and GAs is that while the first operates on only one configuration at a time, a GA maintains a large population of configurations to be optimized simultaneously, say the population goes evolving as a memory gained for experience in the search and increasing this memory in successive iterations.

Random numbers:

Since Genetic Algorithms are stochastic, it is notorious that a big influence in to the underlying algorithm for generation of random numbers.

There are methods for generation of "real" random sequences based on specialized hardware, however for the general user the solution is to implement an algorithm which generates a pseudo-random sequence.

It seems odd to use the most deterministic device (computer) to build a random sequence: any computer program will produce a predetermined output, this way it will not be truly random, for this case we will use the term pseudo-random.

How could we define the vaporous concept of randomness?

A working approach could be that the deterministic program that builds a random series should be different and uncorrelated with the program which uses the sequence. This way two different generators should coincide with the one obtained using a true random device.

¿From a pragmatic point of view an algorithm which is "good" for one application can show its weakness in another one.

There are several tests to prove the goodness of a pseudo random number generating algorithm. Some of them are supported by statistics and some others are based on heuristics: a good pseudorandom number generator should pass all of them. However we should bear in mind that these are necessary conditions, never sufficient. One of the most important theoretical result is the χ^2 criterion of non-correlation, among the heuristical ones we will mention the n -cube distribution for uniform deviates. In recent years, genetic algorithms (GAs) have become more and more robust and easy to use. Current knowledge and many successful experiments suggest that the application of GAs is not limited to easy-to-optimize unimodal functions that contain only one o ptimum. Several results and GA theory give the impression that GAs easily escape from millions of local optima and reliably converge to a single global optimum. However, recent results raise serious doubts that this explanation is always correct.

2 Application on transonic flow optimization

We present the optimization of the pressure around a single airfoil evaluating the best shape with a proposed Genetic Algorithm (GA) in the corresponding minimization problem to solve transonic flow.

The research of an optimal shape in aerodynamics is a problem which has been attracting for a long time engineers, mathematicians, etc.

Now, we have by hand robust procedures for global optimization like Genetic Algorithms (GAs), which allow to minimize more complex functionals like non-convex, high multimodal, nonlinear function and finding solutions close to the optimum.

The problem considered is to reach the airfoil shape which realizes a given p_t pressure distribution on its boundary. This inverse problem consists in finding the shape (denoted λ) of an airfoil which realizes a surface target pressure distribution for a given transonic flow condition.

The problem considered is an inverse problem, it's to say, finding the shape of an airfoil which has a target pressure distribution for a given transonic flow condition.

This problem corresponds to the following formulation:

$$minJ(\lambda)$$

with

$$J(\lambda) = \frac{1}{2} \int_{\lambda} (p_{\lambda} - p_t)^2 \, d\lambda$$

p_t: a given target pressure whose shape is searched and p_{λ} the actual pressure on λ.

Each calculation of the functional J needs to solve the full potential flow equation. We do not describe here the formulation, which is well known and nowaday many solvers are available for the calculations. Thus the computing cost is directly proportional to the number of Computational Fluid Dynamic simulations. We consider finite element for solving the non linear P.D.E.'s of the transonic flow regime.

The population is the set of airfoils, the individual corresponds to the airfoil shape, and the fitness function is a discrete version of J, which includes an additional term that considers the possible singularity point at the end of the airfoil profile.

The natural choice to represent the shape of an airfoil is using points. This has many advantages because generally most airfoils shapes are given in the form of table functions or fixed points. In such a representation one recommendeds the inclusion of some penalization terms to consider the smooth form or the airfoil (only one maximum) and to avoid a possible chaotic distribution of the points.

However, the number of parameters treated by the genetic algorithm could be drastically reduced and the need of the penalization terms avoided, by using an interpretation instead of the point representation.

The Bezier Spline representation is a good alternative due to the fact that a few control points are enough to represent with a very good accuracy any kind of airfoil shape, and also because the new individuals are always candidate solutions (Bezier curves have underlying smoothness properties). Its expression is given by the following equations:

$$\begin{cases} x(t) = \sum_{i=0}^{8} C_8^i.t^i.(1-t)^{8-i}.x_i \\ y(t) = \sum_{i=0}^{8} C_8^i.t^i.(1-t)^{8-i}.y_i \end{cases} \quad C_n^i = \frac{n!}{i!\,(n-i)!}$$

using an order representation with 7 control points and 2 fixed ones, t is the parameter of the curve whose values vary between $[0,1]$, (x_i, y_i) are the coordinates of the control points and the define profile. We define a Bezier curve for the extrados and another Bezier curve for the intrados, the whole airfoil shape being defined by the merge of the two curves. In case of flo w with an incidence angle $\alpha = 0^o$ the results are symmetric and it only is necessary to consider the extrados profile.

We consider the x_i fixed, and the parameters which are coded in the genetic algorithm are only the y_i coordinates or the control points.

The optimization problem is to find the y_i parameters that minimize cost function J.

The fitness function is evaluated from the coding during the GA iteration.

2.1 Test Case

The general parameters of the execution of the test cases are shown above:

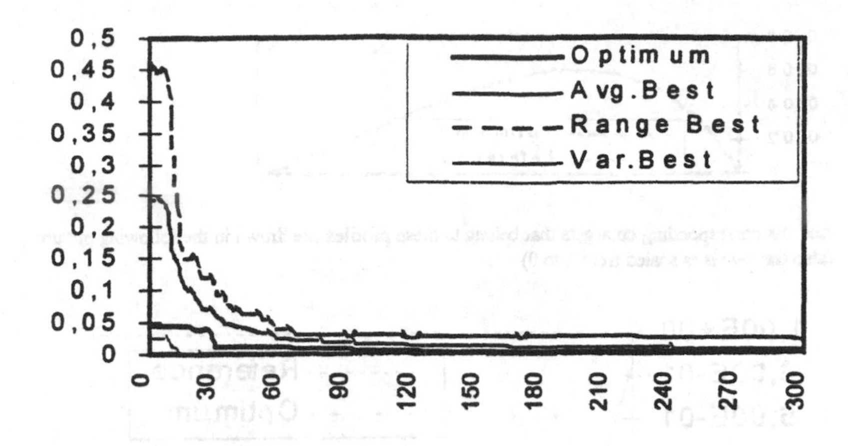

-**Twopoint crossover without GrayCode:**

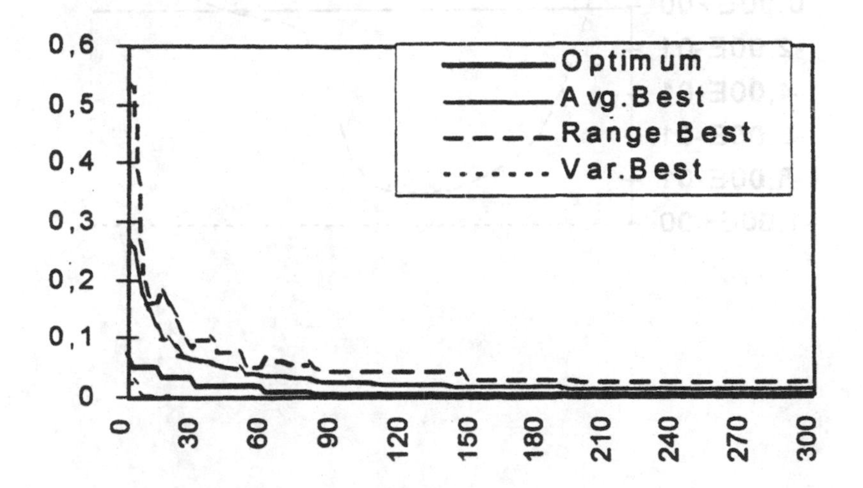

The optimum profile versus the reference of the NACA0012 profile is shown in the next figure (a fitness function of 0.00156); (the y-axis is the y-scale length and the x-axis is scaled from 0 to 1):

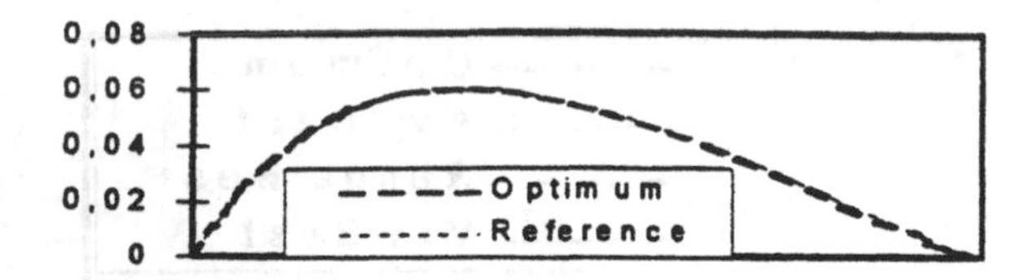

And the corresponding cptargets that belong to these profiles are drown in the following picture: (also the x-axis is scaled from 1 to 0)

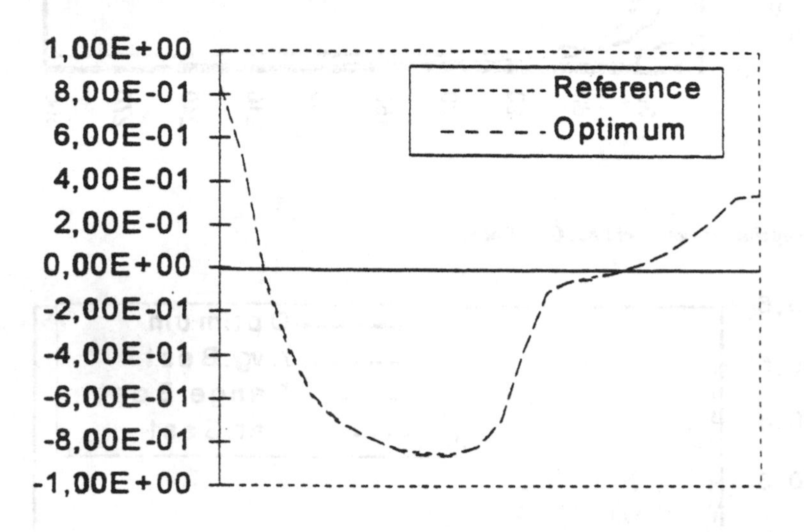

Population replacement: steady state (with 2 springhood members each generation)
Initial Population: duplicates not allowed
Number of runs per case: 10
Random number generation : shuffing
Accuracy: 32 bits per variable (7 control points)
Dimension: 7 (one per each control point)
Population size: 30
Crossover rate: 85%
Mutation rate: 3.8%
Selection strategy : ranking

The Mach number for this test case is 0.8 and $\alpha = 0$. The following figures represent the Optimum (best of all runs), the Avg. Best (arithmetic average of the best in all runs), the Range Best (distance between the optimum and the "worst" best) and the Avg. Var (variance or the best for all runs).

In the y axis the value of the minimization function is shown and the x -axis represents the generation number of the execution.

Acknowledgment: The results described in this paper have been obtained with the collaboration of David Greiner, a promising researcher of CEANI (Centro de Aplicaciones Numéricas en Ingeniería, University of Las Palmas of Gran Canaria)

References

[1] Bäck, T. - Hoffmeister, F. (1991) "*Extended Selection Mechanisms in Genetic Algorithms*", 92-99, Proceedings of the Fourth International Conference on Genetic Algorithms. Morgan Kaufmann Publishers, Los Altos, CA. Belew, R. and Booker, L. (Ed.)

[2] Bäck, T. (1992) "*Self Adaptation in Genetic Algorithms*". Proceedings of the First European conference on Artificial Life, 263-271. The MIT Press, Cambridge, M.A. F.J. Varela and P. Bourgine (Eds.).

[3] Baker, J.E. (1i985) "*Adaptative Selection Methods for Genetic Algorithms*". Proceedings of the First International Conference on Genetic Algorithms and Their Applications, 101-111. Hillsdale, New Jersey. Lawrence Erlbaum Associates. J.J. Grefenstette (Eds.).

[4] Baker, J.E. (1987) "*Reducing Bias Inefficiency in the Selection Algorithm*", 14-21. Proceedings of the Second International Conference on

Genetice Algorithms and Their Applications . Hillsdale, New Jersey. Lawrence Erlbaum Associates, Grefenstette J.J. (Ed.).

[5] Booder, L.B. (1982) "*Intelligent Behaviour as an Adaption to the Task Environment*", Doctoral Disertation , University of Michigan.

[6] Bratley, P., Fox B.L., Schrage L. (1983) "*A guide to Simulation* " Springer-Verlag.

[7] Brindle, A. (1981) "*Genetic Algorihms for Function Optimization*", Doctoral Dissertation, University of Alberta, Edmonton.

[8] Caruana, R.A., Schaffer, J.D. (1988) "*Representation and hidden bias: Gray vs binary coding for genetic algorithms*". Proceedings of the 5th International conference on Machine Liarning. 153-161. Morgan Kauffmann Publishers, San Mateo, California.

[9] Davis, L. (Editor), (1991) "*Handbook of Genetic Algorithms*". Van Nostrand Reinhold, New York.

[10] De Jong, K.A., Sarma J. (1993) "*Generation Gaps Revisited*". Foundation of Genetic Algorithms, 19-27.

[11] De Jong, K.A. (1975) "*An Analysis of the Behaviour of a Class of Genetic Adaptive Systems*". Ph. dissertation, University of Michigan, Ann Arbor, MI.

[12] De Jong, K.A. (1993) "*Genetic Algorithms are not Function Optimizers*". Foundation of Genetic Algorithms, 5-17.

[13] Dihn, Q.V., Glowinski, R., Periaux, J. and Terrasson, G. (1988) "*On The Coupling of the Viscous and Inviscid Models for Incompressible Fluid Flows Via Domain Decomposition*". First International Sysmposium on Domain Decomposition Methods fot PDE, SIAM.

[14] Eshelman, L.J. and Schaffer, J.D. (1992) "*Real-coded Genetic Algorithms and Interval-Shemata*" 187-202, Foundations of Genetic Algorithms, 1, 69-93. Morgan Kaufmann Publishers. G.J.E. Rawlins (Ed.).

[15] Eshelman, L.J. and Schaffer, J.D. (1993) "*Crossover's niche*", 9-14. Proceedings of the Fifth International Conference on Genetic Algorithms. Morgan Kufmann Publishers, Los Altos, CA. Forrest, S. (Ed.)

[16] Eshelman, L.J., Caruana, R.A. and Schaffer, J.D. (1989) "*Biases in the Crossover Landscape*", 10-19. Proceedings of the Third International Conference on Genetic Algorithms and their Applications . Morgan Kaufmann Publishers, San Mateo, CA. Schaffer J.D. (Ed.).

[17] Fogarty, T.C. (1989) "*Varying the probability of mutation in the genetic algorithm*". 104-109. Proceedings of the Third International Conference on Genetic Algorithms and their Applications. Morgan Kaufmann Publishers, San Mateo, CA. Schaffer J.D. (Ed.).

[18] Forsythe G.E., Malcom M.A., Moler C. (1977) "*Computer Methods for Mathematical Computations*" Prentice-Hall.

[19] George, A. (1971) "*Computer implementation of the Finite Element Method*", Report Stan CS-71-208, also Ph. D. thesis Departament of Computer Science, Standford University, Standford, CA.

[20] Glover, F. and Laguna, M. (1993) "*Tabu search. Modern Huristic Techniques for Combinatorial Problems*", Blackwell Scientific Publications, Oxford, 70-141. Reeves C.R. (Ed.).

[21] Goldberg, D.E., Deb, K., Clark, J.H. (1992) "*Genetic Algorithms, noise, and the sizing of populations*". Complex Systems, (6) (332-362).

[22] Goldberg, D.E. Deb, K. (1991) "*A comparative Analysis of Selection Schemes used in Genetic Algortihms*". Foundations of Genetic Algorithms, (1) (69-93). Morgan Kaufmann Publishers. G.J.E. Rawlins (Ed.).

[23] Goldberg, D.E. (1989) "*Genetic Algorithms In Searc, Optimization and Machine Learning*". Reading, Massachusets, Addison-Wesley.

[24] Goldberg, D.E. and Richardson, J. (1987) "*Genetic Algorithms with Sharing for Multimodal Function Optimization*", 41-49. Proceedings of the Second International Conference on Genetic Algorithms and Their Applications. Hillsdale, New Jersey. Lawrence Erlbaum Associates, Grefenstette J.J. (Ed.)

[25] Goldberg, D.E., Deb, K. Thierens, D. (1993) "*Toward a better understanding of mixing in genetic algorithms*". Journal of the Society of Instrument and Control Engineers, 32 (1) (10-16).

[26] Goldberg, D.E., Deb, K., Kargupta, H., Harik, G. (1993) "*Rapid, accurate optimization of difficult problems using fast messy genetic algorithms*". Proceedings of the Fifth Intertional Conference on Genetic Algorithms, 56-54.

[27] Goldberg, D.E., Kelsey M., Tidd, C. (1992) "*Genetic Algorithm: A Bibliography*", (IlIGAL Report n. 92208), University of Illinois at Urbana-Champaign, Illinois Genetic Algorthm Laboratory, Urbana, USA.

[28] Goldberg, D.E., Korb, B., Deb, K. (1991) "*Do not Worry, Be Messy*". Proceedings of the Fourth International Conference on Genetic Aogorithms, 24-30. morgan Kaufmann Publishers, Los Altos, CA. Belew, R., Booker, L. (Ed).

[29] Goldberg, D.E., Korb, B., Deb, K. (1989) "*Messy Genetic Algorithms: Motivation analysis, and first results*". Complex Systems, 3 (5), 493-530.

[30] Grefenstette, J.J., Baker, J.E. (1989) "*How genetic algorithms work: A critical look at implicit parallelism*", 20-27. Proceedings of the Third International Conference on Genetic Algorithms and their Applications Morgan Kaufmann Publishers, San Mateo, CA. Schaffer J.D. (Ed.).

[31] Grefenstette, J.J. (1986) "*Optimization of Control Parameters for Genetic Algorithms*". IEEE Transactions on Systems, Man and Cybernetics SMC-16 (1), 122-128.

[32] Hassan, O. Morgan, K. and peraire, J. (1990) "*An implicit Finite Elemente Method for high speed flows*", 28th. Aerospace Sciences Meeting, Jan 8-11. Reno.Nevada. AIAA(90-0402).

[33] Hesser, J., Manner, R. (1991). "*Towards an optimal mutatuion probability in genetic algorithms*". Parallel Problem Solving from Bature, 23-32. Vol. 496 of Lecture Notes in computer Science, Springer, Berlin. H.P. Schwefel, R. Manner (Eds.).

[34] Holland, H.J. (1975) "*Adaptation in Natural and Artificial Systems*" University of Michigan Press, Ann Arbor, MI.

[35] L'Ecuyer, P. (1988) "*Communications of the ACM*" Vol.31, pp. 742-774.

[36] Lions, J.L. (1968) "*Controle Optimal des systemes gouvernes par des equations aux derivees partielles*". Dunod, Paris.

[37] Marsaglia, G. "*Comments on the perfect uniform random number generator*", Unpublished Notes, Wash.. S.U.

[38] Marsaglia, G. (1994), private e-mail comm.

[39] Metropolis, N., Rosenbluth, A. Rosenbluth, M., Teller, A. and Teller, E. (1953) Journal of Chemical Physics (21) (1087-1092).

[40] Meyer, C., Matyas, S. (1982) "*Cryptography: A New Dimension in Computer Data Security*" John Wiley and Sons.

[41] Michalewicz, Z. (1994) "*Genetic Algorithms + Data Structures = Evolution Programs.*" springer Verla. (Second, Extented Edition). New York.

[42] Miller, B.L. and Goldberg, D.E. (1995) "*Genetic Algorithms, Tournament Selection, and the Effects of Noise*". IllGAL Report no. 95006, July 1995, University of Illinois at Urbana-Champaign, Illinois Genetic Algorithm Laboratory, Urbana, USA.

[43] Mühlenbein, H., Schlierkamp-Voosen, D. (1993) "*Predictive Models for the Breeder Genetic Algorithm*": I-Continuos Parameter Optimization. Evolutionary Computation, (1), (25-49).

[44] Mühlenbein, H., Gorges-scleuter, Kramer O. (1988) "*Evolution Algorithms in Combinatorial Optimization, Parallel computing*", (7), (65-88).

[45] Mühlenbein, H. (1995) "*Genetic Algorithms in Egineering and Compu ter Science* ". Chapter IV, (191-201). John Wiley & Sons. Winter, G., Périaux , J., Galán, M. and Cuesta, P (Eds.).

[46] Naylor, W.C. (1995) private e-mail comm.

[47] Park, S.K. and Miller K.W. (1988) "*Communications of the ACM*", (31), (1192-1201).

[48] Rawlins G.J.e. (1991) "*Introduction in Foundations of Genetic Algorthms*". Morgan Kaufmann Publishers. Rawlins G.J.E. (Ed.).

[49] Schaffer, J.D., Caruana, R.A., Eshelman, L.J., Das, R. (1989) "*A study of control parameters affecting on line performance of genetic algorihms for function optimization*". Proceedings of the Third International Conference on Genetic Algorithms and their Applications, 56-54. Morgan Kaufmann jPublishers, San Mateo, CA. Schaffer, J.D. (Ed.).

[50] Schwefel, H.P. (1981) "*Numerical Optimization for Computer Models*", John Wiley, Chichester, U.K.

[51] Smith, S.F. (1993) "*Adaptively Resizing Populations: An Algorithm and Analysis*" Proceedings of the Fith International Conference on Genetic Algorithms. Morgan Kaufmann Publishers, Los Altos, CA. Forrest, S. (Ed.).

[52] Syswerda, G. (1991)"*A study of Reproduction in Generational and Steady-State Genetic Algorithms*", (1), (100-103). Foundations of Genetic Algorithms. Morgan Kaufmann Publishers. G.J.E. Rawlins (Ed.)

[53] Syswerda, G. (1989) "*Uniform Crossover in Genetic Algorihms*". Proceedings of the Third International Conference on Genetic Algorithms and their Applications, (2-9). Morgan Kaufmann Publishers, San Mateo, CA. Schaffer, J.D. (Ed.)

[54] Thierens, D. and Goldberg, D.E. (1994) "*Convergence models of genetic algorithm selection schems*". Parallel Problem Solving from Nature-PPSN III, (119-129), Berlin, Springer-Verlag. Davidor, Y., Schwefel, H.P., and Manner, R. (Eds.)

[55] Tinney W.F. and Walker, J.W. (1967) "*Direct solutions of sparse network equations by optimally ordered triangular factorization*", Proc. IEEE, (55), (1801-1809).

[56] Vose, M. (1992) "*Modeling Simple Genetic Algorithms*". Foundations of Genetic Algrithms (2), (63-74). Whitley (Ed.), Vail, CO: Morgan Kaufmann.

[57] Whitley, D. (1988) "*GENITOR: A Different Genetic Algorithms*" Proceedings of the Rocky Mountain Conference on Artificial Intelligence, Denver.

[58] Whitley, D. (1989) "*The Genitor algorithm and selection pressure: Why rank-based allocation of reproductive trials is best*". Proceedings of the Third Conference on Genetic Algorithms and their Applications, (116-121). Morgan Kaufmann Publeshers, San Mateo, CA. Schaffer, J.K. (Ed.).

[59] Whitley, D. (1991) "*Fundamental Principles of Decption in Genetic Search*". Foundations of Genetic Algorithms. Morgan Kaufmann Publishers. G.J.E. Rawlins (Ed.).

[60] Whitley, D., Mathias, K. and Fitzhorn, P. (1991) "*Delta Coding: An Iterative Search Strategy for Genetic Algorithms*", (77-84). Proceedings of the Fourth International Conference on Genetic Algorihms, (24-30). Morgan Kaufmann Publishers, Los Altos, CA. Belew, R., Booker, L. (Ed.).

[61] Whitley, D. (1995) "*Genetic Algorithms in Egineering and Computer Science* ". Chapter X, (191-201). John Wiley & Sons. Winter, G., Périaux, J., Galán, M. and Cuesta, P (Eds.).

[62] Winter, G. Montero, G., Cuesta, P., Galán, M. (1994) "*Mesh generation and adaptive rmeshing by genetic algorithms on transonic flow simulation*", (281-287). Proceedings of Computational Fluid Dynamics, 94. John Wiley and Sons Publisher.

[63] Winter, G., Galán, M., Cuesta, P. and Greiner, D. (1995) "*Genetic Algorithms in Engineering and Compu ter Science* ". Chapter XII, (191-201). John Wiley & Sons. Winter, G., Périaux , J., Galán, M. and Cuesta, P (Eds.).

[64] Winter, G., Galán, M., Périaux, J., Montero, G., Greiner, D. and Mantel, B. (1996) "*Computational Methods in Applied Sciences '96*".(273-283). John Wiley & Sons Désidéri, J.A., Hirsch, C., Le Tallec, P., Oñate, E., Pandolfi, M., Périaux, J. and Stein, E. (Eds.).

[65] Wright, A.H. (1991) "*Genetic Algorithms for Real Parameter Optimization*", Foundations of Genetic Algorithms, (205-218). Morgan Kaufmann Publishers. Rawlins, G. (Ed.).

Blackbox And Non-blackbox Optimization: A Common Perspective

Hillol Kargupta[*]
Computational Science Methods Group
Los Alamos National Laboratory
Los Alamos, NM, 87545

Abstract

The SEARCH (Search Envisioned As Relation & Class Hierarchizing) framework developed elsewhere (Kargupta, 1995; Kargupta & Goldberg, 1995) offered an alternate perspective toward blackbox optimization (BBO)—optimization in absense of domain knowledge. This paper argues that the fundamental concepts are also applicable to non-blackbox optimization (NBBO)—optimization in presence of information about the search domain and objective function. The SEARCH framework investigates the conditions essential for transcending the limits of random enumerative search using a framework developed in terms of relations, classes and partial ordering. This paper reviews some of the main results of that work and describes its generality by considering different popular BBO and NBBO algorithms.

1 Introduction

The desire for solving large, high dimensional optimization problems is ever increasing. Developing techniques to solve large linear optimization problems addresses one aspect. There also exists a large class of problems which may not be classified as linear problems. Optimization problems in which sufficient information about the properties of the objective function is available a priori can be called non-blackbox optimization (NBBO). Apart from NBBO problems, there are also many optimization problems in which little knowledge is available about the properties of the search space. Blackbox optimization (BBO) deals with this extreme case in which little domain knowledge about the problem structure is available. Despite the individual advancements of techniques for solving both BBO and NBBO little progress has been been understanding both of them from a common perspective. This paper makes an effort to do that using the SEARCH (Search Envisioned As Relation and Class Hierarchizing) framework introduced elsewhere (Kargupta, 1995).

SEARCH made an effort to capture the fundamental computations in search and optimization in terms of relations, classes and partial ordering. SEARCH is primarily motivated by the observation that searching for optimal solution in a BBO is essentially a pure *inductive process* (Michalski, 1983) and in absence of any relation among the members of the search space, induction is no better than enumeration (Watanabe, 1969). SEARCH decomposed optimization into (1) relation, (2) class, and (3) sample spaces. SEARCH also identified the importance of searching for appropriate relations in BBO. No BBO algorithm can efficiently solve a reasonably general class of problems unless it searches for relations. Kargupta (1995) also showed that the class of *order-k delineable* problems

[*]The author can be reached at, P.O. Box 1663, XCM, Mail Stop F645, Los Alamos National Laboratory, Los Alamos, NM 87545, USA. e-mail: hillol@lanl.gov

G. Winter Althaus and E. Spedicato (eds.), Algorithms for Large Scale Linear Algebraic Systems, 369–385.

can be solved in SEARCH with sample complexity polynomial in problem size, desired quality and reliability of the solution. These results are relevant for any BBO algorithm. Although the analysis is performed for BBO problems we shall see that the underlying concepts remain unchanged for NBBO problems.

In this paper we first present a brief description of the SEARCH effort. We present some of the main results without going through the detailed derivations. Section 2 introduces the fundamental similarities and differences between BBO and NBBO. Section 3 briefly describes the main concepts and the foundation of SEARCH. Section 4 describes what it means to be an optimization algorithm in SEARCH. This is followed by a discussion on problem difficulty in Section 5. Section 6 specializes the results for sequence representation and identifies the class of order-k delineable problems, that can be solved in polynomial sample complexity in SEARCH. Section 7 further explains the implications of order-k delineable problems. Section 8 presents few case studies. Finally, Section 11 concludes this paper.

2 Optimization: Blackbox and Non-blackbox

An optimization problem is comprised of a search domain and an objective function to be optimized. The objective function of an optimization problem can be defined as,

$$\Phi : \mathcal{X} \rightarrow \mathcal{Y} \tag{1}$$

The objective of a maximization problem is to find some $x^* \in \mathcal{X}$ such that $\Phi(x^*) \geq \Phi(x)$ for all $x \in \mathcal{X}$. In blackbox optimization, the objective function is available as a black box, i.e., for a given decision variable, x in the feasible domain, it returns the function value $\Phi(x)$. No local or global information about the function is assumed. For most of the optimization problems with practical importance the size of the search space grows exponentially with the problem size. Since in BBO no problem knowledge is available the only way to do better than enumerative search is to take few samples from the search domain and guess about the possible location of the optimal solution. Therefore, searching for an optimal solution in BBO is essentially an *inductive process* (Michalski, 1983). It has been well known that in absence of any relation among the members of the search space, induction is no better than enumeration (Watanabe, 1969). A relation is a set of ordered pairs. However, since no problem knowledge is available in BBO, we need to search for the appropriate relations.

On the other hand in non-blackbox optimization some knowledge aboout the objective function and/or search domain is available. For example in linear programming problem the objective function is known to be linear; meaning, for any two given points x_1 and x_2 in the search domain, we know that their objective function values are linearly related to each other. If an objective function is convex we know that for some $0 < \lambda < 1$, the convexity condition $\Phi((1-\lambda)x_1 + \lambda x_2) \leq (1-\lambda)\Phi(x_1) + \lambda\Phi(x_2)$ is satisfied. If the local gradient information is available then we know that how the function value changes locally between any two points. As we see, in NBBO we have access to some additional information that relates a set of points to each other. This information about the relative properties of different points can be interpreted as *relations* using the set theoretic terminology. Therefore, we can say that in NBBO the search for relations is made easier depending on the amount of available domain knowledge.

There exists a huge body of literature for NBBO algorithms (Törn & Žilinskas, 1989). Some popular NBBO algorithms are Bayesian methods (Betrò, 1983), Branch and bound methods. Although the amount of work done on BBO algorithms is relatively smaller, many new algorithms like Genetic algorithms (GAs) (Holland, 1975), simulated annealing (SA) (Kirpatrick, Gelatt, & Vecchi, 1983), tabu search (Glover, 1989) became popular recently for solving BBO problems. Despite the

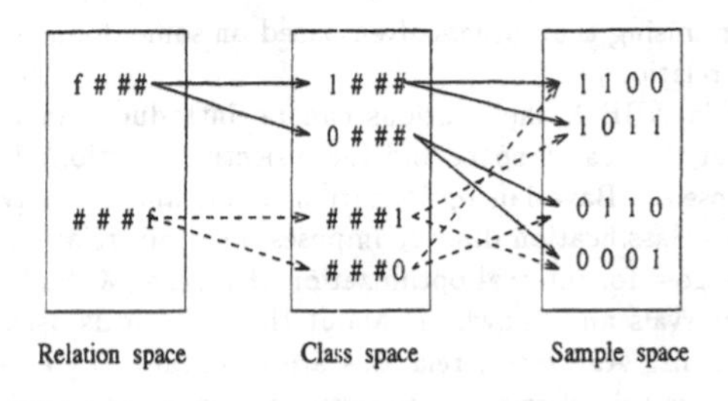

Figure 1: Decomposition of blackbox optimization in SEARCH.

emergence more and more new algorithms, little work has been done to understand the fundamental processes in optimization for both BBO and NBBO from a common ground. The following section introduces the SEARCH framework that takes a small step toward that.

3 The SEARCH Framework

This section presents a brief review of the SEARCH framework. As we noted in the previous section that the main difference between the BBO and NBBO problems is the availability of information about relations among the members of the search space. The search for relations decreases as the availability of such information increases. In order to maintain its generality the SEARCH framework considers relation search as a component of optimization. SEARCH offers a perspective of optimization in a probabilistic and approximate sense in terms of relations, classes and partial ordering. Section 3.1 describes the general concepts. Section 3.2 discusses the bound on sample complexity in SEARCH.

3.1 Foundation

The foundation of SEARCH is laid on a decomposition of the blackbox search problem into relation, class, and sample spaces. As mentioned ealier, a relation is a set of ordered pairs. For example, in a set of cubes, some white and some black, the color of the cubes defines a relation that divides the set of cubes into two subsets—set of white cubes and set of black cubes. Consider a 4-bit binary sequence. There are 2^4 such binary sequences. This set can be divided into two classes using the equivalence relation[1] $f\#\#\#$, where f denotes position of equivalence; the # character matches with any binary value. This equivalence relation divides up the complete set into two equivalence classes, 1### and 0###. The class 1### contains all the sequences with 1 in the leftmost position and 0### contains those with a 0 in that position. The total number of classes defined by a relation is called its index. The order of a relation is the logarithm of its index with some chosen base. In a BBO problem, relations among the search space members are often introduced through different means, such as representation, operators, heuristics, and others. The above example of relations in binary sequence can be viewed as an example of relation in the sequence representation. In a sequence space of length ℓ, there are 2^ℓ different equivalence relations. The search operators also define a set of relations by introducing a notion of neighborhood. For a given member in the search space, the search operators define a set of members that can be reached by one or several application of the operators. This introduces relations among the members. Heuristics identifies a subset of the

[1] An equivalence relation is a relation that is reflexive, symmetric, and transitive.

search space as more promising than others often based on some domain specific knowledge. Clearly this can be a source of relations.

On the other hand in NBBO, the relations can be introduced in a more direct way by using domain knowledge about the search space and the objective function. For example, Perttunen and Stuckman (1990) proposed a Bayesian optimization algorithm that divides the search space into Delaunay triangles. This classification directly imposes a certain relation among the members of the search space. The same goes for interval optimization (Ratschek & Voller, 1991), where the domain is divided into many intervals and knowledge about the problem is used to compute the likelihood of success in those intervals. As we see, relations are introduced by every search algorithm, either implicitly or explicitly. The role of relations in optimization is very fundamental and important.

Relations divide the search space into different classes and the objective of sampling based non-enumerative optimization is to detect those classes that are most likely to contain the optimal solutions. To do so requires constructing a partial ordering among the classes defined by a relation. The classes are evaluated using samples from the search domain and a *class comparison statistic* is used for comparing different classes. For a given *class comparison statistic* $\leq_{\mathcal{T}}$ and some number M, a relation is said to *properly delineate* the search space if the class containing the optimal solution is within the top M classes, when the set of all classes defined by the relation are ordered using $\leq_{\mathcal{T}}$. This basically means that if a relation satisfies the delineation constraint then, given sufficient samples, the relation will pick up the class containing the optimal solution within the top M ranked classes. If a relation does not satisfy this, then the relation leads to wrong decision and as a result success in finding the optimal solution is very unlikely.

A particular relation may not satisfy the delineation constraint for different problems, different class comparison statistics, and different values of M. One relation may work for a particular case and may fail to do so for a different setting. Therefore, any algorithm that aspires to be applicable for a reasonably general class of problems, must search for appropriate relations. Determining whether or not a relation satisfies this delineation constraint requires decision making in absence of complete knowledge. For a given relation space Ψ_r, an optimization algorithm must identify the relations that properly delineate the search space with certain degree of reliability and accuracy. This requires comparing one relation with another using a *relation comparison statistic* and constructing a partial ordering among them.

An optimization algorithm in SEARCH cannot be efficient if it needs to consider relations that divide the search space in classes, with the total number of classes growing exponentially with the problem dimension. For example, in an ℓ-bit sequence representation, if there is a class of problems which requires considering the equivalence relations with $(\ell - 1)$ fixed bits then there is a major problem. This relation divides the search space into $2^{\ell-1}$ classes and we cannot solve this problem in complexity polynomial in ℓ. However, in optimization the ultimate objective is to identify the optimal solution which basically defines a singleton class. The smaller the cardinality of the individual classes, the larger the index of the corresponding relation. So we need the higher order relations for finally identifying the optimal solution, but we cannot directly evaluate them since their index is large. The solution is to limit our capability and realize that we can only solve those problems which can be addressed using low order relations and when high order relations are decomposable to those low order relations. This means that the information about low order relations can be used to evaluate the higher order relations. Consider the following example. Let r_0 be a relation that is logically equivalent to $r_1 \wedge r_2$, where r_1 and r_2 are two different relations; the sign $\wedge$ denotes logical AND operation. If either of r_1 or r_2 was earlier found to properly delineate the search space, then the information about the classes that are found to be bad earlier can be used to eliminate some classes in r_0 from further consideration. This process in SEARCH is called *resolution*. Resolution basically evaluates the relations of higher order using the information gathered by direct evaluation of lower

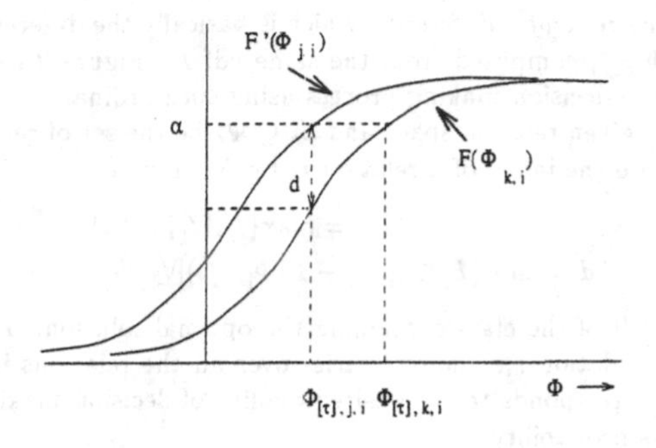

Figure 2: Fitness distribution function of two classes $C_{j,i}$ and $C_{k,i}$.

order relations.

The above description gives a brief informal overview of the SEARCH framework. As we saw, SEARCH addresses optimization on three distinct grounds: (1) relation space, (2) class space, and (3) sample space. Figure 1 shows this fundamental decomposition in SEARCH. The major components of SEARCH can be summarized as follows:

1. classification of the search space using relations;
2. sampling;
3. evaluation, ordering, and selection of better classes;
4. evaluation, ordering, and selection of better relations;
5. resolution.

A detailed description of each of these processes can be found elsewhere (Kargupta, 1995). In the following part of this section we consider the expression for sample complexity in SEARCH derived elsewhere (Kargupta, 1995) and define the class of order-k delineable problems that can be efficiently solved in SEARCH.

3.2 Sample complexity in SEARCH

For a given relation space, and an algorithm in SEARCH, it is possible to derive the bound on sample complexity for the desired quality and reliability of decision making. Defining an algorithm in SEARCH first requires specifying class and relation comparison statistics. Although, most of the existing optimization algorithms do not explicitly define them, SEARCH does so in order to quantify and understand the role of decision making in the relation and class spaces. Kargupta (1995) considered distribution free ordinal comparison statistics. In an ordinal comparison statistic, two distributions are compared on the basis of some chosen percentile. Figure 2 shows the cumulative distribution function F' and F of two arbitrary subsets $C_{j,i}$ and $C_{k,i}$, respectively. Indices j, k represent the two classes defined by some relation r_i. When these two classes are compared on the basis of the α quantile, then we say $C_{j,i} \leq_\alpha C_{k,i}$, since $\Phi_{[\tau],j,i} \leq \Phi_{[\tau],k,i}$; $\Phi_{[\tau],j,i}$ and $\Phi_{[\tau],k,i}$ are the solutions of $F'(\Phi_{j,i}) = \alpha$ and $F(\Phi_{k,i}) = \alpha$, respectively. Let us define

$$d = F(\Phi_{[\tau],k,i}) - F(\Phi_{[\tau],j,i}).$$

The variable d defines the *zone of indifference*, which is basically the difference in the percentile value of $\Phi_{[\tau],k,i}$ and that of $\Phi_{[\tau],j,i}$ computed from the same cdf F. Figure 2 clearly explains this definition. We can quantify the decision making process using such ordinal class and relation comparison statistics. Let Ψ_r be the given relation space and $S_r \subseteq \Psi_r$ be the set of relations needed to solve the given problem. We denote the index of a relation r_i by N_i. Define,

$$N_{\max} = \max\{N_i | \forall r_i \in S_r\}$$
$$d' = \min\{F(\Phi_{[\tau],*,i}) - F(\Phi_{[\tau],j,i}) | \forall j, \forall i\}.$$

where $F(\Phi_{[\tau],*,i})$ is the cdf of the class containing the optimal solution. The index j varies over all the classes defined by a relation r_i. Index i varies over all the relations in Ψ_r. If d^* is a constant such that $d' \geq d^*$, that corresponds to the desired quality of decision making in the class space, the bound on overall success probability is,

$$[(1 - 2^{nH(\alpha)}(\alpha - d^*)^{\alpha n})^{(N_{\max} - M_{\min})}]^{\|S_r\|} q_r \geq q \tag{2}$$

where $H(\alpha)$ is the binary entropy function, $H(\alpha) = -\alpha \log_2 \alpha - (1-\alpha)\log_2(1-\alpha)$. From this we can derive the bound on overall sample complexity,

$$\text{SC} \leq \frac{N_{\max}\|S_r\| \log(1 - \left(\frac{q}{q_r}\right)^{\frac{1}{\|S_r\|(N_{\max} - M_{\min})}})}{-d^*}. \tag{3}$$

where q is the overall desired success probability in SEARCH and q_r is the desired success probability in the relation space. $M_{\min}$ is a constant that depends on the memory used by the algorithm. The success probability in the relation space depends on the appropriate decision making and the statistic defined for comparing relations with each other. Similar expression for the computational complexity in the relation space can be derived as shown elsewhere (Kargupta, 1995).

Inequality 3 presents the upper bound on the sample complexity in optimization. As we increase the success probability in the relation space, the overall success probability in the combined relation and class spaces increases. The sample complexity should therefore decrease as success probability in the relation space increases. This also shows that SC decreases with increase in q_r. Note that the ratio $\left(\frac{q}{q_r}\right)^{\frac{1}{\|S_r\|(N_{\max} - M_{\min})}}$ approaches 1 in the limit as $\|S_r\|(N_{\max} - M_{\min})$ approaches infinity. Therefore, SC grows at most linearly with the maximum index value $N_{\max}$ and the cardinality of the set S_r. Recall that d^* defines the desired region of indifference; in other words, it defines a region in terms of percentile within which any solution will be acceptable. The sample complexity decreases as the d^* increases. Kargupta (1995) also showed that when no relations are considered, this expression points out that the sample complexity will be of the order of the size of search space; in other words search will be no better than enumeration. For a given relation space and a class of problems that can be solved considering a bounded number of relations from that space, inequality 3 gives the bound on sample complexity for desired quality and reliability of the decision making. This expression is valid for any finite blackbox search space. The following section identifies the different components of an optimization algorithm in SEARCH.

4 An Optimization Algorithm In SEARCH

The SEARCH framework decomposed optimization in a given search domain into different components such as the relation space, class space, and the sample space. Searching in these spaces requires some fundamental tools, such as some comparison statistics and a perturbation operator for generating samples. In this section, we list them together and project a complete picture of what it means to be an optimization algorithm in SEARCH.

1. SEARCH views the solution domain through relations, classes, and samples. An algorithm in SEARCH should be provided with a set of relations Ψ_r. Representation in genetic algorithms (GAs), perturbation operators of simulated annealing (SA), and the neighborhood heuristic in k-opt algorithm are some examples of different sources of relations.

2. SEARCH also needs explicit storage for processing relations, classes, and samples. The maximum possible values of M_r[2] and M_i determine the size of the memory for storing relations and classes respectively. In a genetic algorithm the population serves as the memory for all three of these spaces. In SA the evaluation of different classes defined using the perturbation operators is distributed over time and only one sample is taken at a time. The state of the SA algorithm serves as the memory for the sample space.

3. Two statistic measures for comparing classes and relations are required. A selection operator is used for comparing classes in the simple GA. The simple GA does not really search for better relations. On the other hand, in simulated annealing, the Metropolis criterion is used for comparing two states; this can be viewed as a class comparison statistic.

4. A perturbation operator, $\mathcal{P}$ is required for generating samples. In GA, crossover and mutation generate new samples. SA makes use of a neighborhood generator for sample generation.

5. Accepting criterion of success probability, q, and q_r are necessary. Almost every practical application of GA and SA either implicitly or explicitly makes use of an acceptance criterion for success. Neither simple GA nor SA actually searches for better relations. Neither of them has any explicit criterion like q_r.

6. Required precision in solution quality, d^* is required. Again, in practice, both GA and SA somehow introduce the factor controlling the desired solution quality. In SEARCH, this is introduced in an non parametric way. Examples of parametric approaches may be found elsewhere Goldberg, Deb, and Clark (1993, Holland (1975).

SEARCH provides a common ground for developing new sampling based adaptive optimization algorithms in the future. Regardless of the motivation and background, any blackbox optimization algorithm should clearly define each of the above listed components. An optimization algorithm should define how it processes relations, classes, and samples. It should state its relation and class comparison statistics. The following section addresses this important aspect of optimization—problem difficulty.

5 Problem Difficulty

SEARCH presents an alternate perspective of problem difficulty in optimization. In this section we identify the main dimensions of problem difficulty in SEARCH and precisely define a characterization of difficult problems in SEARCH.

The expression for the sample complexity developed in the previous section immediately leads to identifying different facets of problem difficulty in SEARCH. As we saw from Inequality 3 the sample complexity grows linearly with the size of the set of relations considered to solve the problem, S_r. Often this size depends on the "size" of the problem; the word "size" defines a parameter ℓ that bounds the search domain. In a sequence representation with constant alphabet size, the length of

[2] M_r, the relation space memory comes into the picture when the memory is not large enough to process all the relations in Ψ_r

the sequences needed to represent the search space may be an example of such a size parameter. This finally sets the stage for introducing problem difficulty in SEARCH.

Definition 1 (Problem difficulty in SEARCH) *Given an optimization function* $\Phi : X \rightarrow \Re$ *and a set of relations* Ψ_r*, we call a problem difficult for an algorithm if the total number of samples needed to find the globally optimal solution grows exponentially with* ℓ, q, q_r, $1/d^*$, *and* $1/d_r^*$.

The size of the problem is represented by ℓ; q denotes the bound in the overall decision success probability in choosing the right classes; $1/d^*$ defines the quality of the desired solution.Both q and $1/d^*$ together can be viewed as representing the overall accuracy and the quality of the solution found; q_r is the bound in success probability in choosing the right relations, and $1/d_r^*$ represents the desired quality of the relations.

The above definition of problem difficulty in SEARCH can be physically interpreted into the following items:

1. growth of the search space along problem dimension;

2. inadequate source of relations and decision making in relation space;

3. inaccurate decision making in choosing classes;

4. quality of the desired solution and relations.

This gives a general description of the SEARCH perspective of problem difficulty. The following section brings us closer to the ground by specializing the framework for sequence representation. We identify a class of problems in sequence representation that can be solved in polynomial sample complexity in SEARCH.

6 Sequence representation and the class of order-k delineable problems

Sequence representation is used in many evolutionary optimization algorithms. We therefore choose this for exploring the class of problems that can be efficiently solved.

A sequence representation can be defined as $I : X \rightarrow \Lambda^{\ell}$, where Λ is the alphabet set. This sequence representation induces a set of equivalence relations, $\Psi_r = \{f, \#\}^{\ell}$, where f indicates values that must match for equivalence and # is a wild character that matches any value. The cardinality of the set of all such equivalence relations $\|\Psi_r\| = 2^{\ell}$.

Definition 2 (Order k delineable problems) *Let us define a subset of* Ψ_r *containing every order-k relation as follows:* $\Psi_{\{o(r) \leq k\}} = \{r_i : o(r_i) \leq k \;\&\; r_i \in \Psi_r\}$. *For a given class comparison statistic* $\leq_{\mathcal{T}_i}$, *a problem is order-k delineable if there exists a subset* $\Psi' \subseteq \Psi_{\{o(r) \leq k\}}$ *and at least one member of* Ψ' *has an order equal to* k, *such that its every member* r_i *satisfies the delineation constraint with memory size* M_i *and the size of the intersection set,*

$$\mathcal{G} = \bigcup_{a_1, a_2, \cdots a_k} \bigcap C_{[a_1],i} C_{[a_2],i} \cdots C_{[a_k],i},$$

is bounded by a polynomial of ℓ, $\rho(\ell)$. *The indices* $a_1, a_2, \ldots a_k$ *can take any value in between* 1 *and* M_i.

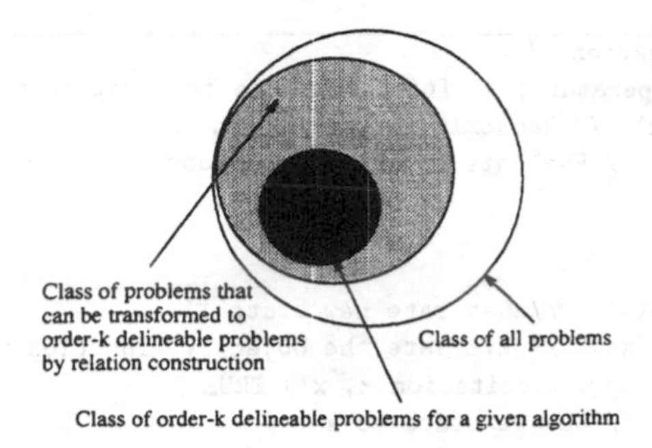

Figure 3: Optimization problems from the delineability perspective.

It has been shown elsewhere (Kargupta, 1995) that this class of problems can be solved in sample complexity polynomial in q, q_r, $1/d^*$, $1/d^*_r$, and the problem size ℓ. To achieve an overall success probability of q, the required sample complexity is,

$$SC \leq \|\Lambda\|^k k \frac{\log(1 - \left(\frac{q}{q_r}\right)^{\frac{1}{(\ell-k+1)(\|\Lambda\|^k - M_{\min})}})}{-d^*} + \rho(\ell). \tag{4}$$

When $q/q_r << 1$, this can be approximated as,

$$SC \leq \|\Lambda\|^k k \frac{\left(\frac{q}{q_r}\right)^{\frac{1}{(\ell-k+1)(\|\Lambda\|^k - M_{\min})}}}{d^*} + \rho(\ell)$$

This basically says that the problems that can be solved using a polynomially bounded number of relations can be efficiently solved in SEARCH. Note that this class of problems is fundamentally defined in terms of relation space. Note that, a problem may be order k-delineable in one relation space but fail to be for another relation space. Therefore when the relation space is already chosen by fixing the sequence representation, an optimization problem can be solved efficiently if it is order-k delineable in that relation space.

Before we conclude this review on SEARCH, let us revisit the issue of order-k delineability in order to clear up our current objectives and future directions for designing optimization algorithms.

7 Implications Of Order-k delineability

As we saw in the previous section, the notion of order-k delineability presents a picture of the general class of optimization problems from the perspective of an algorithm. In SEARCH, defining an optimization algorithm requires specifying the relation space, class comparison statistic, and the constant M that defines how many "top" classes will be picked up. Therefore, by definition an algorithm in SEARCH specifies the class of order-k delineable problems. For a chosen class comparison statistic and M, the relation space restricts the class of order-k delineable problems for an algorithm. Changing the relation space by constructing new relations may convert a non-order-k delineable problem to an order-k delineable one. For some problems finding such transformation by constructing new relations may be possible in sample complexity, polynomial in problem size, reliability, and accuracy of the solution. Clearly, there may exist a class of non-order-k delineable

```
/* Initialization */
T = High_temperature; // Initialize the temperature to a high value
Initialize(x); // Randomly initialize the state
Evaluate(x); // Evaluate the objective function value
{
Repeat
{
    Generate(x'); // Generate new state
    Evaluate(x'); // Evaluate the objective function value
    If ( Metropolis_criterion(x, x') TRUE )
      x = x' // Change state to x'
}
Until (Equilibrium is reached)
  Decrease(T); // Decrease the temperature
}
Until ( T < T_min Or (termination criterion TRUE) )
```

Figure 4: A pseudo-code for simulated annealing.

problems, that can be transformed to order-k delineable problems in polynomial sample complexity. Figure 3 shows a schematic description of this classification of optimization problems.

It is important to note that, membership of a problem in the class of order-k delineable problems does not necessarily guarantee that the algorithm will solve that problem. It only says that the problem is " efficiently solvable" in the chosen relation space, class comparison statistic, and M. The algorithm needs to perform adequate sampling and make decisions with high confidence in the relation and class spaces in order to find the desired quality solution. Therefore, the first step of an algorithm should be to make sure it can solve its own order-k delineable class of problems. That will define the first milestone. The next step should be to introduce mechanism for new relation construction and investigate what kind of problems can be dynamically transformed to order-k delineable problems efficiently. Unfortunately, there hardly exists any algorithm that can adequately guarantee the capability of solving even its order-k delineable problems. More work is needed to develop optimization algorithms that follow the systematic decomposition of SEARCH.

In the following sections we shall consider some popular BBO and NBBO algorithms and describe their components from the SEARCH perspective to demonstrate the generality of this approach.

8 SEARCH And Simulated Annealing

Like many other algorithms, simulated annealing (SA) algorithm does not explicitly consider the relations. Therefore, the projection of SA into the SEARCH framework depends on our perspective toward SA as well. Since relations can be defined in many ways, when the relation space is not explicitly specified, identifying it leaves room for speculation. The original version of SA does not emphasize representation. Moreover, the random neighborhood generation operator does not pay enough consideration to the relations and classes defined by the chosen representation.

In this section, we therefore choose to view SA as a processor of relations and classes defined by the neighborhood generation operator. The following part of this section briefly discusses different counterparts of SEARCH in the SA.

- **Relation space**: A state x_i and the neighborhood generation operator ($\mathcal{P}$) are the two ingre-

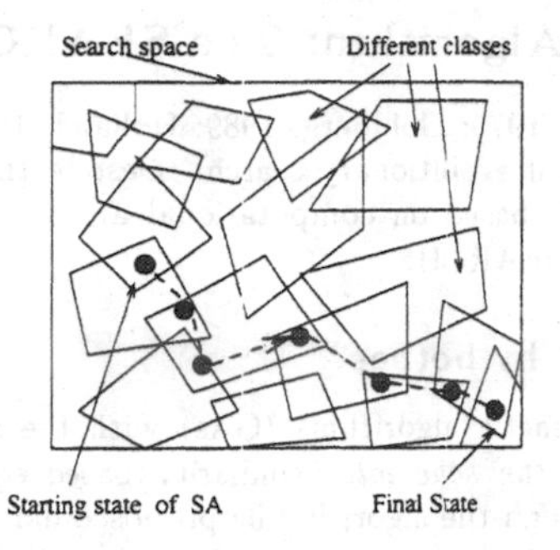

Figure 5: The SEARCH perspective of SA.

dients of the relations processed by the SA. For a given state x_i, the neighborhood generation operator defines a set of states that can be reached in certain number of steps (s) from x_i. This defines a relation among a certain subset of the search space. Therefore, a relation r_i in SA can be specified by $(x_i, \mathcal{P}, s)$.

- **Class space**: The relation $(x_i, \mathcal{P}, s)$ divides the search space into two classes—(1) the set of states that can be reached from x by applying $\mathcal{P}$ for s number of times and (2) the rest of the search space. This defines the class space for a given relation. Let us denote the first class by $C_{1,i}$ and the second by $C_{2,i}$.
- **Sample space**: The SA processes only one sample at a time. The sample represents the state of the algorithm.

Searching for the optimal solution in SEARCH also requires different comparison statistics and resolution for combining the features of different classes from different relations. The following discussion points out their counterpart in SA.

- **Relation and class comparison statistics**: Since SA does not explicitly defines the relations and classes, only one statistic, defined by the *Metropolis criterion*, is used for serving both purposes. This comparison statistic varies as the *temperature* changes.

- **Resolution**: Consider the two relations $(x_1, \mathcal{P}, s)$ and $(x_2, \mathcal{P}, s)$, where x_1 and x_2 are two arbitrary states from the search space. Let us denote the set of states that can be reached from x_1 and x_2 by applying $\mathcal{P}$ for s times by $C_{1,1}$ and $C_{1,2}$, respectively. Let x_i be the current state of SA and x_{i+1} be the next state. Now if x_1 and x_2 are such that the $x_i \in C_{1,1}$ and $x_{i+1} \in C_{1,2}$, then the next state, x_{i+1}, is basically a sample from the intersection set of the two classes $C_{1,1}$ and $C_{1,2}$. Generating samples from the intersection set of classes is essentially what resolution does.

The above discussion presents a perspective of SA in the light of SEARCH. Figure 5 pictorially depicts this perspective of SA. This figure schematically shows the trajectory of SA within the overlapping classes. As we mentioned earlier, this presents only one possible way to define classes and relations in SA. Since SA does not explicitly define them, different possibilities may be speculated. The following section considers another popular BBO algorithm, namely the genetic algorithms and describe the underlying computations in the light of SEARCH.

9 Simple Genetic Algorithm: The SEARCH Perspective

Genetic algorithms (De Jong, 1975; Goldberg, 1989; Holland, 1975) are BBO algorithms designed primarily motivated by natural evolutionary search. Despite their biological motivation, Holland (1975) presented the concepts based on computational arguments. In this section we revisit the relevant issues in the light of SEARCH.

9.1 Schema analysis: Why bother?

Holland (1975) introduced genetic algorithms (GAs) with the spirit of searching by equivalence class processing. He defined the *schemata*—similarity based equivalence classes in the sequence space (Radcliffe, 1991). Although the algorithm he proposed did not explicitly consider equivalence classes, he visioned GAs as an implicit processor of schemata at an abstract level. Since, relations play a critical role in making BBO efficient, it is important to study the processing of classes in any BBO algorithm. Clearly, Holland's approach using schema analysis complied with the emphasis on the relations defined by the representation. During the last decade as GAs gained popularity many different models of SGA (Srinivas & Patnaik, 1993; Nix & Vose, 1992) came up. Although these models can present an exact picture of the dynamics of the population members from generations to generations, the notion of class processing needs to be incorporated in order to use these models for designing efficient GAs. Jones (1995) offered a different kind of relation source in GAs. He defined a landscape based on crossover perturbation. Although it does not use any schema analysis, counterparts of schemata in the so called *crossover landscape* needs to be introduced for comparing the performance of GAs with that of random enumerative search. As we see, since relations and classes are what make a BBO algorithm transcend the limits of random search, any model of GAs that aspire to provide insight for designing efficient GAs must project a picture of the class processing. The following section discusses another rule of thumb believed by a portion of the GA research community and study its utility in the light of SEARCH.

9.2 Building block hypothesis: The bottom-line

The schema theorem clearly says that low order, short, above average equivalence classes are expected to be explored more. The term "short" captures the idea that one point crossover can efficiently generate samples and compute set intersections only for those equivalence relations, in which the fixed(f) bits are closer to each other. Schema theorem has been extrapolated to hypothesize that the success of a GA may depend on the increase in the proportion of members in low order, short, better classes (termed *building blocks* in (Goldberg, 1989)) and the computation of the intersection of these good classes using the crossover operator. This is often called as *building block hypothesis*. Again, this hypothesis does not necessarily describe the behavior of a GA, since it is partially based on the schema theorem, which is a simple first order bound. However, we need to pay little more attention to appreciate the rationale behind this hypothesis. It has been shown elsewhere (Kargupta, 1995) that the bottom-up organization of the blackbox search has computational significance. Although algorithms, which do not satisfy this criterion, do not necessarily have to be of exponential complexity, but they will have a large constant in front of the bounding polynomials. This desirable feature of the polynomial complexity search exactly resonates with the essence of building block hypothesis. The bottom-up organization requirement of the blackbox search says that the low order equivalence relations need to be evaluated first, which is also suggested by building block hypothesis. Therefore, the building block hypothesis should be viewed as a desirable feature for a polynomial complexity GA rather than something that describes the behavior of all GAs.

Since GAs work from a population of samples, relations can be evaluated in parallel. The following section points out that the relation, class, and sample spaces are defined by a single population in GA.

9.3 Implicit definition of relation, class and sample spaces

In simple GA the relation, class, and the sample spaces are defined together in a single population. As we noted earlier during the development of SEARCH, relation and class spaces require distinct decision makings and interference among them may cause decision error. The SGA uses a single selection operator for making decisions in each of these spaces. As a result decision making is very noisy in SGA. This is one among the major bottlenecks of the SGA.

9.4 Crossover as resolution of SEARCH

In SGA, crossover exchanges parts of parent strings. If two partially good solutions can be detected and combined to produce a better solution crossover can be effective in finding good solutions. However, the condition of detecting partially good solution can only be satisfied when the optimization problem is delineable at a certain order in the chosen representation. The resolution operation of SEARCH that computes intersection of different classes can also be viewed as the process of combining class features. Therefore crossover can be fundamentally viewed as a way to implement resolution. However, in simple GA the relations and classes are processed implicitly and crossover is randomized with no direct bias toward good classes. The bias toward better classes is introduced through selection and random operation of crossover probabilistically hopes for accomplishing the right resolution among better class features.

9.5 Linkage: Search for relations

As we saw earlier, search for proper relations that satisfy the delineation requirement plays an important role in the blackbox search.

The simple GA considers only a small fraction of relations defined by the representation. A simple GA with one-point crossover (De Jong, 1975) favors those relations in which positions in sequence space defining equivalence are closer to each other and neglects those relations that contain equivalence defining positions far apart. One-point crossover also fails to generate samples for the intersection set of two equivalence classes in which fixed bits are widely separated. For example, in a 20-bit problem, single-point crossover is very unlikely to generate a sample from the intersection set of $1\#\#\cdots\#$ (first bit is fixed) and $\#\cdots\#1$ (last bit is fixed). In biological jargon, this is called the *linkage problem*. Unfortunately, this is a major bottleneck of SGA. Although Holland (1975) realized the importance of solving this problem and suggested use of the *inversion* operator (Holland, 1975), it has been shown elsewhere (Goldberg & Lingle, 1985) that inversion is very slow and unlikely to solve this problem efficiently. One-point crossover is not the only type to suffer from this problem. Uniform crossover is another kind of crossover (Syswerda, 1989) often used in the simple GA. In uniform crossover, the exchange of bit values among the two parent strings takes place based on a randomly generated binary mask string. If the value of this mask string at a particular locus is 1, the corresponding bits in the parent strings get exchanged; otherwise they do not. Unlike one-point crossover, uniform crossover does not have any preference bias toward the closely spaced partitions. Since the relation space and the sample space are combined, random perturbation of the sample strings also result in disrupting proper evaluations of the relations. Uniform crossover should also fail to accomplish proper search in the relation space. In fact, this is exactly what Thierens and Goldberg (1993) reported. Their analysis and experimental results showed that the sample complexity grows exponentially with the problem size for solving bounded deceptive problems (Thierens & Goldberg,

1993) using a simple GA with uniform crossover. This discussion clearly points out that the search in the relation space is very poor in the case of a simple GA with either one-point or uniform crossover.

9.6 Discussion

The simple GA has many interesting features. The first important aspect is that GA emphasizes the role of representation in search and representation is one possible way to define a rich source of relations. Searching by constructing and ordering equivalence relations is very natural in GA. Apart from this, crossover provides an interesting tool to implement sample generation process from the intersection set of different equivalence classes. Probably one of the most interesting observations that Holland made in his book (1975) is the idea of implicit parallelism. Although this conjecture was not thoroughly laid in terms of computational arguments, the SEARCH framework confirms this observation in a quantitative manner. Despite these interesting features, the simple GA has several major problems. The main bottlenecks of the simple GA are listed below:

- Poor search for relations;
- Lack of precise mechanism for implicit parallelism;
- Relation, class, and sample spaces are combined.

Now that we have discussed the computational aspects of two BBO algorithms like SA and SGA in the light of SEARCH, it will be interesting to consider an NBBO in the light of SEARCH. The following section presents the SEARCH perspective of Bayesian optimization algorithms.

10 SEARCH and Bayesian Optimization

Bayesian algorithms try to solve an optimization problem by gradually developing an approximation of the objective function in an implicit way. The approximation of the objective function is captured by using a random variable. Like any other adaptive sampling-based search techniques Bayesian algorithms make use of sample points from the search space. These algorithms can also be viewed in the light of SEARCH and the computational results derived in this section are also valid for this approach. In the following part of this section we discuss this possibility.

Although different Bayesian algorithms exists (Betrò, 1983) in the literature, I shall confine myself to the method by Kushner (1963) for illustrating the fundamental characteristics of this class of algorithms, following Jones and Stuckman (1992). Kushner's (1963) univariate method divide the entire search space into different line segments. It then tries to capture the distribution of the objective function values within these line segments using the mean and variance of a random variable $X(s)$. The objective function is evaluated at the end points of a segment and a point within the segment, s' is chosen that maximizes the probability of improving the current maximum. Computation of this point is based on second order approximation and requires the mean $E[X]$ and the variance $\mathrm{Var}[X]$. For example, if $[a, b]$ defines a segment, then the probability that $X(s)$ will exceed the current maximum, $\Phi_{\max}$ by an amount ϵ can be expressed as (Jones & Stuckman, 1992),

$$Pr[X(s) \geq \Phi_{\max} + \epsilon] = 1 - \Xi\left(\frac{\Phi_{\max} + \epsilon - E[X(s)]}{\sqrt{\mathrm{Var}[X(s)]}}\right)$$

where $s \in [a, b]$ and $\Xi(\cdot)$ is a monotonically increasing function. Once a point is chosen that maximizes this probability, the segment can be further divided based on this new point. The process is continued in the same way, each iteration including a new point and further subdividing the segment. Following (Jones & Stuckman, 1992) we summarize this process in Figure 6.

```
/* Initialize */
Set the complete search space to a single segment.
Repeat
{
1.   Evaluate the objective function at the endpoints of the segment.
2.   Compute the point, s' that maximizes the probability of increasing
     the objective function value by more than ε.
3.   Evaluate Φ(s') and update the estimate Φ_max.
4.   Define the new set of segments based on previous end points and s'.
}
Until (termination criterion TRUE)
```

Figure 6: A pseudo-code for Kushner's Bayesian optimization algorithm.

In the following part of this section we identify the main correspondence between Bayesian optimization with SEARCH.

- **Relation space:** The relation space depends on how the algorithm chooses to classify the search space. Kushner's (1963) method chooses to construct the line segments using the point s' that maximizes the probability of increasing the objective function value. Perttunen and Stuckman (1990) extended Kushner's (1963) method to multi-dimension case and suggested *Delaunay triangulation* of the search space for classification. Again, Delaunay triangulation defines the relation space for the algorithm by Perttunen and Stuckman (1990).

- **Class space:** The class space is defined by the segments and Delaunay triangles in case of Perttunen's algorithm.

- **Sample space:** Kushner's algorithm evaluates the objective function at the end points of the segments. This defines the sample space

- **Relation and Class comparison statistics:** The classes in Bayesian optimization are compared based on the probability of increase in objective function value. The maximum probability values are computed and the class with the highest value is selected. The search for relations is not very much emphasized. Segmentation of the search space into lines or Delaunay triangles provide the fundamental source of relations. Like many other algorithms, search for relations is loosely defined and combined with class selection.

- **Resolution:** In Bayesian optimization previously eliminated classes are explicitly rejected from future considerations. considerations in Bayesian optimization.

The following section concludes this paper.

11 Conclusion

The SEARCH framework provided insightful quantitative results relevant to any BBO algorithms. This paper showed that the underlying concepts are also applicable to NBBO algorithms that make use of domain knowledge. The availability of domain knowledge in NBBO contributes to defining the relation space and class comparison statistic. Given sufficient domain knowledge the search for relations may be trivial and the class comparison statistic may become accurate. On the other hand a BBO algorithm may require more involved search for appropriate relations. Unfortunately, most of the popular BBO algorithms like simulated annealing, genetic algorithms do not properly search for relations. Therefore, we need algorithms that are more constructively designed following the decomposition offered by SEARCH. Hopefully this work will also provide a systematic approach to introduce domain knowledge in optimization algorithms.

12 Acknowledgment

This work was supported by US. Department of Energy and AFSOR Grant F49620-94-1-0103. The author acknowledges many insightful suggestions, discussions, and help from David E. Goldberg.

References

Betrò, B. (1983). A bayesian nonparametric approach to global optimization. In Stähly, P. (Ed.), *Methods of Operations Research* (pp. 45–47). Atenäum Verlag.

De Jong, K. A. (1975). An analysis of the behavior of a class of genetic adaptive systems. *Dissertation Abstracts International*, *36*(10), 5140B. (University Microfilms No. 76-9381).

Forrest, S. (Ed.) (1993). *Proceedings of the Fifth International Conference on Genetic Algorithms.* San Mateo, CA: Morgan Kaufmann.

Glover, F. (1989). Tabu search - part I. *ORSA Journal on Computing*, *1*, 190–206.

Goldberg, D. E. (1989). *Genetic algorithms in search, optimization, and machine learning.* New York: Addison-Wesley.

Goldberg, D. E., Deb, K., & Clark, J. H. (1993). Accounting for noise in the sizing of populations. In Whitley, L. D. (Ed.), *Foundations of Genetic Algorithms 2* (pp. 127–140). San Mateo, CA: Morgan Kaufmann.

Goldberg, D. E., & Lingle, R. (1985). Alleles, loci, and the traveling salesman problem. In Grefenstette, J. J. (Ed.), *Proceedings of an International Conference on Genetic Algorithms and Their Applications* (pp. 154–159). Hillsdale, NJ: Lawrence Erlbaum Associates.

Holland, J. H. (1975). *Adaptation in natural and artificial systems.* Ann Arbor: University of Michigan Press.

Jones, D. R., & Stuckman, B. E. (1992). Genetic algorithms and the Bayesian approach to global optimization. *Proceedings of the 1992 International Fuzzy Systems and Intelligent Control Conference*, 217–235.

Jones, T. (1995). *Evolutionary algorithms, fitness landscapes and search.* Doctoral dissertation, Department of Computer Science, University of New Mexico, Albuquerque, NM.

Kargupta, H. (1995, October). *SEARCH, Polynomial Complexity, and The Fast Messy Genetic Algorithm.* Doctoral dissertation, Department of Computer Science, University of Illinois at Urbana-Champaign, Urbana, IL 61801, USA. Also available as IlliGAL Report 95008.

Kargupta, H., & Goldberg, D. E. (1995, October). *SEARCH: An alternate perspective toward blackbox optimization.* In Commmunication.

Kirpatrick, S., Gelatt, C. D., & Vecchi, M. P. (1983). Optimization by simulated annealing. *Science*, *220*(4598), 671–680.

Kushner, H. (1963). A new method of locating the maximum of an arbitrary multipeak curve in the presence of noise. In *Proceedings of the Joint Automatic Control Conference* (pp. 100–111). American Society of Mechanical Engineers.

Michalski, R. S. (1983). Theory and methodology of inductive learning. In Michalski, R. S., Carbonell, J. G., & Mitchell, T. M. (Eds.), *Machine learning: An artificial intelligence approach* (pp. 323–348). Tioga Publishing Co.

Nix, A. E., & Vose, M. D. (1992). Modeling genetic algorithms with Markov chains. *Annals of Mathematics and Artificial Intelligence*, *5*, 79–88.

Perttunen, C., & Stuckman, B. (1990). The rank transformation applied to a multi-univariate method of global optimization. *IEEE Transactions on System, Man, and Cybernetics*, *20*, 1216–1220.

Radcliffe, N. J. (1991). Equivalence class analysis of genetic algorithms. *Complex Systems*, 5(2), 183–206.

Ratschek, H., & Voller, R. L. (1991). What can interval analysis do for global optimization? *Journal of Global Optimization*, *1*, 111–130.

Srinivas, M., & Patnaik, L. (1993). Binomially distributed populations for modelling GAs. See Forrest (1993), pp. 138–145.

Syswerda, G. (1989). Uniform crossover in genetic algorithms. In Schaffer, J. D. (Ed.), *Proceedings of the Third International Conference on Genetic Algorithms* (pp. 2–9).

Thierens, D., & Goldberg, D. (1993). Mixing in genetic algorithms. See Forrest (1993), pp. 38–45.

Törn, A., & Žilinskas, A. (1989). *Global optimization.* Berlin: Springer-Verlag.

Watanabe, S. (1969). *Knowing and guessing - A formal and quantitative study.* New York: John Wiley & Sons, Inc.

Messy Genetic Algorithms: Recent Developments

Hillol Kargupta*
Computational Science Methods Group
Los Alamos National Laboratory
Los Alamos, NM, 87545

Abstract

Messy genetic algorithms define a rare class of algorithms that realize the need for detecting appropriate relations among members of the search domain in optimization. This paper reviews earlier works in messy genetic algorithms and describes some recent developments. It also describes the gene expression messy GA (GEMGA)—an $O(\Lambda^k(\ell^2+k))$ sample complexity algorithm for the class of order-k delineable problems (Kargupta, 1995) (problems that can be solved by considering no higher than order-k relations) of size ℓ and alphabet size Λ. Experimental results are presented to demonstrate the scalability of the GEMGA.

1 Introduction

One of the most important task of a "general purpose" blackbox optimization (BBO) algorithm is to search for appropriate relations among the members of the search space. In genetic algorithms (GAs) the representation is the main source of relations. Unfortunately, like many other BBO algorithms GAs do not properly search for relations. Messy genetic algorithms (mGAs) (Goldberg, Korb, & Deb, 1989) realize the need for adequate search for relations and make an effort to do that.
The main objective of this paper is to present a brief review of the earlier works on messy GAs, describe the problems and accomplishments, and present some recent works. Section 2 presents the motivation behind messy GAs and identify the main distinguishing features of mGAs. Section 3 reviews previous efforts in messy GA research and also points out the main accomplishments and shortcomings of the previous efforts. Section 4 presents the gene expression messy GA (GEMGA)—a recent development in messy GA research. The scalability of GEMGA for order-k delineable problems are supported by experimental results in Section 5. Finally, Section 6 concludes this paper.

2 Messy GAs: Why, What, And All That

Since most of the realistic optimization problems do not come with sufficient knowledge about the properties of objective function, optimization algorithms like genetic algorithms (Holland, 1975), simulated annealing (Kirpatrick, Gelatt, & Vecchi, 1983) that do not require much of problem information, gained popularity among practitioners. Unfortunately this has resulted in a race for developing apparently different, new algorithms with each claiming superiority from others. Some algorithms are quite general, while some of them are for special purposes. Some emphasize the role of representation and some do not. The nature of the search operators varies widely between algorithms. Given this situation, an introduction of mGAs first demand theoretical justification and motivation behind the effort. Section 2.1 presents this. Section 2.2 briefly describes the main feature of messy GAs.

2.1 Why messy GAs?

Since messy GAs are a class of genetic algorithms, it is natural to first discuss the strengths and weaknesses of simple GA (De Jong, 1975; Goldberg, 1989) in order to justify the development of mGAs. Doing that

*The author can be reached at, P.O. Box 1663, XCM, Mail Stop F645, Los Alamos National Laboratory, Los Alamos, NM 87545, USA. e-mail: hillol@lanl.gov

G. Winter Althaus and E. Spedicato (eds.), Algorithms for Large Scale Linear Algebraic Systems, 387–400.

rigorously requires first defining what it means to be a good BBO algorithm. The SEARCH framework introduced elsewhere (Kargupta, 1995) provided a general framework for BBO by decomposing blackbox search in relation, class, and sample spaces. SEARCH also explored the conditions for polynomial complexity search on analytical ground. In the remaining portion of this paper we shall use SEARCH as the foundation for making qualitative remarks about the strengths and weaknesses of an algorithm.
The simple GA has many interesting features. The first important aspect is that it emphasizes the role of representation in search. Representation is one possible way to define a rich source of relations. Searching by constructing and ordering equivalence relations is very natural in GA. Apart from that, crossover provides an interesting tool to implement sample generation process from the intersection set of different equivalence classes. Probably one of the most interesting observations that Holland made in his book (1975) is the idea of implicit parallelism. Although this conjecture was not thoroughly laid in terms of computational arguments, the SEARCH framework confirms this observation in a quantitative manner. Moreover, the population based approach also makes GA highly suitable for parallel implementations. Despite these interesting features, the simple GA has several major problems. The main bottlenecks of the simple GA are listed below:

1. **relation, class, and sample spaces are all combined together:** Almost every paradigm of evolutionary algorithms including simple GA uses a single population of samples. This essentially means that the relation, class, and sample spaces are combined and as a result the decision making processes in each of them can affect others in an undesirable way. Only one selection operator is used for deciding in both relation and class spaces together.

2. **poor search for relations:** Simple GA does not emphasize the role of search for relations. Although Holland (1975) designed GAs along the right direction using the perspective schemata and partitions, unfortunately simple GA fails to process relations in a reasonable way. SGA with single point crossover does not exploit the complete relation space defined by the representation. It evaluates and processes only those relations that are defined over positions close to one another. Uniform crossover (Syswerda, 1989) does not let proper evaluations of relations unless selection produces more copies. Since the relation space and the sample space are combined, random perturbation of the sample strings also result in disrupting proper evaluations of the relations. Uniform crossover should also fail to accomplish proper search in the relation space. In fact, this is exactly what Thierens and Goldberg (1993) reported. Their analysis and experimental results showed that the sample complexity grows exponentially with the problem size for solving bounded deceptive problems (Thierens & Goldberg, 1993) using a simple GA with uniform crossover.

3. **Lack of precise mechanism for implicit parallelism:** Parallel evaluation of relations using the same sample set is often termed as implicit parallelism. Unfortunately, simple GA does not have a mechanism to precisely evaluate relations. This may result in decision error as a price for implicit parallelism.

Clearly, these are very fundamental bottlenecks and evolutionary algorithms cannot enjoy the taste of success for wide classes of problems unless these bottlenecks are eliminated. The main objective of messy GAs is to eliminate these problems of simple GA.

2.2 What is a messy GA?

The messy GAs have many unique differences from simple GAs. Although some of them like variable-length strings, local search template are implementation specific differences, there are several fundamental distinctions as listed in the following:

1. mGAs emphasize explicit processing of relations, classes, and samples;

2. Since in absence of relations blackbox search cannot be any better than random enumerative search, mGAs do not search for optimal solutions until appropriate relations are detected with desired accuracy and reliability. The search process is distinctly divided into two stages, namely:

    - Primordial stage: Detects appropriate relations and preserves better classes;
    - Juxtapositional stage: Computes intersection among the better classes to find the optimal solutions. This stage corresponds to the *resolution* operation of SEARCH.

3. Adequate decision making in relation and class spaces satisfying desired accuracy and reliability.

4. Messy GAs have been traditionally designed for solving quasi-decomposable problems with bounded inappropriateness of representation. In this paper, however, we shall use the recently proposed class of order-k delineable problems—problems that can be solved by searching for a polynomially bounded order of relations—(Kargupta, 1995) as the scope of messy GAs.

The following section presents a brief review of previous works on messy GAs.

3 Previous versions of messy GAs

The work on mGAs was initiated by Goldberg, Korb, and Deb (1989). The originally proposed version of messy GA (Deb, 1991; Goldberg, Korb, & Deb, 1989) took at least two important steps:

1. separated the relation and class spaces from the sample space.

2. deterministically processed all order-k relations and classes.

Original messy GA used a population that contained all (deterministically enumerated) order-k classes defined by the chosen representation. This population defined the relation and class spaces together. These classes were evaluated using a template string (a locally optimal solution), which defined the sample space. In other words, in mGA the sample space was comprised of the locally optimal solutions, found by the algorithm in different iterations. In a particular iteration the sample space was however defined by only one locally optimal solution, called template. This decomposition and the emphasis on search for relations made decision making in mGA more accurate compared to many other evolutionary algorithms.
Different versions of messy GAs studied different aspects of BBO by decomposing blackbox search along different dimensions. These investigations have directly influenced the development of SEARCH and the design of the GEMGA. Another interesting aspect of their work was the class of problems they wanted to solve. The class of bounded deceptive problems captures the essence of order-k delineability developed in SEARCH. An order-k bounded deceptive problem is order-k delineable with respect to class average comparison statistics. It has been shown elsewhere that order-k delineable problems can be solved in polynomial sample complexity. Messy GAs tries to solve this class of problems efficiently. The original version of messy GA was $O(|\Lambda|^k \ell^k)$ complexity algorithm, where ℓ is the problem length and Λ is the alphabet size of the representation.
Merkle (1992) developed a parallel implementation of original messy GA. Merkle and Lemont (1993) addressed data distribution strategies for the parallel implementation of mGA. Although the population size in primordial stage grows polynomially with problem size ℓ, $O(\ell^k)$ is a fairly large number for any reasonable value of k. Plevyak (1992) investigated the possibility of smaller population size in the primordial stage. Messy GAs have also been reported to be used for solving some application problem. Mohan (1993) applied mGAs for clustering. A hierarchical controller based on messy GAs is reported elsewhere (Hoffmann & Pfister, 1995).
The original messy GA had many problems. Some of them are listed in the following:

1. **Combined relation and class spaces:** In mGA the relation space is implicitly defined together with the class space. The SEARCH framework pointed out that these two different spaces require different decision makings and they should be processed separately in order to avoid undesirable errors in decision making.

2. **Sample space comprised of one template:** A locally optimal solution, the template, defines the sample space. Evaluating and comparing different classes on the basis of one member make not be appropriate.

3. **Lack of implicit parallelism:** The explicit enumeration of all order-k classes is very expensive ($O(|\Lambda|^k \ell^k)$). SEARCH pointed out that the same sample set can be used for evaluating different relations, which may be viewed as a quantification of the benefits of implicit parallelism. The mGA lacked such computational benefits.

The fast messy GA (fmGA) proposed elsewhere (Goldberg, Deb, Kargupta, & Harik, 1993; Kargupta, 1995) made an effort to reduce the cost of deterministic initialization by using the so called *probabilistically complete*

Table 1: Counterparts of different components of SEARCH in natural evolution.

SEARCH	Natural evolution
Relation space	gene regulatory mechanism
Class space	amino acid sequence in protein
Sample space	DNA space

initialization (PCI) and *building-block filtering* (BBF) techniques. Instead of explicitly generating all the order-k classes, PCI technique initialized the population with order-ℓ classes. This is followed by the BBF process in which the good order-k classes are gradually detected using random gene deletion and thresholding selection. The fmGA indeed reduced the cost of initialization but it succumbed to a fundamental problem that all the versions of mGAs had—implicit definition of the relation space in the class space. The primary objective of the thresholding selection of messy GAs was to select good classes defined by the same relation. However, thresholding selection was also implicitly responsible for selecting good relations. This was simply because the chosen value of thresholding parameter was always less than the string length. The main problem of the fmGA was that thresholding selection could not satisfactorily maintain the growth of strings which are instances of good classes. Undesirable cross-competition among classes from different relations usually eliminated some of the good classes and as a result the algorithm required several expensive iterations for solving large problems (Kargupta, 1995). Apart from this problem, the fmGA also faced the same questions regarding the use of single local template. Another problem of fmGA was it was slow because of the thresholding selection and it required order ℓ population sizing since selection took place only at the string or sample level.

Despite several practical applications as noted earlier, different versions of the messy GAs primarily studied different fundamental issues and were used for verifying the theoretical observations. Nevertheless, messy GAs should get the credit for taking the following conceptual leaps: (1) realizing the need for searching for relations, (2) decomposing the search space into sample space and class space (with implicit definition of relation space), and (3) paying careful attention to the decision making in genetic algorithms.

In the following section, I shall present a new generation of messy GA, called the *gene expression messy GA* (GEMGA) that eliminates many of the above mentioned problems of messy GAs by taking the inspiration from the lessons of SEARCH and the alternate perspective of natural evolution offered by SEARCH.

4 The Gene Expression Messy GA

The main objective of the gene expression messy GA (GEMGA) is to eliminate the computational shortcomings of the previous versions of messy GA. Apart from the conceptual motivations, the GEMGA also has a strong biological motivation. The implementation of GEMGA is motivated by the intra-cellular expression genetic information often known as gene expression. Section 4.1 presents that. Section 4.2 discusses the representation in GEMGA. Section 4.3 explains the population sizing in GEMGA. This is followed by Section 4.4 that describes the main operators, transcription, selection, and recombination. Section 4.5 presents of the overall mechanisms.

4.1 The biological motivation

The GEMGA also has a strong biological foundation, which is based on a mapping from the relation, class, and sample spaces to different components of natural evolutionary search space. Table 1 summarizes this correspondence. It also hypothesized a perspective of the search for appropriate relations in evolution through *gene expression*.

Unfortunately, many of the existing computational models of evolution address only the extracellular flow of genetic information. Existing models of evolutionary computation like genetic algorithms (Holland, 1975), evolutionary strategie (Rechenberg, 1973), and evolutionary algorithms (Fogel, Owens, & Walsh, 1966) are some examples. These existing perspectives of evolutionary computation do not assign any computational role to the nonlinear mechanism for transforming the information in DNA into proteins. The same DNA is used for different kinds of proteins in different cells of living beings. The development of different expression control mechanisms and their evolutionary objectives are hardly addressed in these models. They primarily emphasize the extra-cellular flow. The main difference among these models seems to be the emphasis on crossover compared to mutation or vice versa. The GEMGA emphasizes the computational role of gene expression in evolution. It makes use of transcription like operators for detecting relations and classes. A more detailed description of the biological motivations behind GEMGA can be found elsewhere (Kargupta, 1996a).

4.2 Representation

The GEMGA uses a sequence representation. Each sequence is called a *chromosome*. Every member of this sequence is called a *gene*. A gene is a data structure, which contains the *locus*, *value*, *weight* and a dynamic list of integers, called the *linkage set*. The *locus* determines the position of the member in the sequence. The locus does not necessarily have to be the same as the physical position of the gene in the chromosome. For example, the gene with locus i, may not be at the i-th position of the chromosome. When the chromosome is evaluated, however the gene with locus i gets the i-th slot. This positional independence in coding was introduced elsewhere (Deb, 1991; Goldberg, Korb, & Deb, 1989) to enforce the proper consideration for all relations defined by the representation. The GEMGA does not depend on the particular sequence of coding. For a given ℓ bit representation, the genes can be placed in arbitrary sequence. A gene also contains the *value*, which determines the value of the gene, which could be any member of the alphabet set, Λ. The weights associated with every gene take a positive real number. The weight space over all the genes define the class space of the GEMGA. The linkage set of a gene is a list of integers defining the set of genes related with it. If the genes with loci, $\{1, 5, 10, 15\}$ are related to each other then the gene with locus 1 will have the linkage set $\{5, 10, 15\}$. Similarly, the gene with locus 5 will have the linkage set $\{1, 10, 15\}$. The *linkage set* space over all genes defines the relation space of the GEMGA. No two genes with the same locus are allowed in the sequence. In other words, unlike the original messy GA (Deb, 1991; Goldberg, Korb, & Deb, 1989) no under or overspecifictions are allowed. A population in GEMGA is a collection of such chromosomes.

4.3 Population sizing

The GEMGA requires at least one instance of the optimal order-k class in the population. For a sequence representation with alphabet Λ, a randomly generated population of size Λ^k is expected to contain one instance of an optimal order-k class. The population size in GEMGA is therefore, $n = c\Lambda^k$, where c is a constant. When the signal from the relation space is clear, a small value for c should be sufficient. However, if the relation comparison statistic produces a noisy signal, this constant should statistically take care the sampling noise from the classes defined by any order-k relation. Since the proposed version of GEMGA uses sequence representation, the relation space contains total 2^ℓ relations. However, GEMGA processes only those relations with order bounded by a constant, k. In practice, the order of delineability (Kargupta, 1995) is often unknown. Therefore, the choice of population size in turn determines what order of relations will be processed. For a population size of n, the order of relations processed by GEMGA is, $k = \log(n/c)/log|\Lambda|$. If the problem is order-k delineable (Kargupta, 1995) with respect to the chosen representation and class comparison statistics then GEMGA will solve the problem otherwise not. If GEMGA cannot solve the problem for a given population size, a higher population size should be used to address possible higher order delineability.

4.4 Operators

GEMGA has four primary operators, namely: (1) *transcription*, (2) *class selection*, (3) *string selection*, and (4) *recombination*. Each of them is described in the following.

4.4.1 Transcription

As mentioned before, the weight space of the proposed version of the GEMGA chromosomes represents the class space. On the other hand the relation space is defined by the linkage set associated with every gene. The transcription operator detects the appropriate order-k relations. The transcription phase I operator determines the instances of genes contributing to the locally optimal classes. The transcription phase II operator determines the clusters of genes precisely defining the relations among those instances of genes.

```
// pick is the currently considered gene
TranscriptionPhaseI(CHROMOSOME chrom,
    int pick)
{
  double phi, delta;
  int dummy;
  double dwt;

  dwt = chrom[pick].Weight();
    phi = chrom.Fitness();
    dummy = chrom[pick].Value();
    // Change the value randomly
    chrom[pick].PerturbValue();
    // Compute new fitness
    chrom[pick].EvaluateFitness();
    // Compute the change in fitness
    delta = chrom[pick].Fitness() - phi;
    // For minimization problem
    if(delta < 0.0)
      delta = 0.0;
    // Set the weight
    if(dwt < delta OR delta == 0.0)
      chrom[pick].SetWeight(delta);
    // Set the value to the original value
    chrom[pick].SetValue(dummy);
    // Set the original fitness
    chrom[pick].SetFitness(phi);
}
```

Figure 1: Transcription Phase I operator for minimization problem. For maximization problem, if **delta**< 0 absolute value of **delta** is taken and otherwise **delta** is set to 0.

Comparing relations requires a relation comparison statistics. The GEMGA does not process the relations in a centralized fashion; instead it evaluates relations locally in a distributed manner. Every chromosome tries to determine whether or not it has an instance of a good class belonging to some relation. The transcription phase I operator considers one gene at a time. The value of the gene is randomly flipped to note the change in fitness. For a *minimization problem*, if that change cause an improvement in the fitness (i.e. fitness decreases) then the original instance of the gene certainly do not belong to the instance of the best class of a relation, since fitness can be further improved. Transcription sets the corresponding weight of the gene to zero. On the other hand if the fitness worsens (i.e. fitness increases) then the original gene may belong to a good class; at least that observation does not say it otherwise. The corresponding weight of the gene is set to the absolute value of the change in fitness. Finally, the value of that gene is set to the original value and the fitness of the chromosome is set to the original fitness. In other words, ultimately transcription phase I does not change anything in a chromosome except the weights. For a maximization problem the conditions for the weight change are just reversed. The same process is continued deterministically for all the ℓ genes in every chromosome of the population. Figure 1 shows the Transcription phase I operator. Transcription phase II identifies the exact relations among the genes and constructs the linkage set of every gene in a choromosme. This operator performs pairwise consideration of genes. The objective is to identify the set of genes that are related with any given gene from the chromosome. Among the $\binom{\ell}{2}$ possible pair of choices only those pairs are considered in which both the genes have non-zero weights. In other words if a gene is identified as a possible contributor to an instance of locally optimal set of genes then its dependencies on other such genes in that chromosome are tested using the transcription phase II operator. For every gene with non-zero weight the linkage set is constructed and the real weights are replaced by boolean weights. Figure 2 shows the pseudo-code for this operator, where `pick1` and `pick2` define the loci of the pair of genes.

For genes with higher cardinality alphabet set (Λ) this process is repeated for some constant $C < |\Lambda|$ times. The following section describes the two kinds of selection operators used in GEMGA, which correspond to the selective pressures in protein and DNA spaces of natural evolution described elsewhere (Kargupta, 1996b).

4.4.2 Selection

Once the relations are identified, selection operator is applied to make more instances of better classes. GEMGA uses two kinds of selections—(1) class selection and (2) string selection. Each of them is described in the following:

- **Class Selection:** The class selection operator is responsible for selecting individual classes from the chromosomes. Better classes detected by the transcription operator are explicitly chosen and given more copies at the expense of bad classes in other chromosomes. Figure 3 describes the operator. Two chromosomes are randomly picked; A set of genes with non-zero weights are chosen from one of them, `chrom1`; those genes with cardinality of their `LinkageSet` strictly greater than those of their counterparts in the other participating chromosome are collected in a list called `SelectSet`. Then the genes of the chromosome `chrom1` corresponding to `SelectSet` are copied on the corresponding genes of `chrom2`.
- **String Selection:** This selection operator gives more copies of the chromosomes. A standard binary tournament selection operator (Brindle, 1981; Goldberg, Korb, & Deb, 1989) is used. Binary tournament selection randomly picks up two chromosomes from the population, compares their objective function values, and gives one additional copy of the winner to the population at the expense of the looser chromosome.

The following section describes the recombination operator in GEMGA.

4.4.3 Recombination

Figure 4 shows the mechanism of the recombination operator in GEMGA. It randomly picks up two chromosomes from the population and considers all the genes in the chromosomes for possible swapping. It randomly marks one among them. Just like the `ClassSelection` operator `Recombination` selects a set of genes called the `ExchangeSet`. Genes of `chrom1` and `chrom2` corresponding to the members of `ExchangeSet` are exchanged.

The following section describes the overall mechanism of the algorithm.

4.5 The algorithm

GEMGA has two distinct phases: (1) primordial stage and (2) juxtapositional stage. The primordial stage first applies the transcription phase I operator for ℓ generations, deterministically considering every gene in each generation. This is followed by the application of the transcription phase II operator for each pair of genes with non-zero weights. During this stage the population of chromosomes remains unchanged, except that the weights of the genes change and the linkage sets get constructed. This is followed by the juxtapositional stage, in which the string selection, class selection, and recombination operators are applied iteratively. Figure 5 shows the overall algorithm. The length of the transcription phase I application is ℓ. The length of the application of the transcription phase II application is $\ell^2 - \ell$ in the worst case. The length of the juxtapositional stage can be roughly estimated as follows. If t be the total number of generations in juxtapositional stage, then for binary tournament selection, every chromosome of the population will converge to same instance of classes when $2^t = n$, i.e. $t = \log n / \log 2$. Substituting $n = c|\Lambda|^k$, we get,$t = \frac{\log c + k \log|\Lambda|}{\log 2}$. A constant factor of t is recommended for actual practice. Clearly the number of generations in juxtapositional stage is $O(k)$. Let us now compute the overall sample complexity of GEMGA. Since the population size is $O(|\Lambda|^k)$ and the primordial stage continues for $C\ell = O(\ell)$ generations, the overall sample complexity,

$$\begin{aligned} SC &= O(|\Lambda|^k(\ell + \ell^2 - \ell + k)) \\ &= O(|\Lambda|^k(\ell^2 + k)) \end{aligned}$$

Note that the transcription phase II operator is applied on those pair of genes that have non-zero weights. Therefore, the complexity of this operation is quadratic only in the worst case when all the genes in a chromosome have non-zero weights.

GEMGA is a direct realization of the lessons from the SEARCH framework. Now that we described the algorithm let us look back to our SEARCH framework and see how the different components of GEMGA maps onto the SEARCH.

- **Relation, class, and sample spaces:** The linkage set defines the relation space; the weights define the class space and the (value, locus) pair define the sample space.
- **Class comparison statistics:** The class comparison statistic in GEMGA is defined in two stages. First of all, the transcription phase I operator identifies the locally optimal set of genes by bitwise perturbation. Once the transcription phase II operation is performed to identify the relations among these set of genes, all the genes defining locally optimal classes are assigned a non-zero weight 1. This makes sure that the classes are not given any undeserved bias based on their local evaluation. The earlier version of GEMGA had this undue bias. The local evaluation was used to compare classes at a global level.
- **Relation comparison statistics:** Transcription phase II identifies the linkage set of the locally optimal set of genes. The relation and class comparison statistics are mutually dependent in GEMGA.
- **Constant M:** The value of M is controlled in a distributed manner in GEMGA. The class selection operator controls the value of M in the GEMGA. As we increase the class selection probability the M value is decreased and the vice versa.
- **Recombination:** This defines the resolution operation of SEARCH.

The following section presents the test results.

5 Test Results

Designing a test set up requires careful consideration. An ideal set up should contain problems with different dimensions of problem difficulty, such as multi-modality, bounded inappropriateness of relation space, problem size, noisy objective function. In this paper, we present the performance of GEMGA for problems with varying degree of difficulties along the first three dimensions. The following sections describe the test functions and present the experimental results.

5.1 Experimental design

A test function is constructed by concatenating multiple numbers of order-5 trap functions (Ackley, 1987). Each of these subfunctions is an order-5 trap function. The particular version of the deceptive trap function used can be defined as follows:

$$
\begin{aligned}
f(x) &= \ell \text{ if } u = \ell \\
&= \ell - 1 - u \text{ otherwise,}
\end{aligned}
$$

where u is the number of 1-s in the string x and ℓ is the string length. If we carefully observe this trap function, we shall note that it has two peaks. One of them corresponds to the string with all 1-s and the other is the string with all 0-s. For $\ell = 200$, the overall function contains 40 subfunctions; therefore, an order-5 bounded 200-bit problem has 2^{40} local optima, and among them, only one is globally optimal. As the problem length increases the number of local optima exponentially increases. This class of problems is order-5 delineable with respect to the class average comparison statistics (i.e. when classes are compared with respect to the distribution means). This problem in order-5 deceptive representation has only $\ell/5$ proper relations among the $\binom{\ell}{5}$ order-5 relations. Therefore, searching for the appropriate relations is not a trivial job in this class of problems. This is the primary reason behind the failure of most of the existing blackbox optimization algorithms for such problems. Clearly this class of problems are massively multimodal and has bounded inappropriateness of the relation space, defined by the representation. The following section presents the test results.

5.2 Results

The GEMGA is tested against order-5 deceptive problems of different sizes. Figure 6 shows the average number of sample evaluations from six independent runs needed to find the globally optimal solution for different problem sizes. The population size is 500, chosen as described earlier in this paper. As we see, the sample complexity linearly depends on the problem size.

Figure 7 show the gradual detection of the relations during the primordial and juxtapositional stages for a 30-bit order-5 deceptive problem. Each figure represent the relation space of the whole population at a certain generation. The x-axis denotes the weights in the genes, ordered on the basis of the *locus* of the gene. In other words the values along the x-axis correspond to the actual value of the locus of a gene in a chromosome. The y-axis corresponds to the different members in the population. The z-axis, perpendicular to the page denotes the weights of the corresponding gene in the corresponding chromosome. Since the test function is comprised of order-5 trap functions, for any particular gene in a chromosome, there are only 4 other genes that are related with it. The complete relation space has a cardinality of 2^{30}. Among $\binom{30}{5}$ order-5 relations there are only 6 relations that correctly correspond to the actual dependencies defined by the problem. GEMGA needs to detect the relations that relate genes with loci ranging from 0 to 4 together, from 5 to 9 together and so on. These relations are gradually detected in different chromosomes that contain good classes from those relations. More instances of good classes are produced by selection and they are exchanged among different strings to create higher order relations that finally lead to the optimal solution. The following section concludes this paper.

6 Conclusion

Research on messy GAs contributed in understanding several important issues in blackbox optimization. The most important one is the search for appropriate relations from the relation space defined by the representation. Decomposition of search into relation, class, & sample spaces and decision theoretic construction of ordering among classes & relations are examples other important contributions. In this paper we also reviewed the shortcomings of earlier version of messy GAs. The GEMGA eliminated many of these problems. The main accomplishments of GEMGA are,

1. explicit processing of relations and classes,
2. eliminating the need for a template solution,
3. reducing the population size from $O(\Lambda^k \ell)$ to $O(\Lambda^k)$ for order-k delineable problems in sequence representation of length ℓ,
4. introducing high degree of parallelism (even more than simple GA), and
5. reducing the running time by a large factor.

Experimental results clearly showed that GEMGA can detect appropriate relations efficiently for different classes of problems. However, the transcription phase I and phase II requires more work. For example, the current version of transcription phase II will fail to work satisfactorily for some functions in which the overall objective function value is the product of the function values of individual subfunctions. Moreover, currently GEMGA can at most solve problems that are order-k delineable in the chosen representation. One future possibility is transforming non order-k delineable problems to delineable problems by representation construction.

7 Acknowledgment

This work was supported by US. Department of Energy. The author also acknowledges many useful discussions with Professor David E. Goldberg and Liwei Wang.

```
// pick1, pick2 are the indices of a pair of genes
TranscriptionPhaseII(CHROMOSOME chrom,
    int pick1, int pick2)
{
  double phi, delta;
  int dummy1, dummy2;

  if(chrom[pick1].Weight() > 0) {
    dummy1 = chrom[pick1].Value();
    phi = chrom.Fitness();
    chrom[pick1].PerturbValue();
    chrom.EvaluateFitness();
    if(chrom[pick2].Weight() > 0.0) {
      dummy2 = chrom[pick2].Value();
      chrom[pick2].PerturbValue();
      chrom.EvaluateFitness();
      delta = chrom.Fitness() - phi;
      // For minimization problem
      if(delta < 0.0)
        delta = 0.0;
      if(delta != chrom[pick2].Weight()) {
        chrom[pick1].AddLinkageSet(pick2);
        chrom[pick2].AddLinkageSet(pick1);
        chrom[pick1].SetWeight(1.0);
        chrom[pick2].SetWeight(1.0);
      }
      chrom[pick2].SetValue(dummy2);
    }
    chrom[pick1].SetWeight(1.0);
    // Set the value to the original value
    chrom[pick1].SetValue(dummy1);
    // Set the original fitness
    chrom.SetFitness(phi);
  }
}
```

Figure 2: Transcription Phase II operator for minimization problem.

```
ClassSelection(chrom1, chrom2)
CHROMOSOME chrom1, chrom2;
{
int i;

for(i=0; i<Problem_length; i++) {
  if(Rnd()<0.5 AND chrom1[i].Weight()>0) {
    if(chrom1[i].LinkageSet.Length() >
      chrom2[i].LinkageSet.Length()) {
      // Collect linkage sets of chosen genes
      SelectSet.Collect[LinkageSet[i]]; }
  }
}
for(i=0; i<SelectSet.Length(); i++)
  chrom2[SelectSet[i]]=chrom1[SelectSet[i]];
}
```

Figure 3: Class selection operator in GEMGA. A consistent coding (where *chrom*1[*i*] and *chrom*2[*i*] has common *locus*) is used in place of messy coding for the sake of illustration. `Rnd()` generates a random number in between 0 and 1.

```
Recombination(chrom1, chrom2)
CHROMOSOME chrom1, chrom2;
{
int i;
GENE dummy;

for(i=0; i<Problem_length; i++){
  if(Rnd()<0.5 AND chrom1[i].Weight()>0) {
    if(chrom1[i].LinkageSet.Length() >
      chrom2[i].LinkageSet.Length()) {
      // Collect linkage sets of chosen genes
      ExchangeSet.Collect[LinkageSet[i]]; }
  }
}
for(i=0; i<ExchangeSet.Length(); i++) {
  dummy=chrom1[ExchangeSet[i]];
  chrom1[ExchangeSet[i]]=chrom2[ExchangeSet[i]];
  chrom2[ExchangeSet[i]]=dummy;
}
}
```

Figure 4: Recombination operator in GEMGA. A consistent coding (where `chrom1[i]` and `chrom2[i]` has common *locus*) is used in place of messy coding for the sake of illustration. `Rnd()` generates a random number in between 0 and 1.

```
void GEMGA() {
POPULATION Pop;
int i, j, k, C, k_max;

// Initialize the population at random
Initialize(Pop);
i = 0;
// Primordial stage
While(i < C) { // C is a constant
  j = 0;
  Repeat {
    // Identify better relations
    TranscriptionPhaseI(Pop, j);
    // Increment generation counter
    j = j + 1;
  } Until(j == Problem_length)
  i = i + 1;
}
TranscriptionPhaseII(Pop);
k = 0;
// Juxtapositional stage
  Repeat {
    // Select better strings
    Selection(Pop);
    // Select better classes
    ClassSelection(Pop);
    // Produce offspring
    Recombination(Pop);
    Evaluate(Pop); // Evaluate fitness
    // Increment generation counter
    k = k + 1;
  // k_max is of O(log(Problem_length))
  } Until ( k > k_max )
}
```

Figure 5: Pseudo-code of GEMGA. The constant C ¡ $|\Lambda|$, where $|\Lambda|$ is the cardinality of the alphabet set.

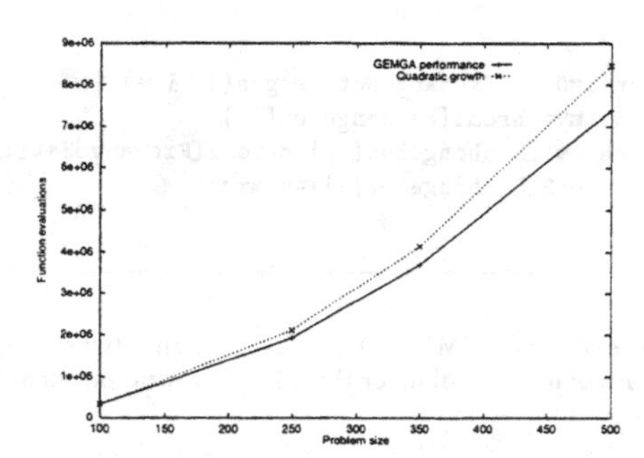

Figure 6: Growth of the number of function evaluations with problem size.

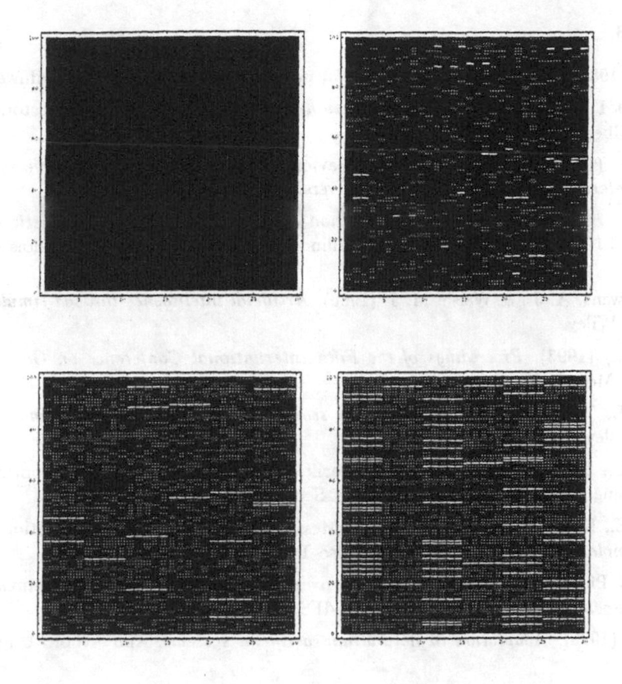

Figure 7: The relation space during primordial generation 1, 10, and juxtapositional generations 1, 4 (from top to bottom).

References

Ackley, D. H. (1987). *A connectionist machine for genetic hill climbing.* Boston: Kluwer Academic Publ.

Brindle, A. (1981). *Genetic algorithms for function optimization.* Unpublished doctoral dissertation, University of Alberta, Edmonton, Canada.

De Jong, K. A. (1975). An analysis of the behavior of a class of genetic adaptive systems. *Dissertation Abstracts International, 36*(10), 5140B. (University Microfilms No. 76-9381).

Deb, K. (1991). *Binary and floating-point function optimization using messy genetic algorithms* (IlliGAL Report No. 91004). Urbana: University of Illinois at Urbana-Champaign, Illinois Genetic Algorithms Laboratory.

Fogel, L. J., Owens, A. J., & Walsh, M. J. (1966). *Artificial intelligence through simulated evolution.* New York: John Wiley.

Forrest, S. (Ed.) (1993). *Proceedings of the Fifth International Conference on Genetic Algorithms.* San Mateo, CA: Morgan Kaufmann.

Goldberg, D. E. (1989). *Genetic algorithms in search, optimization, and machine learning.* New York: Addison-Wesley.

Goldberg, D. E., Deb, K., Kargupta, H., & Harik, G. (1993). Rapid, accurate optimizaiton of difficult problems using fast messy genetic algorithms. See Forrest (1993), pp. 56–64.

Goldberg, D. E., Korb, B., & Deb, K. (1989). Messy genetic algorithms: Motivation, analysis, and first results. *Complex Systems, 3*(5), 493–530. (Also TCGA Report 89003).

Hoffmann, F., & Pfister, G. (1995, July). *A new learning method for the design of hierarchical fuzzy controller using messy genetic algorithms.* Presented on IFSA'95, Sao Paulo.

Holland, J. H. (1975). *Adaptation in natural and artificial systems.* Ann Arbor: University of Michigan Press.

Kargupta, H. (1995, October). *SEARCH, Polynomial Complexity, and The Fast Messy Genetic Algorithm.* Doctoral dissertation, Department of Computer Science, University of Illinois at Urbana-Champaign, Urbana, IL 61801, USA. Also available as IlliGAL Report 95008.

Kargupta, H. (1996a). SEARCH and a Computational Perspective of Evolution. In *Proceedings of the Artificial Life V* (pp. 56–63).

Kargupta, H. (1996b, January). *SEARCH, evolution, and the gene expression messy genetic algorithm.* Los Alamos Unclassified Report LA-UR-96-60.

Kirpatrick, S., Gelatt, C. D., & Vecchi, M. P. (1983). Optimization by simulated annealing. *Science, 220*(4598), 671–680.

Merkle, L. D. (1992, December). *Generalization nd parallelization of messy genetic algorithms and communication in parallel genetic algorithms.* Master's thesis. Air Force Institute Of Technology, WPAFB OH 45433.

Merkle, L. D., & Lemont, G. B. (1993). Comparison of parallel messy genetic algorithm data distribution strategies. See Forrest (1993), pp. 191–205.

Mohan, C. K. (1993). A messy genetic algorithm for clustering. In Dagli, C. H., Burke, L. I., Fernádez, & Ghosh, J. (Eds.), *Intelligent Engineering Systems Through Artificial Neural Networks* (pp. 831–836). New York: ASME Press.

Plevyak, J. (1992). *A messy GA with small primordial population.*

Rechenberg, I. (1973). Bionik, evolution und optimierung. *Naturwissenschaftliche Rundschau, 26*, 465–472.

Syswerda, G. (1989). Uniform crossover in genetic algorithms. In Schaffer, J. D. (Ed.), *Proceedings of the Third International Conference on Genetic Algorithms* (pp. 2–9).

Thierens, D., & Goldberg, D. (1993). Mixing in genetic algorithms. See Forrest (1993), pp. 38–45.

LIST OF CONTRIBUTORS:

Prof. Pedro Almeida Benítez
Dpto. de Matemáticas
Edfc. de Informática y Matemáticas
Campus Universitario de Tafira
35017 Las Palmas de G.C., Spain
Fax: +34 (9)28 - 458811

Prof. Claude Brezinski
Université des Sciences et Technologies de Lille
Laboratoire d' Analyse Numérique
et d'Optimisation
UFR IEEA M3 59655
Villeneuve d'Ascq Cedex, France
Fax: (33) 20 43 68 69

Prof. C. G. Broyden
Università di Bologna,
Facoltà di Scienze MM NN FF,
via Sacchi N.3
47023 Cesena, Italy
Fax: 39-547-610100

Prof. David J. Evans
Department of Computer Studies
University of Technology Loghborough
Leicestershire LE11 3TU, UK

Prof. Luis Ferragut Canals
Departamento de Matemática Pura y Aplicada
Facultad de Ciencias.
Universidad de Salamanca
Pza. de la Merced 1-4 37008, Salamanca, Spain
Fax: (34)(23)294583

Prof. Alan George
Dept. of Computer Science
University of Waterloo
Waterloo, Ontario N2L 3G1, Canada
Fax: 519/885-1208

Prof. Hillol Kargupta
Computational Methods Group,
X Division,
Los Alamos National Laboratory
University of California
P.O. Box 1663, Mail Stop F645
Los Alamos National Laboratory
Los Alamos, NM 87545, USA
Fax: (505) 665-4479

Prof. Gustavo Montero García
Dpto. de Matemáticas
Edfc. de Informática y Matemáticas
Campus Universitario de Tafira
35017 Las Palmas de G.C., Spain
Fax: +34 (9)28-458811

Prof. Victor Pan
Lehman College
The City University of New York
Department of Mathematics and Computer
Science (718) 960-8117
250 Bedford Park Boulevard West
Bronx, NY 10468-1589, USA
Fax: (718) 960-8969

Prof. Yousef Saad
Department of Computer Science
4-192 EE/CSci Building
200 Union Street S.E.
Minneapolis, MN 55455, USA
Fax: (612) 625 0572

Prof. Emilio Spedicato
Università di Bergamo
Dipartimento di Matematica, Statistica,
Informatica ed Applicazioni
241[illegible] Bergamo, Piazza Rosate 2, Italy
Fax: 39 35 249598

Prof. Zdenek Strakos
Institute of Computer Science
Academy of Science
Pod Vodareunkou Vezi
2182 07 Praha 8, Czech Republic

Prof. Henk A. van der Vorst
Department of Mathematics
University of Utrecht
P.O. Box 80.010
3508 TA Utrecht, The Netherlands
Fax: +31-30-518394

Prof. Gabriel Winter Althaus
University of Las Palmas
de Gran Canaria
Campus de Tafira Baja
35017 Las Palmas de G.C., Spain
Fax: +34-(9)28-451921

Index